With this cutting-edge work, Rob DeLeo  tion to both the theory and practice of publ (seemingly) straightforward question, one ment: Why can't public officials anticipate and take steps to avert or to lessen their in literature on agenda-setting as well as policy typologies and the implementation process, DeLeo addresses this question by means of a sophisticated analysis that is sensitive to the behavior of political and administrative institutions while also considering critical substantive differences across such issue areas as nanotechnology, public health pandemics, and climate change. The result is a wealth of insights into the distinctive dynamics of anticipation in the public arena, including suggestions for the right—and wrong—ways to improve performance in our age of "preparedness."

David A. Rochefort, *Northeastern University*

Most studies of policy change cite reactions to recent events as prime motivators for policy change. Rob DeLeo turns this process on its head, and considers how the policy process can address problems before they become damaging, attention-grabbing events. The book is firmly grounded in policy theory, and contains a number of examples of effective anticipatory policymaking that resulted in the mitigation of risk and prevented a bad outcome. DeLeo's painstaking and clear presentation shows that policymakers are not always reactive, that they have the capacity to anticipate and mitigate risks, and that this anticipatory policymaking is characterized by patterns of behavior among actors and institutions that deserve our careful attention. This book is a fine example of contemporary policy research, and a useful corrective to the notion that the policy process simply waits for a disaster before rousing itself to action. The process is more complex and subtle, and DeLeo helps us see past this oversimplification.

Thomas A. Birkland, *North Carolina State University*

# Anticipatory Policymaking

Public policy analysts and political pundits alike tend to describe the policymaking process as a reactive sequence in which government develops solutions for clearly evident and identifiable problems. While this depiction holds true in many cases, it fails to account for instances in which public policy is enacted in anticipation of a potential or future problem. Whereas traditional policy concerns manifest themselves through ongoing harms, "anticipatory problems" are projected to occur sometime in the future, and it is the prospect of their potentially catastrophic impact that generates intense speculation and concern in the present.

*Anticipatory Policymaking: When Government Acts to Prevent Problems and Why It Is So Difficult* provides an in-depth examination of the complex process through which United States government institutions anticipate emerging hazards. Using contemporary debates over the risks associated with nanotechnology, pandemic influenza, and global warming as case study material, Rob A. DeLeo highlights the distinctive features of proactive governance. By challenging the pervasive assumption of reactive policymaking, DeLeo provides a dynamic approach for conceptualizing the political dimensions of anticipatory policy change.

**Rob A. DeLeo** is an Assistant Professor of Public Policy at Bentley University.

# Routledge Research in Public Administration and Public Policy

# Anticipatory Policymaking

When Government Acts to Prevent
Problems and Why It Is So Difficult

Rob A. DeLeo

Routledge
Taylor & Francis Group
New York London

First published 2016
by Routledge
711 Third Avenue, New York, NY 10017, USA

and by Routledge
2 Park Square, Milton Park, Abingdon, Oxon OX14 4RN

First issued in paperback 2017

*Routledge is an imprint of the Taylor & Francis Group, an informa business*

© 2016 Taylor & Francis

The right of Rob A. DeLeo to be identified as author of this work has
been asserted by him in accordance with sections 77 and 78 of the
Copyright, Designs and Patents Act 1988.

*Library of Congress Cataloging-in-Publication Data*
DeLeo, Rob A.
Anticipatory policymaking : when government acts to prevent
problems and why it is so difficult / Rob A. DeLeo.
    pages cm. – (Routledge research in public administration and
    public policy ; 14)
    Includes bibliographical references and index.
    1. Public administration–United States. 2. Policy sciences–United
    States. 3. Policy sciences–Decision making. 4. Emergency
    management–United States. 5. United States–Politics and
    government. I. Title.
    JK421.D44 2015
    361.6'10973–dc23                                        2015013384

ISBN 13: 978-1-138-30747-6 (pbk)
ISBN 13: 978-1-138-81342-7 (hbk)

Typeset in Sabon
by Wearset Ltd, Boldon, Tyne and Wear

# Contents

# Illustrations

## Figures

## Table

# Acknowledgments

This book is the culmination of a lifetime's worth of support and guidance. The number of friends, family, and colleagues worthy of acknowledgment is far too great to possibly list. This said, a number of individuals were particularly instrumental in helping me complete this project. I am greatly indebted to David Rochefort who invested an enormous amount of time and energy into virtually every stage of this book. I thank him for his continued support and mentorship. This book would not exist without him.

I also owe a great deal of gratitude to Bentley University for providing me with the resources necessary to complete this project, including generous summer grant money and funding for research assistantships. I am grateful to my colleagues in the 2014–2015 Valente Center Humanities Seminar, whose interdisciplinary perspective was a great source of inspiration. A special thank you is owed to Juliet Gainsborough for providing a detailed review of my conceptual framework. I also want to thank my colleagues in the Department of Global Studies for providing much needed encouragement throughout this project. I am particularly grateful for the support of my Department Chair, Joni Seager, who has been an unwavering advocate of my work both inside and outside the classroom.

I want to thank Christopher Bosso and John Portz for providing extensive feedback on earlier versions of this manuscript as well as Thomas Birkland for his useful insights into the dynamics of risk and disaster policymaking. I was also fortunate to have the support of a number of brilliant and hard working research assistants, specifically Mohamed Jaafar, Shannen Amicangioli, and Geoffrey Steber.

I want to thank Natalja Mortensen of Routledge for her support of this project and for guiding me through the process of academic publishing. I also want to thank the anonymous reviewers for their detailed commentary and insights. And to the rest of the team at Routledge, thank you for promptly—and patiently—responding to my inordinate number of inquiries about formatting requirements, permission standards, and all things "production."

Finally, I am extremely grateful to my family for their love and support. To my mother, thank you for instilling in me the intellectual curiosity that continues to fuel my passion for academic work. More importantly, thank you for helping me endure life's many trials-and-tribulations. This book would not have been possible without your love. To my father, thank you for teaching me the value of hard work. Your example has shown me that anything is possible, so long as I put my mind to it. What is more, few political scientists are lucky enough to have a father who knows how politics *really* works. Your insights into the political world are invaluable. To my sister, I continue to marvel at your compassion and commitment to making this world a better place. I truly cherish our relationship and appreciate your unfettered love and support. To Olga, you are unquestionably one of my biggest fans and perhaps the only one in our small family who truly understands the academic lifestyle. Thank you for assuaging the anxieties of the son you never wanted. And to Gram, you truly are the light of my life. I love you more than words can possibly express and hope this book makes you proud.

This book is dedicated to Jenny, who, for the last eight years, has been my greatest source of strength, inspiration, and love. Your sacrifices have not gone unnoticed and I will be forever grateful for your support of this "grumpy writer." I am truly blessed to have you in my life.

# 1    Anticipation and Public Policy

*December 31, 1999.* While the rest of the world flocked to the streets and congregated in friends' homes in celebration of the coming year, John Koskinen and a group of 400 highly skilled computer technicians anxiously watched the countdown to midnight in a $50 million federal government facility located just blocks away from the White House in Washington, D.C. Koskinen, the White House's so-called Y2K Czar, and his team had little time for champagne, midnight embraces, and "Auld Lang Syne." Instead, they were busy overseeing the fate of humanity, as the transition from "1999" to "2000" threatened a catastrophic crash of computer systems around the globe.[1]

The Y2K problem or Year 2000 bug, which was first discovered in the 1980s, stemmed from a programming practice dating back to the 1960s that used a two-digit abbreviation code to account for four digit years. The year "1983," for example, was abbreviated as "83," the year "1997," as "97." By using a two, as opposed to a four, digit-coding scheme, programmers saved critical storage space and memory, scarce commodities at this time. However, in the mid-1980s technicians realized the dawn of the twenty-first century would create a problem in that the year "2000" would be indistinguishable from the year "1900"—both abbreviated to "00." In turn, computing systems across the globe were expected to revert back to the year 1900 on January 1, 2000.[2,3] Because dates and times are critical to a computer's operating system, the Y2K problem was anticipated to create widespread IT failures in the banking, security, medical, transportation, utility, and a host of other sectors. Dire predictions abounded. Many feared entire hospitals and medical devices would suddenly stop working, critical national security information would be compromised, banking systems would crash, and planes would, literally, fall from the sky.[4]

In anticipation of this potentially catastrophic event, governments around the world enacted sweeping Y2K preparedness plans. In October 1998, the United States Congress passed the Year 2000 Information and Readiness Disclosure Act, which limited liability for private companies in order to facilitate greater and more transparent information sharing. Prior to the law, many companies withheld critical information on the status of their Y2K readiness plans out of fear that the government might sue them

2 Anticipation and Public Policy

for failing to properly prepare or for disclosing potentially controversial information. Congress was indeed highly involved in the preparedness effort, holding more than 100 hearings in the four years leading up to 2000.[5]

Then-President Bill Clinton established the President's Council on Year 2000 Conversion, which was headed by Koskinen. The Council convened a series of "Year 2000" summits in order to help states prepare. It also encouraged international cooperation and greater information sharing between business and government. The Office of Management and Budget coordinated activities across the executive branch, monitoring individual agency readiness, designating lead agencies for overseeing the government's "high impact" federal programs (i.e., food stamps, Medicare, and federal electric power generation and delivery), and helping private businesses develop contingency programs and plans.[6] Commenting on the government's response, President Clinton stated: "We've worked hard to be ready. I set a government-wide goal of full compliance by March of 1999. The American people have a right to expect uninterrupted service from government, and I expect them to deliver."[7]

The federal government spent close to $9 billion preparing for Y2K. The private sector, state governments, and other stakeholders spent another $90 billion. Save some minor glitches, Y2K caused few disruptions and the government's efforts were heralded as a smashing success.[8] In February 2000, former U.S. Senator Bob Bennett (R-UT) rose on the Senate floor and declared, "The record is fairly clear that had we as a nation not focused on this issue and dealt with it, we would have had very significant problems."[9] Reflecting on the legacy of Y2K, Farhad Manjoo of *Slate* argued the case is emblematic of something much more important—much more profound—than simply an example of successful crisis management and political leadership. Instead, the Y2K case challenges one of the most pervasive assumptions about the American system of government: that it is inherently *reactive*. According to Manjoo:

> [I]t's the only recent example of something exceedingly rare in America—an occasion when we spent massive amounts of time and money to improve national infrastructure to *prevent* a disaster. Typically, we write checks and make plans after a catastrophe has taken place, as we did for 9/11 and Hurricane Katrina.[10]

The reactive assumption, like so many others about our system of government and the problems and issues it deals with, is often communicated by means of a narrative or story. This "reactivity narrative" holds that government tends to wait for policy issues to develop into large scale—and often costly—problems before addressing them. A more logical approach, critics argue, would be for policymakers to act quickly with countermeasures or, better yet, anticipate or prevent problems before they occur, saving both precious resources and, in some instances, lives. Government should therefore embrace a proactive approach to policymaking, as opposed to responding to issues after they have emerged and blossomed into significant ills. This notion that a proactive approach is somehow superior is only reinforced by the high esteem Americans hold for private sector enterprises, which are perpetually "looking ahead" as a way to achieve a strategic advantage. Government, it seems, is only adept at cleaning up society's problems and, even in this regard, Americans are skeptical about its abilities.

A simple Internet search of the terms "reactive government," "reactive policymaking," or "reactive Congress" suggests the pervasiveness of this narrative in popular depictions of government. Scholars, journalists, and government officials have all lamented government's reactive disposition. Often, the situation is portrayed as evidence of a stifling or prohibitive trait that hinders government's very ability to solve complex policy problems. Perhaps no event did more to elevate this concern than the attacks of September 11, 2001. Although the federal government was criticized on a number of grounds, one of the most prominent critiques held that intelligence agencies failed to anticipate this national tragedy despite an abundance of indicators that an attack was imminent. As Donald Kettl points out, the proverbial "take away" lesson of September 11 for policymakers looking to construct a twenty-first century homeland security infrastructure is that security would require both prevention—future attacks would have to be anticipated and stopped—and response—in the event that the uncertain did occur, government would have to react swiftly and effectively.[11] Writes Kettl:

> [The Bush Administration] sought *prevention*: to do everything possible to ensure that those who might launch such attacks were stopped before they could try. But they also needed to strengthen *response*: to do everything possible, should an attack occur, to minimize injuries, the loss of life, and damage to property. Administration officials knew that although any attack was unacceptable, total protection was impossible. The terrorists had proved that they were cunning strategists who worked hard to identify and exploit points of vulnerability. Officials also were aware that they needed to strengthen the system's response. But that would matter only if the prevention strategy failed, and they did not want to talk publicly about the possibility. Officials thus needed to maximize their ability to respond while doing everything possible to prevent attacks in the first place.[12]

The federal government's capacity to respond to crisis certainly requires strengthening (these deficiencies were again highlighted by Hurricane Katrina) and much of the post-9/11 debate has focused on providing government with the means to anticipate "all-hazards," be they man-made or naturally occurring.[13] Indeed, anticipation sometimes seems a novel construct that only truly became the centerpiece of government reform efforts due to national tragedy. Calls for greater "prevention," "anticipation," "foresight," and "proactiveness" abounded in the wake of 9/11. For example, in presenting a menu of options for Congress to consider in its revision of aviation security policy, the Congressional Research Service (CRS) explicitly integrated anticipation into the methodology it used to evaluate various policy options. One of CRS's five evaluative criteria asked: "Is the national strategy for aviation security forward-looking, or does it perpetuate a reactive approach to strategic security planning in the aviation domain?"[14] Summarizing the claims of those who criticized the current aviation security standard, CRS writes: "Critics argue that reacting to single events is near-sighted and goes against the very purpose of developing strategies and plans in the first place, which is to be proactive in assessing threats and directing resources to mitigate associated risks."[15] Here, then, is a clear normative preference toward anticipation, reflecting an assumption that the existing aviation security domain is reactive and thus not prepared to deal with future terrorist threats.

Countless variations of this same narrative now exist in other domains. In "Making Data Breach Prevention a Matter of Policy," Chris Sullivan, vice president of customer solutions at Courion, an identity access and management company, lamented the "reactive model" of data breach prevention adopted by many organizations and perpetuated by government policy. Sullivan summarized this faulty model as entailing three broad steps: "Step 1—Wait for a breach to occur, hoping it doesn't happen to your organization; Step 2—Get breached and notify consumers; Step 3—Get money and get focused post-breached, and try to catch up and fix the problem."[16] While data-breach prevention is in many ways a problem for individual organizations to deal with, Sullivan noted that the breach notification laws adopted by government are a far cry from preventive and actually serve to reinforce the reactive model. And, in fact, he reports that recent studies demonstrate notification statutes have not reduced identity theft.

A June, 2010, report by the Institute of Medicine charged that the Food and Drug Administration is "too often caught flat-footed when problems arise" and the agency should devote greater resources toward preventing illness, as opposed to simply responding to outbreaks. Although the panel proposed a number of structural reforms that might facilitate a more prospective approach, such a change would ultimately require "a cultural change, a different way of doing business," stated the study's main author, Dr. Robert Wallace of the University of Iowa's College of Public Health.[17]

The assumption that government is solely reactive has permeated academic circles as well.[18] Colebach implicitly summarized this preconception within the study of public policy, writing:

> The dominant paradigm in texts on policy practice sees the policy process as an exercise in informed problem-solving: a problem is identified, data

is collected, the problem is analyzed and advice is given to the policy-maker, who makes a decision which is then implemented.[19]

He adds that the "textbook account" of the policy process assumes policy-making is a process leading to a "known and intended outcome: it is a collective attempt to construct a policy in order to address *some evident problem*" (emphasis added).[20] The policy cycle implies that government reacts to evident—clearly seen, obvious, or understood—problems, which cause tangible harm to human life, property, or economy.

But what about those problems that do not cause immediate or "evident" harm but nonetheless threaten significant future impacts? Anticipation or "the act of looking forward" deals with events, situations, changes, and other phenomena that have yet to occur.[21] The concept of anticipation has received little explicit consideration within the public policy literature. Where policy scholars have considered anticipation it has been presented as little more than contextual sidelight or an outgrowth of particularistic problems, like resource planning,[22] technology research and development,[23] public health,[24] biotechnology,[25] and environmental issues.[26] A handful of studies have also explored the political dynamics of anticipation during the policy design and adoption process, including policymaking in anticipation of future constraints (e.g., budgetary short-falls, swings in public opinion, changes in the composition of government),[27] the capacity of state legislatures to preempt court decisions,[28] as well as the inherent challenges associated with drawing lessons from problems that have yet to occur.[29] Yet these studies rarely devote more than a paragraph to topic of anticipation.

Broadly speaking, this book focuses on the anticipation of risks and hazards. A hazard is a source of potential damage or harm.[30] Risk refers to the likelihood of harm that is associated with some physical phenomenon, human activity, or technology.[31] *Risk* is thus the likelihood of incurring harm if exposed to a *hazard*. All of us encounter an innumerable number of private risks in our personal lives. Poor diet, smoking, even reckless driving are but a sampling of the multitude of private risks encountered by countless Americans everyday. However, the risks examined in this book threaten to produce significant and aggregated harms across a broad cross-section of society. These risks can have disastrous, if not catastrophic consequences, for *many* people, causing a significant loss of human life and economy in a relatively short period of time. In turn, it is generally assumed that government is responsible for protecting citizens against these types of risks.[32]

The management and governance of large scale hazards and risk involves an array of policy areas, including public health policy, natural disaster policy, environmental policy, counterterrorism policy, and science and technology policy. Broad differences aside, these fields tend to break the process of hazard and disaster management into a series of distinct but overlapping stages. Even a cursory review of these risk management and preparedness "cycles" reveals many risks span entire time horizons, the past, the present, and, most important for this work, the future. Risk management cycles devote entire stages to preventing, mitigating, and preparing for emerging problems before they become full blown disasters.[33] Some fields, most notably public health, advocate for the sustained surveillance and identification of both routine hazards and unknown threats, like novel diseases.[34] All of these approaches build from the assumption that anticipation and proactive governance is absolutely critical to the successful management of crises, regardless of their source.

Of course, even the best-laid plans cannot prevent all hazards. Once disaster strikes, response strategies are activated to help manage crises as they occur, a dynamic process that requires a blend of reflexivity, resilience, and individual leadership.[35] And in the wake of disaster, a sustained period of recovery and relief is needed to help impacted communities and regions return to normalcy.[36] Ideally, the recovery process will include a period of reflection and "lesson learning" in order to inform future preparedness, prevention, and mitigation efforts.[37]

Yet even in the most proactive of policy domains, a "reactive bias" still exists. Disaster relief spending dwarfs preparedness spending.[38] Healy and Malhotra note that between 1985 and 2004, the U.S. government spent an annual average of $3.05 billion on disaster relief and a mere $195 million on disaster preparedness.[39] Sainz-Santamaria and Anderson estimate that between 1985 and 2008 the federal government spent more than $82 billion on relief programs but only $7.5 billion on preparedness.[40] Similar disparities have been noted in non-disaster fields. Faust estimates medical treatment spending outpaces preventative medicine spending by a 1:99 cents ratio.[41]

This gulf in spending has been attributed to an array of political, institutional, and cultural factors. At its most basic level, relief spending addresses highly salient problems with an immediate and very tangible impact on voters, whereas preparedness spending is far less visible and often addresses problems that have yet to occur, let alone those with an immediate impact on the electorate.[42] Activities like rebuilding homes, repairing roadways, and revitalizing economies are probably the *highest priority* items in disaster stricken communities. In fact, studies have shown political candidates who support relief spending receive a much higher vote share than those who support preparedness spending, a pattern that only reinforces the so-called reactive bias.[43] And because elected officials are expected to deliver results every four to six years, long term investments in planning and preparedness for problems that might take months or even years to fully manifest (if at all) is rarely seen as a politically expedient strategy.[44] Others have observed that many elected officials are philosophically reluctant to commit to long-term investments and protective measures against "low probability, high consequence" events, a phenomenon aptly named the "commitment conundrum."[45]

This bias might be institutionally induced as well. Birkland shows constitutional arrangements have limited the national government's capacity to engage in proactive policymaking, a testament to the fact that preparedness and disaster response policy is technically a responsibility of the states. National policymakers cannot mandate that states engage in preparedness planning or mitigation activities and are thus limited to offering "inducements."[46] This same arrangement also exists in the public health arena, where the federal government is reduced to a supportive role in emergency response.[47]

Despite these disincentives, policymakers—even at the federal level—can and often do engage in anticipatory policymaking. Activities, like preparedness, resilience, planning, and mitigation simply cannot be divorced from the political process, as evidenced in the Y2K problem, the aforementioned examples, and the three case studies examined in this book. What is more, anticipatory measures are cost effective and can reduce future damage, further incentives for proactive government.[48] Rose et al. show that $1 in preparedness spending offsets roughly $4 in relief spending.[49] Healy and Malhotra offered an even more extreme estimate, noting that every $1 spent on preparedness results in roughly $7 in immediate savings but close to $15 in future damage reductions.[50] Moreover, anticipatory action is critical to establishing well-prepared and resilient communities, capable of both absorbing and recovering from a range of natural and man-made threats.[51]

Government is unquestionably predisposed to reactive policymaking. But the notion that government is solely reactive is a misnomer. As noted above, anticipation factors prominently in various stages of the risk management cycle. Nor is anticipatory policymaking purely a technical exercise. In fact, Gerber and Neeley show that the same citizens who reward elected officials for relief spending can demand anticipatory policy, especially when they feel threatened and have sufficient information about actual risks.[52]

Relatively little is known about how and under what conditions policymakers, especially elected officials, create policy in anticipation of potentially hazardous and even catastrophic events. This book establishes a distinctive "type" of policy problem called "anticipatory." Unlike many disasters, which tend to crop-up with very little forewarning, anticipatory problems are foreseeable and can be identified months if not years in advance. This distinctive feature in turn allows policymakers to proactively fashion a policy response. Yet despite the fact that they provide ample forewarning, anticipatory problems, like all risks and hazards, are marked by tremendous uncertainty and it is impossible to conclusively determine how, when, if, and/or to what extent they will occur. In some instances, these problems are so novel that policymakers struggle to even determine their signs and symptoms. However, because a failure to act can have dire consequences, there exists a tremendous onus on government to at least entertain the development of policy—even in the face of these uncertainties. This book thus explores policy change that *precedes* the onset and culmination of a hazardous event.

To illustrate this pattern of policymaking, I will review three case studies. In each case, policy change is pursued in anticipation of a potential or likely problem. My investigation begins with nanotechnology. Nanotechnology, which refers to the manipulation of matter at a subatomic level, has been heralded as arguably the most revolutionary technology of the twenty-first century. Nanotechnology is a broad designation that denotes a wide array of technologies in computing, energy production and transmission, health care, agriculture, national security, materials, physics, and other fields. It promises to provide mankind with a plethora of new and improved products, such as improved drug delivery systems, stronger building materials, longer lasting batteries, and lighter but more protective athletic equipment, to name a few. Yet the hopes of scientists and policymakers alike are tempered by concern that this potentially revolutionary technology may also pose unforeseen and unexpected risks. At present, there is little consensus as to what those risks may be, but recent studies imply nanotechnology might have deleterious effects on human health and the environment. Therefore, a concerted movement is afoot to identify and understand these dangers *before* they have even occurred and to develop regulatory regimes capable of mitigating them.

My second example of anticipatory policymaking will be the H5N1 avian influenza case. H5N1 avian influenza, commonly referred to as "bird flu," was, as implied by its name, an avian-borne influenza virus that first emerged in 1997 in fowl populations in the Guangdong Province, China. Less than a year after the Guangdong outbreak, H5N1 influenza infected and sickened 18 Chinese citizens. Six cases were fatal. At least within the international public health community, the presence of a potentially lethal virus capable of infecting both human and animal populations represented a viable public policy problem, and indeed one worthy of government attention and action. However, from a public policy perspective, and from the perspective of American policymakers, the true source of anxiety—the "real" problem at hand—was not simply the sporadic infection of citizens, but the possibility this virus might undergo a mutation making it capable of airborne transmissibility. This would mean H5N1 could be spread via a sneeze or cough, like any other seasonal influenza virus. Such a mutation, it was feared, would spark a global pandemic infecting and perhaps even killing millions of people worldwide. These extraordinarily dire predictions reflected the fact that H5N1 was essentially a novel virus for which humans had no natural immunity. In the face of these concerns, federal policymakers enacted sweeping preparedness plans, allocating billions of dollars toward readying the nation for pandemic H5N1 influenza.

The final case in this book will explore the policy dynamics associated with climate change. As early as 1904, when Swedish scientist Svante Arrhenius suggested that minor changes in atmospheric temperature might be caused by Western industries, scientists have raised concerns about the potential implications of climate change. Beginning in the 1980s, a political movement crystallized aiming at both increasing awareness of climate change and advocating for substantive government regulation of greenhouse gases, the alleged cause of the rise in the Earth's temperature. Since then, the global warming debate has become a mainstay on the American political landscape, with both major parties staking out positions on this controversial policy issue. This study focuses on both climate change mitigation and adaptation efforts. Whereas mitigation policies work to prevent further warming, particularly by reducing greenhouse gas emissions, adaptation concedes that certain warming effects are inevitable (e.g., extreme weather, rising tides, heat waves) and therefore works to reduce existing vulnerabilities.

The organization of my book is as follows. In Chapter 2, a framework for studying anticipatory problems and policymaking is presented. This book melds two streams of public policy literature. The first stream encompasses the literature on policy typologies, a conceptual device used to classify and distinguish different categories of public policies. Policy typologies are founded on the assumption that policy issues can be qualitatively different from one another and engender distinctive patterns of political development and resolution. Most importantly, a number of these studies have examined the concept of anticipation. Drawing from these seminal works, this section of my literature review closes by presenting a definition of the anticipatory policy type. The second stream examines the literature on policy change, which describes the various institutional and social forces that induce—or prevent—policymaking. This book conceptualizes policy change as a multifaceted process, beginning with the identification and definition of social problems through to the implementation of substantive government outputs. Although the vast literature on policy change has readily explored policymaking in the wake of disaster, far less is known about when and under what conditions policy change can *precede* the onset of a problematic event.

With this conceptual structure in place, Chapters 3, 4, and 5 cover the three cases of nanotechnology, avian influenza, and climate change. All of the cases represent classic examples of anticipatory problems in that they examine instances wherein the underlying source of concern—the underlying policy "problem"—has yet to occur. Thus, in each instance, policymakers are compelled to proactively fashion legislation in order to stave off or at least prepare for a future threat.

Chapter 6 closes by highlighting the distinctive features of the anticipatory policy type, paying particularly close attention to its implications for contemporary theories of policy change. To what extent does anticipatory policymaking differ from reactive policymaking? Does the fact that problems have yet to occur create special policymaking patterns throughout the process of policy change? Does temporality represent a distinguishing characteristic, specifically with regard policy problems and policymaking in general? What does anticipatory policymaking reveal about extant models of policy change? These are the kinds of overarching questions that will be formally addressed based on my case studies.

Contrary to popular belief and scholarly assumption, policymaking can and often is a proactive process. Environmental threats, new technologies, and emerging diseases are but a sampling of the countless public problems that we can anticipate. Scores of contemporary society's most pressing problems dictate a proactive and anticipatory policymaking process. With this in mind, the following study represents a modest effort to illustrate the distinctive political and policymaking dynamics associated with anticipation.

## Notes

1 John Simons. 1999. "Clinton's Y2K Czar Won't Have Time for Champagne." *Wall Street Journal*, December 30. Available at www.wsj.com/articles/ SB94651122974624 6519 (accessed March 27, 2015).
2 Robert Sam Anson. 1999. "The Y2K Nightmare." *Vanity Fair* 80.
3 Some programmers argue Y2K was not a programming error. They purport large computer companies required programmers use a two-digit coding scheme, which saved storage and ultimately money.
4 Simons 1999; Anson 1999; Michael S. Hyatt. 1999. *The Y2K Personal Survival Guide: Everything You Need to Know to Get From This Side of the Crisis to the Other*. Washington, D.C.: Regnery Publishing.
5 General Accounting Office. 2000. *Year 2000 Computing Challenge: Lessons Learned Can Be Applied to Other Management Challenges*. Washington, D.C. Available at www.gao.gov/assets/240/230628.pdf (accessed March 27, 2015).
6 Ibid.
7 "Clinton Urges Americans to Act on Y2K Problem." 1998. CNN, July 14. Available at www.cnn.com/ALLPOLITICS/1998/07/14/clinton.y2k/ (accessed March 27, 2015).
8 Farhad Manjoo. 2009. "Apocalypse Then." *Slate*, November 11. Available at www.slate.com/articles/technology/technology/features/2009/apocalypse_then/ was_y2k_a_waste.html. (accessed March 27, 2015); On the other hand, a number of detractors argued other countries spent considerably less than the U.S. and did not encounter any problems.
9 Manjoo 2009.
10 Ibid.
11 Donald F. Kettl. 2007. *System Under Stress: Homeland Security and American Politics, 2nd Edition*. Washington, D.C.: CQ Press.
12 Ibid., 8.
13 Peter J. May, Ashley E. Jochim, and Joshua Sapotichne. 2011. "Constructing Homeland Security: An Anemic Policy Regime." *Policy Studies Journal* 39(2): 285–307.
14 Bart Elias. 2008. "National Aviation Security Policy, Strategy, and Mode-Specific Plans: Background and Considerations for Congress." Congressional Research Service, January 2: 21. Available at www.au.af.mil/au/awc/awcgate/ crs/rl34302.pdf. (accessed March 27, 2015).
15 Ibid., 21.
16 Chris Sullivan. 2008. "Making Data Breach Prevention a Matter of Policy." *SC Magazine*, July 9. Available at www.scmagazine.com/making-data-breach-prevention-a-matter-of-policy/article/112213/ (accessed March 27, 2015).
17 Andrew Zajac. 2010. "FDA Urged to be Proactive, Not Reactive, in Preventing Food Safety Problems." *Los Angeles Times*, June 8. Available at http:// articles.latimes.com/2010/jun/08/nation/la-na-fda-20100609 (accessed March 27, 2015).

18 Anticipation is not entirely foreign to the social sciences and a number of fields have considered the difficulties associated with anticipating emerging events. Notable examples include organization studies and the work on High Reliability Organizations (HROs); international relations; decision theory, especially game theoretic and rational choice models; disaster management; and science, technology, and society studies. It has received far less attention, however, within the subfield of public policy.

19 Hal K. Colebach. 2006. "What Work Makes Policy?" *Policy Sciences* 39(4): 309–321, 309.

20 Ibid., 311.

21 Cynthia Selin. 2010. "Anticipation." In *Encyclopedia of Nanoscience and Society*, ed. David H. Guston. Thousand Oaks: Sage. Available at http://know ledge.sagepub.com/view/nanoscience/n10.xml (accessed March 27, 2015).

22 Gila Menaham. 1998. "Policy Paradigms, Policy Networks and Water Policy in Israel." *Journal of Public Policy* 18(3): 283–310.

23 Henry W. Lambright and Jane A. Heckley. 1985. "Policymaking for Emerging Technology: The Case of Earthquake Prediction." *Policy Sciences* 18(3): 227–240.

24 Christopher H. Foreman, Jr. 1994. *Plagues, Products, & Politics: Emergent Public Health Hazards and National Policymaking*. Washington, D.C: The Brookings Institution Press.

25 Richard P. Hiskes. 1988. "Emergent Risks and Convergent Interests: Democratic Policymaking for Biotechnology." *Policy Studies Journal* 17(1): 73–82; Noah Zerbe. 2007. "Risking Regulation, Regulating Risk: Lessons from the Transatlantic Biotech Dispute." *Review of Policy Research* 24(5): 407–423.

26 Renee J. Johnson and Michael J. Scicchitano. 2000. "Uncertainty, Risk, Trust, and Information: Public Perceptions of Environmental Issues and Willingness to Take Action." *Policy Studies Journal* 28(3): 633–647.

27 John W. Kingdon. 2003. *Agendas, Alternatives, and Public Policies, 2nd Ed.* New York: Addison-Wesley Educational Publishers.

28 Laura Langer and Paul Brace. 2005. "The Preemptive Power of State Supreme Courts: Adoption of Abortion and Death Penalty Legislation." *Policy Studies Journal* 33(3): 317–340.

29 Richard Rose. 1993. *Lesson-Drawing in Public Policy: A Guide to Learning Across Time and Space*. Chatham: Chatham House Publishers.

30 Arjen Boin, Louise K. Comfort, and Chris C. Demchak. 2010. "The Rise of Resilience." In *Designing Resilience: Preparing for Extreme Events*, eds. Louise K. Comfort, Arjen Boin, and Chris C. Demchak. Pittsburgh: University of Pittsburgh Press: 1–12.

31 Peter J. May and Chris Koski. 2013. "Addressing Public Risks: Extreme Events and Critical Infrastructures." *Review of Policy Research* 30(2): 139–159, 139.

32 Peter W. Huber. 1986. "The Bhopalization of American Tort Law." *Hazards: Technology and Fairness*. Washington, D.C.: National Academies Press: 89–110, 90. See also: Peter J. May. 1991. "Addressing Public Risks: Federal Earthquake Policy Design." *Journal of Policy Analysis and Management* 10(2): 263–285; May and Koski 2013.

33 Boin, Comfort, and Demchak 2010; Charles F. Parker and Eric K. Stern. 2005. "Bolt From the Blue or Avoidable Failure? Revisiting September 11 and the Origins of Strategic Surprise." *Foreign Policy Analysis* 10(3): 301–331.

34 Rob A. DeLeo. 2014. "Centers for Disease Control and Prevention: Anticipatory Action in the Face of Uncertainty (1946–Present)." In *Guide to U.S. Health and Health Care Policy*, ed. Thomas R. Oliver. Washington, D.C.: CQ Press: 51–64; Foreman 1994; Elizabeth W. Etheridge. 1992. *Sentinel for Health: A History of the Centers for Disease Control*. Berkeley: University of California Press.

35 Boin, Comfort, and Demchak 2010; Ashley D. Ross. 2014. *Local Disaster Resilience: Administrative and Political Perspectives.* New York: Routledge; Arjen Boin, Paul 't Hart, Eric Stern, and Bengt Sundelius. 2005. *The Politics of Crisis Management: Public Leadership under Pressure.* Cambridge: Cambridge University Press.

36 Jamie Sainz-Snatamaria and Sarah E. Anderson. 2013. "The Electoral Politics of Disaster Preparedness." *Risks, Hazards & Crisis in Public Policy* 4(4): 234–249; Boin, Comfort, and Demchak 2010; Andrew Healy and Neil Malhotra. 2009. "Myopic Voters and Natural Disaster Policy." *American Political Science Review* 103(3): 387–406.

37 Thomas A. Birkland. 2006. *Lessons of Disaster: Policy Change After Catastrophic Events.* Washington, D.C.: Georgetown University Press.

38 George Haddow, Jane Bullock, and Damon P. Coppola. 2007. *Introduction to Emergency Management, 3rd Edition.* Burlington: Butterworth-Heinemann; Dennis Mileti. 1999. *Disasters by Design: A Reassessment of Natural Hazards in the United States.* Washington, D.C.: Joseph Henry Press.

39 Healy and Malhotra 2009.

40 Sainz-Santamaria and Anderson 2013.

41 Halley S. Faust. "Prevention vs. Cure: Which Takes Precedence?" 2005. *Medscape Public Health & Prevention,* 3 (May).

42 Howard Kunreuther. 2006. "Has the Time Come for Comprehensive Natural Disaster Insurance?" In *On Risk and Disaster: Lessons from Hurricane Katrina,* eds. Ronald J. Daniels, Donald F. Kettle, and Howard Kunreuther. Philadelphia: University of Pennsylvania Press: 175–202; Healy and Malhotra 2009; Thomas A. Birkland. 1997. *After Disaster: Agenda Setting, Public Policy, and Focusing Events.* Washington, D.C.: Georgetown University Press.

43 Healy and Malhotra 2009.

44 Kunreuther 2006.

45 May and Koski 2013; Philip R. Berke and Timothy Beatley. 1992. *Planning for Earthquakes: Risks, Politics, and Policy.* Baltimore: The Johns Hopkins University Press; Raymond Burby and Peter J. May. 1998. "Intergovernmental Environmental Planning: Addressing the Commitment Conundrum. *Journal of Environmental Planning and Management* 41(1): 95–110. Peter J. May and Thomas A. Birkland. 1994. "Earthquake Risk Reduction: An Examination of Local Regulatory Efforts." *Environmental Management* 18(6): 923–937; Sammy Zahran, Samuel D. Brody, Arnold Vedlitz, Himanshu Grover, and Caitlyn Miller. 2008. "Vulnerability and Capacity: Explaining Local Commitment to Climate-Change Policy." *Government and Policy* 26(3): 544–562.

46 Thomas A. Birkland. 2010. "Federal Disaster Policy: Learning, Priorities, and Prospects for Resilience." In *Designing Resilience: Preparing for Extreme Events,* eds. Louise K. Comfort, Arjen Boin, and Chris C. Demchak. Pittsburgh: University of Pittsburgh Press: 106–128.

47 Rebecca Katz. 2011. *Essentials of Public Health Preparedness.* Sudbury: Jones & Bartlett Learning.

48 Conversely, Sainz-Santamaria and Anderson (2013) show election concerns can undermine this efficiency. Politicians often invest preparedness money in contested electoral districts as opposed to "high risk" areas. Politicians calculatedly allocate preparedness money in home districts, not disaster prone areas, in order to shore up constituent support.

49 Adam Rose, Keith Porter, Nicole Dash, Jawhar Bouabid, Charles Huyck, John Whitehead, Douglass Shaw, Ronald Eguchi, Craig Taylor, Thomas McLane, L. Thomas Tobin, Philip T. Ganderton, David Godschalk, Anna S. Kiremidijian, Kathleen Tierney, and Carol Taylor West. 2007. "Benefit-Cost Analysis of FEMA Hazard Mitigation Grants." *Natural Hazards Review* 8(4): 97–111. See also: David Godschalk, Adam Rose, Elliot Mittler, Keith Porter, and Carol

Taylor West. 2009. "Estimating the Value of Foresight: Aggregate Analysis of Natural Hazard Mitigation Benefits and Costs." *Journal of Environmental Planning and Management* 52(6): 739–756.

50 Healy and Malhotra 2009; Immediate savings refers to savings over a four-year period or election cycle.

51 However, Gerber argues the federal government magnifies disaster losses by promoting mitigation practices in hazard-prone areas, instead of asking citizens to avoid settling and building in high risk areas. See: Brian J. Gerber. 2007. "Disaster Management in the United States: Examining Key Political and Policy Challenges." *Policy Studies Journal* 35(2): 227–238; See also: Kathleen Tierney. 2014. *The Social Roots of Risk: Producing Disasters, Promoting Resilience.* Redwood City: Stanford University Press.

52 Brian J. Gerber and Grant W. Neeley. 2005. "Perceived Risk and Citizen Preferences for Governmental Management of Routine Hazards." *Policy Studies Journal* 33(3): 395–418.

# 2   Toward an Anticipatory Policy Type

The truth is hard to face. For decades, we have acted as if American primacy was the natural order of things rather than a legacy built on the vision and the sacrifices of our predecessors. We have been encouraged to think of ourselves as fortune's favored children, and the sad consequences of that are all too apparent. We must now learn to govern ourselves more intelligently. The first step is to accept that, in a complex universe, the only true constants are surprise and change. Success goes to those who anticipate.[1]

—Leon Fuerth, Research Professor of International Affairs at George Washington University and former National Security Advisor to Vice President Albert Gore

The wisdom of Fuerth notwithstanding, there are considerable institutional, political, and cultural barriers to anticipatory policymaking. From electoral concerns to policymaker reluctance to make long-term investments, it is no surprise that spending on *reactive* policies, like disaster recovery, far outpaces spending on *proactive* measures, like preparedness and planning. Yet, as evidenced in the Y2K and other examples described in Chapter 1, government can—and in fact does—produce polices in anticipation of emerging problems. But what are the factors and forces that catalyze certain anticipatory problems onto the crowded governmental agenda, allowing them to overcome the previously described reactive bias? Why does government pursue anticipatory policy in some cases but not others? To what extent does anticipatory policymaking differ from the traditional pattern of reactive policymaking associated with most disaster domains? These questions, as well as others, will be investigated in the pages ahead.

This is certainly not the first study to examine government and its capacity to anticipate. Fuerth, for example, has devoted most of his academic career to developing institutional mechanisms that promote "forward engagement," a term which broadly denotes the ability of government to identify, address, and adapt to emerging threats.[2] Famed futurist Alvin Toffler once described "anticipatory democracy" as a logical response to the "decisional overload" brought about by the political technologies of the industrial era. Anticipatory governance, he argued, requires a blend of foresight and sustained citizen participation in politics and government.[3] And, as described below, anticipation continues to factor in many of the most important theories of political decision making and, more specifically, organizational behavior.[4]

However, as already noted, the concept of anticipation has received surprisingly little attention to date from public policy scholars. A proper remedy of this problem must begin by delving into two distinct strands of public policy literature focusing on policy typologies and policy change. Policy typologies assume that certain public issues entail features that make them objectively discernible from other issues. Such features refer not only to the descriptive characteristics of the issue, but also the very nature or pattern of political conflict that arises during decision making. Because they categorize policies and issues based on their core features and characteristics, existing typologies offer the conceptual foundation for the addition of a new anticipatory category.

The literature on policy change describes the institutional and social factors that provoke public action on new issues or prompt policymakers to revisit and alter existing policy approaches. Certain scholars have become increasingly concerned with the management and governance of hazards, risks, and disasters. Much of this work scrutinizes the enactment of public policies *after* disaster or in the wake of large-scale "focusing" events.[5] Indeed, a number of studies have even highlighted the difficulties associated with planning for disasters or "public risks," which threaten significant and aggregated harms, and can emerge at a moment's notice.[6] Anticipatory problems also threaten serious harms to a large cross-section of society, but, unlike public risks and many other natural disasters, they provide a degree of forewarning and time to prepare. This study draws on the existing literature on policy change to consider where it is and is not relevant for understanding anticipation as a political and governmental process.

## Policy Typologies and the Anticipation of Risk

The notion of a type of policy we can usefully designate as "anticipatory" suggests that certain problems and their policymaking styles differ profoundly from other types of problems and policymaking styles, and that such differences matter. In making this claim, the present study draws from the tradition of devising policy typologies as analytical schemas that classify public policies into different categories of action. These categories or "policy types" correspond not only to distinctive policy outputs and goals, but also identify different patterns of interest group participation and mobilization, degrees of public salience, levels of conflict, and institutional locations.[7]

By reducing the seemingly infinite number of policies adopted by government to a handful of categories, policy typologies hugely simplify and clarify the policymaking process, bringing attention to the institutions, actors, and interests most relevant to a policy debate. When accurate in their causal assumptions and empirical observations, they can also have predictive power with regard to political behavior and policy outcomes.[8] The literature on policy typologies stems from the seminal works of Lowi in the 1960s and 1970s.[9] Lowi's typology was founded on two assumptions. First, Lowi argued that policy causes politics, meaning that different types of policy produce different types of political conflict. Second, Lowi assumed that coercive power, or the ability to force individuals and groups into certain activities, is the most important characteristic of government. He constructs a two-dimensional typology where one dimension represents the likelihood of government applying its coercive powers (ranging from immediate to remote) while the other represents the target of coercion (the individual or the environment of conduct).[10]

Lowi established four policy types—distributive, redistributive, regulatory, and constituent. Distributive policy focuses on the distribution of new resources. In these cases where everyone is said to win, the likelihood of coercion is remote, and any that occurs is directly applied to individuals. Distributive policy is characterized by a political process with very little visibility, little political conflict, and with decision-making authority resting chiefly in the hands of entrenched political actors ("iron triangles"). Redistributive policy deliberately moves resources from one group to another. Redistributive policies are difficult to enact because they are marked by high public visibility and high levels of political conflict, largely due to the reality of clear winners and losers. Regulatory policy controls certain private activities. Regulatory policies have low visibility and conflict is generally isolated to industry groups. Additionally, the pressure put on government tends to be very organized given the elite status of the groups involved. Finally, constituent policies "set up" new, or reorganize existing, institutions. Constituent policies tend to be noticeably "top-down" and dominated by elected officials and administrative agencies.[11]

Policy typologies are often heralded as one of the most important analytical constructs in all of political science.[12] Critics counter that typologies and their associated policy types are vague, overlapping, and tell us little about the actual implications of specific policy proposals.[13] Detractors aside, scores of studies have built on Lowi's original typology, adding new categories and expanding the dimensions of his policy types.[14] Others have applied Lowi's typology to the "stages model" of public policy, rendering predictions about the nature of political mobilization during the agenda setting, policy formulation, and implementation stages.[15] Rival typologies have introduced totally new policy types, including substantive and procedural policies;[16] public and private goods;[17] and principle, lumpy, cuts/redistribution, and increase issues to name a few.[18]

*Anticipation and Typologies*

The concept of anticipation has been integrated into a number of typological schemes. Richardson, Gustafsson, and Jordan categorized national policy styles based on two features: (1) the government's approach to problem solving, which ranges from anticipatory to reactive; and (2) the government's relationship to other actors involved in the policymaking process, which ranges from consensus seeking to impositional. Based on these dimensions, they offer four types of policy styles: (1) anticipatory-consensual; (2) reactive-consensual; (3) reactive-impositional; and (4) anticipatory-impositional. In the case of the anticipatory-consensual style, the government tends to conduct a great deal of policy planning and illicit considerable input from policy communities. The reactive-consensual style also strives for consensus, but instead of proactive planning the government will incrementally react to existing problems. Incrementalism and reactive governance are also trademarks of the reactive-impositional style. However, in this case, the government will impose its decisions, as opposed to seeking consensus. Finally, with regard to the anticipatory-impositional style, government will plan without seeking consensus from others.[19] Richardson, Gustafsson, and Jordan's typology has been applied to a number of Western European countries,[20] Canada,[21] Japan,[22] and the state of California.[23] Research has also questioned whether increased economic integration, international legal norms, and the transfer of technological knowledge have diminished these differences in national policy styles, a trend known as "policy convergence."[24]

The distinction between anticipatory and reactive policymaking is prominent in the interdisciplinary study of risks, hazards, and disasters as well. Much of this work stems from Wildavsky's *Searching for Safety*, which introduced two rival patterns for managing uncertainty.[25] In the first pattern, policymakers try to anticipate possible risks in order to avoid or eliminate them in advance. Should these risks occur, the anticipatory pattern calls for damage control, meaning decision makers should minimize any negative effects as much as possible. Because some problems are highly uncertain and volatile, Wildavsky offers a second pattern or the "resilient" approach. Resilience refers to the "capacity to cope with unanticipated dangers after they have come manifest, learning to bounce back."[26] Resilience strives for adaptability and flexibility in the face of extreme events. Wildavsky advocates for a blend of resilience and anticipation, with policymakers moving from one end of the spectrum to the other depending on the degree of predictability and amount of knowledge associated with each problem. He concludes that, although anticipation is at times preferable, society should strive for resilience, as many problems are not preventable.[27]

To be sure, Wildavsky describes two dominant management approaches, not theoretically derived policy types.[28] However, he demonstrates that the practice of anticipation—and especially with respect to risks and hazards—is different from many other governmental activities and warrants special analysis. Wildavsky's work inspired a bevy of risk research. "High Reliability Organizations" (HROs) theory, for example, presents a dynamic model of anticipation and resilience in organizational settings. HROs are expected to perform at a very high level despite having to manage complex and "tightly coupled" systems—organizations where failure in one part of a system often results in failure across the entirety of the system. Examples include power plants, naval aircraft carriers, air traffic control centers, and hospital emergency rooms.[29] HRO theory stresses the importance of cultivating cultures that aggressively track small failures, clues, and other indicators implying an emerging problem. Like Wildavsky, HRO theory construes anticipation as the optimal, but not always attainable, management technique. In turn, these organizations must develop resilient structures that can rapidly respond to crisis and quickly return to normal operating protocols.[30]

There has also been uptick in interest in resilience in recent years. Boin, Comfort, and Demchak define resilience as "the capacity of a social system (e.g., an organization, city, or society) to proactively adapt to and recover from disturbances that are perceived within the system to fall outside the range of normal and expected disturbance."[31] Proponents of this approach argue that increased global interconnectedness coupled with the enormous uncertainty associated with many contemporary problems (e.g., terrorism, emerging technologies, and climate change) demand that government institutions and communities ready for events that are incredibly difficult to manage, let alone prevent. Resilience calls for long-term commitments to reducing vulnerabilities, strengthening physical infrastructures, cultivating responsive and flexible disaster management networks, and, most importantly, bolstering the capacity to learn and adapt both during and after disaster.[32] Strong social ties or "social capital" is another important determinant of resilience and regions with stronger community networks are usually better positioned to recover from disaster.[33]

Some have construed the increased interest in resilience as evidence of a waning interest in anticipation.[34] Although Wildavsky presented anticipation and resilience as distinct activities, the contemporary literature on resilience is, for the most part, committed to both ends.[35] Resilience spans virtually every stage in the disaster management cycle, including the identification and surveillance of emerging threats, development of preparedness plans, and even implementation of mitigation strategies. Resilience does not encourage managers to passively wait for the next disaster to emerge, but to acknowledge the fact that even foreseeable events can have unpredictable consequences. Comfort advocates for a dynamic interplay between anticipation and resilience wherein managers strive to achieve both goals as part of a comprehensive risk management portfolio.[36]

The notion of an anticipatory policy type is thus well grounded in existing scholarly literature. The present work, however, differs from the above-described studies in at least two respects. First, neither Wildavsky, nor Richardson, Gustafsson, and Jordan, considered the concept of anticipation in relation to policy change. Whereas Wildavsky's work broadly described two patterns for managing uncertainty, the policy styles concept uses the "nation state" as a unit of analysis.[37] Indeed, Parsons critiqued the work on policy styles arguing that the concept reveals very little about what governments actually do and, more importantly, "with what effects and outcome."[38] This book explicitly considers the intricacies of the policy-making process—it explicitly considers what governments actually do and how they do it.

Second, unlike Wildavsky, this book does not use the term "anticipatory" to denote a special managerial approach that is somehow different from resilience, preparedness, mitigation, prevention, or surveillance—concepts with special meanings in disaster management.[39] The *outcome* of the anticipatory policymaking process described in this book, can, as a point of fact, result in greater resilience, a higher level of preparedness, better surveillance, more robust mitigation efforts, and maybe even prevention. This book simply describes *how* policymaking occurs in the face of anticipatory problems.

## The Anticipatory Policy Problem

This is not the first study to focus on a specific type of policy issue, as opposed to developing a multidimensional typology. Schulman, for example, examined the policymaking pattern associated with "large-scale policies."[40] He argued that large-scale policies have distinguishing categorical properties, most notably their largeness of scale. Unlike many policies, which can be broken into separate parts or units, large-scale policies entail "the provision of *indivisible* payoffs by indivisible means."[41]

Large-scale policies require "go/no-go," "all or nothing" decisions, meaning they must be delivered in a large lump sum otherwise they cannot be provided at all. Examples include space exploration, the war on cancer, and the war on poverty. They require significant and long-term political commitments and therefore foreclose the possibility of incremental adjustments over time. Unlike most policies, which can eventually achieve a relative state of equilibrium, support, and acceptance, proponents must perpetually mobilize support—even after enactment—for large-scale policies because policymakers are reluctant to make long-term commitments.

Schulman's work is important to this study for two reasons. First, it shows the need, at times, to look past broadly comprehensive policy schemas in favor of a more detailed focus on a single category of policy action. Second, it spotlighted a policy type based not on theoretical criteria, but rather empirical properties—"largeness" of scale.[42] Writes Schulman:

> Perhaps no physical characteristic can as conclusively condition an organism's relationship with it environment as those pertaining to scale. The scale of an organism can directly affect its environmental demands and at the same time condition its response and defense

capacities. The relative scale of environmental object can, in turn, affect the behavior of the organism by triggering or eluding its attention, signaling danger or suggesting vulnerability.[43]

Size represents a far more practical designation than, for example, "regulatory," "distributive," and "redistributive," terms that ultimately derive from theoretical constructs. The framework in this book revolves around another kind of recognizable practical attribute, that of timeframe. Problems are distinguished by their timing or the fact that they are projected to occur in the future. Time is inherent to the moment-to-moment human condition—it is not a construct of policy theory.

Drawing from the above-described literatures, I can now offer a definition of anticipatory policy problems:

Anticipatory policy problems are risks that become known to policymakers, stakeholder interests, the media, and the public prior to the full emergence of a significantly hazardous, if not disastrous, culminating event. Anticipatory problems are marked by acute uncertainty regarding the foreshadowing of the problem, its perceived seriousness, and level and type of intervention required within the policy process. Further, the implementation stage of policy is characterized by special complexity due to the fact that these problems have yet to fully emerge and assume all the attributes by which they will eventually come to be known.

With this definition in mind, we can now turn to the literature on policy change, which provides an essential, if incomplete, framework for analyzing the challenge of anticipation and policymakers' response to it. In recent years, scholars added tremendously to our understanding of the interactions between political language and agenda setting, the process of policy design and selection of policy instruments, and the critical role of subsystems before, during, and after policy enactment. The task in this book, then, will be to draw on these insights while adapting them to the special circumstances of anticipation.

## The Dynamics of Policy Change

Agenda setting is the process through which issues are identified and selected for consideration by a decision making body—be it a legislature, bureaucratic agency, court, or chief executive. There are two types of agendas.[44] The systemic agenda consists of all issues currently garnering public interest and media attention. The institutional agenda refers to all issues that government decision makers are actively considering.[45]

Policy change broadly denotes the adoption of new laws or rendering of decisions by government. There are various types or "degrees" of policy change. Policy maintenance refers to minor alterations or adjustments that ensure a policy continues to meet its goals, thereby keeping things "on track."[46] Policy termination refers to a policy that is "abandoned, wound down, and public expenditure on that policy is cut."[47] Most research, however, focuses on policy innovation, which refers to government involvement in a new area.[48]

But policy change—and especially innovation—encompasses actions and processes much more complex than simply the passage of a law. Radical policy change triggers a paradigmatic shift within existing policy subsystems, which are more or less established areas of policy that establish boundaries for policymaking.[49] Because policymaking often includes interactions across multiple actors, institutions, and levels of government, scholars use subsystems as a unit of analysis. Conceptualizing policy change as subsystem change demands that researchers look beyond procedural end points, such as voting results or a single institution, to consider the myriad of institutional, social, and political challenges associated with the formulation, enactment, and implementation of policies across levels and layers of government.[50]

Subsystems are comprised of networks of actors. Often working through a handful of institutions, these actors fixate on a relatively narrow set of ideas or policy frames, which refer to the ways in which a problem is categorized, defined, understood, and measured. Frames further specify the menu of acceptable policy solutions, legitimizing certain proposals while banishing others. Innovation alters power dynamics across an entire policy domain, giving rise to new laws or decisions, policy frames, prevailing institutions, stakeholders, and, most importantly, a new class of dominant actors.[51]

This book relies primarily on two models of policy change, Kingdon's "streams metaphor" and Baumgartner and Jones' punctuated equilibrium model.[52] The streams metaphor portrays policy change as resulting from the convergence of problem, political, and policy streams. First, the problem stream refers to the various problems government works to solve as well as their attributes. Problems can reveal themselves through indicators (measures or metrics of a policy problem), focusing events (large scale disasters and other "attention grabbing" events), or feedback (information generated during the implementation of public policy). Policymakers and other stakeholders use problem definition to convince others that change in these empirical features are symbolic of a larger social or political problem. The political stream encompasses public opinion as well as the balance of political power within government (e.g., its partisan composition, the capacity for consensus building, and the organization of interest groups and other political forces). The policy stream includes specified solutions for a recognized problem. The policy stream integrates input from communities of experts, advocates, interest groups, elected officials, and bureaucrats who introduce and vet competing policy solutions. When the three streams converge, it can open a "window of opportunity" for policy change. Proponents of change must seize on this opportunity and "push" their pet solutions through the window.[53]

Kingdon's streams model has been criticized for focusing primarily on moments of change.[54] Baumgartner and Jones' punctuated equilibrium theory, by contrast, examines both periods of policy change (*punctuations*) and periods of stasis or stability (*equilibrium*). Punctuated equilibrium theory maintains that subsystems are drivers of policy change. During periods of stability, select subsystems or "policy monopolies" dominate the ways in which policy is framed as well as the range of acceptable solutions. Policy monopolies work within a carefully selected set of policy venues or institutional locations with decision-making authority (e.g., a committee, bureaucratic agency, or a specific court). Favorable venues work to enact policies that expand and further entrench the monopoly's dominance.[55]

Because policy monopolies resist new ideas, periods of stasis only accommodate minor and non-disruptive changes. Major change therefore requires the forceful introduction of new ideas, problem attributes, and issue frames. Like streams theory, punctuated equilibrium assumes that changes in the empirical features of a problem through external perturbations (focusing events), information (indicators), or feedback can heighten policymaker, public, and media interest in a given problem. Changes in the political environment (e.g., the composition of a legislature or shifts in public opinion) help facilitate change as well. Proponents of change must dislodge the issue from the policy venues currently supporting the existing policy monopoly and move it to a new venue that is receptive to different policy images. Punctuation—policy change—results from a shift in dominant policy frames as well as the incorporation of new policy venues. In the wake of these upheavals, the system resets and a new monopoly assumes the reins of power and eventually settles into a period a stasis.[56]

Both theories stress the importance of policy entrepreneurs or individuals who are willing to invest a great deal of personal resources in order to procure a future policy return. Beyond promoting pet solutions, policy entrepreneurs work to define problems and create an environment favorable to policy change.[57] In this respect, entrepreneurs play a crucial role in mobilizing action across larger policy subsystems. In the absence of these individuals, policy change is difficult.

Equally important, both theories build from the assumption that policy change is the product of political mobilization or the extent to which groups collectively organize for or against an issue. Schattschneider's classic model of group mobilization assumes the advantaged party will work to contain the scope of conflict to its original participants and institutions. Advantaged groups are, in some respects, similar to policy monopolies in that they dominate the problem definition process and enjoy the support of government institutions. As such, they tend to reject definitions and proposals that challenge their authority.[58]

Given this arrangement, disadvantaged groups work to expand the scope of conflict beyond its original participants by "going public." Going public requires these groups to offer new definitions challenging the status quo. This tactic allows disadvantaged groups to appeal to previously ambivalent or inactive stakeholders who can help tip the scales of power in their favor. Changes in venue are, again, inextricably linked to this process since opponents need to appeal to new institutions that will be receptive to their dissenting claims.[59]

Anticipatory problems raise important questions with respect to both theories. Most importantly, can policy change precede the onset of a culminating event? To what extent can political mobilization and the development of policy solutions take place when the problem is so speculative? What does it take for punctuation to occur in the absence of an existing problem? The remainder of this section investigates the various components of each of these theories in greater detail.

### Problem Identification: Focusing Events and Public Risks

Risks, disasters, and hazards challenge existing policy monopolies by introducing new problem attributes and altering dominant perceptions about an existing problem. Although some observers believe a fully developed theory of crisis policymaking does not exist, extant theory widely recognizes the capacity of disaster events to promote policy change.[60] Birkland's focusing event model of policy change provides a particularly comprehensive analysis. Birkland defines a potential focusing event as one that is:

> sudden, relatively rare, can be reasonably defined as harmful or revealing the possibility of future harms, inflicts harms or suggests potential harms that are or could be concentrated on a definable geographical area or community of interest, and that is known to policy makers and the public virtually simultaneously.[61]

His study considers "domains prone to disaster," or policy domains where change usually only occurs in the wake of disaster events.[62] Focusing events can catalyze problems simultaneously onto the public and institutional agendas, often allowing them to leapfrog other items.[63] The power of these events is derived not only from the objective features of the disaster itself, but also from *how* it is interpreted. A disaster must come to be seen as an important symbol of programmatic failure in order to be considered a focusing event. In some cases, policy change will only occur in the wake of multiple events, which collectively symbolize a problem.[64] On other occasions, events may be sufficiently large or "significant" to induce change across a variety of unrelated domains, although such occurrences are exceedingly rare.[65]

If focusing events direct our attention to policymaking after disaster, the concept of public risks describes how governments plan and prepare for disaster. Public risks are temporally remote and broadly distributed threats that fall outside the bounds of decision-maker control.[66] They include things like earthquakes, nuclear power plant accidents, oil spills, and hurricanes, all of which occur with very little forewarning.[67] Still, because many of these hazards occur with relative frequency, they necessitate that governments plan and prepare for the next—and often inevitable—event. May observed that public risk domains tend to lack an organized public, meaning interest groups do not readily mobilize for or against preparedness measures.[68] The groups and individuals that do participate are often well trained bureaucrats or experts, many of whom accumulate a great deal of specialized knowledge and can thus exert a considerable influence over the policymaking process.[69]

However, unlike public risks or even focusing events, anticipatory problems *do* provide forewarning—months if not years in some cases—thereby allowing for the fashioning of legislation *prior* to the full emergence of a culminating event. Does this distinct temporal dimension—the ability to "foresee" these problems—induce a different policymaking dynamic than in the case of focusing events or public risks? To what extent does advanced forewarning make policymakers more likely to involve themselves in the preparedness and planning process? How does this anticipatory dynamic alter mobilization patterns? What role, if any, can focusing events play in an anticipatory agenda setting process?

*Problem Identification: Indicators*

Indicators are more or less objective metrics of a problem. They include quantitative measures, like the number of cases of a disease, traffic accidents, and annual home sales. At times these measures alone are sufficient to capture policymaker attention. Other times, they are compiled in written reports outlining and interpreting the dimensions of a particular problem in a new or at least more concentrated way.[70] Indicators help describe the magnitude or severity of a problem as well as its rate of change. Of course, given policymakers' limited information processing capacity, indicators may represent little more than "noise," especially if they percolate in the media and public discourse for an extended period of time.[71] Therefore, it can take sudden changes (upticks or downticks) in indicators to capture policymaker attention. Abrupt change signifies a departure from the status quo. On occasion, indicators will not capture policymaker attention until and unless they breach some imaginary or even predetermined threshold of acceptability.

Quantitative indicators are afforded a tremendous amount of legitimacy in the political arena. Policymakers, the media, and the public often assume "countable" items are the most significant forms of data possible. Although the political world is saturated with different problem measures, policymakers tend to "lock-in" on one or a handful of trusted metrics, which come to represent trusted symbols of a problem.[72] The selection of appropriate indicators is therefore a highly contested process. Policy debates hinge not only on the relevance and accuracy of a particular indicator, but often the very methodologies used to derive these measures.[73]

Indicators allow policymakers to identify and, theoretically, prepare for emerging problems. Birkland alludes to this important distinction when he contrasts "domains prone to disaster," the topic of his book, and "domains not prone to disaster." In domains not prone to disaster, a focusing event is not required to produce policy change because "problems become known slowly, as *indicators* of problems accumulate and become more evident."[74] Birkland uses the example of the avian influenza, one of the three cases examined in this book, to illustrate this dynamic:

> In 2005, for example, the problem of the H5N1 strain of bird flu influenza gained worldwide attention, and its transmission to humans in Turkey and Europe in early 2006 has increased concern about pandemic flu, and in particular about the possibility of its transmission from person to person rather than from birds to people. But a global flu pandemic is a different kind of disaster from the type described in this book because it can be anticipated before the pandemic occurs.[75]

This accumulation of indicators is important because it theoretically allows policy change to *precede* the onset of the actual event. But is this dynamic representative of *all* areas of anticipatory policymaking? Are indicators alone enough to generate concern about emerging problems? When and under what conditions will policymakers begin to pay attention to these indicators?

## The Social Construction of Problems

Problem definition is the process through which policymakers, interest groups, advocates, and other stakeholders assign meaning to and interpret problems.[76] Problem definition weaves complex narratives, specifying a problem's scope, severity, novelty, cause, the characteristics of the population it impacts (e.g., favorable or unfavorable, sympathetic or unsympathetic, worthy of government attention or unworthy), and the extent to which the problem impinges on personal interests (its proximity). Problem definition also offers solutions to the identified problem. Problem definition is an important element of policy change, as new narratives are critical to upending monopolies.[77]

Hazards, disasters, and risks do not typically crack the institutional agenda unless they are defined as a crisis—a dire circumstance where immediate correction is needed.[78] Logically, focusing events represent an important symbol of crisis in that they allow problem definers to reference a dire event marked by aggregated physical harm, economic loss, destruction, and, in many cases, death. In the absence of such an event, descriptions of risk or impending crisis are bound to be "conjectural" and organized interests will need to strategically conjecture or speculate as to the likely implications of action or inaction.[79] Birkland introduced the concept of conjectural policymaking during his examination of policymaking in the wake of the Three Mile Island (TMI) nuclear meltdown. Unlike the other disasters considered in his book, TMI did not result in a catastrophic loss of life or economy. It was a near—albeit highly disconcerting—miss event. In turn, ensuing policy debates were marked by conjecture about the possible implications of a large-scale reactor meltdown, as opposed to an analysis of post-event recovery and fallout.[80]

Just as focusing events require interpretation, so too do indicators. Innes shows that indicators and other numerical measures are often packaged as part of larger policy "myths" which communicate narratives describing why—or why not—changes in a particular measure are cause for concern. Because policymakers are perpetually searching for ways to simplify the political process, myths offer "a useful shorthand for leaders to both define an issue and to justify public concern."[81] Myths are often more powerful than the indicators themselves in that they allow policymakers to forgo the rigors of empirical analysis and instead speak in subjective and politically strategic anecdotes. Equally important, the so-called "measurement discourse" associated with most indicators can give highly uncertain problems a tangible quality.[82] Stone writes:

> to count something is to assert that it is an identifiable entity with clear boundaries. No one could believe in a count of something that can't be identified, so to offer a count is to ask your audience to believe the thing is *countable*.[83]

As such, the interpretation of metrics can help substantiate claims of crisis, even in the absence of a "manifest" event.

The use of analogy, metaphor, and other forms of implied comparison is yet another important feature of the problem definition process. Drawing "lessons from the past" allows policymakers to fill present gaps in knowledge. Most analogical reasoning, of course, draws comparisons between two similar events. Analogies allow problem definers to make sweeping assertions about the dangers of inaction or action while accenting specific features of a problem (e.g., its scope, severity, cause, etc.). And, like all problem definition narratives, analogy brings attention to certain policy solutions while disregarding others.[84]

Although all issues entail some level of uncertainty, the uncertainty associated with anticipatory problems is different and indeed much more inherent than in many other cases. With respect to anticipatory problems uncertainty abounds as to when, how, and to what extent these problems will occur. In some instances, there is reason to believe, despite forewarning, they might not occur at all. How can problem definers assign meaning to problems that do not exist? Are there strategic advantages or disadvantages to asserting claims about the future? What sort of discursive strategies can be used to help minimize reluctance to address problems that might never come to fruition?

*Linking Ideas to Institutions*

Policy venues are institutional locations with decision-making authority.[85] Venues have distinct rules, constituencies, incentives, decision-making procedures, and cultures that shape access to their respective policy jurisdictions as well as the ways in which they process political issues.[86] Each venue will be more or less receptive to different ideas, definitions, and policy solutions.[87] For example, legislative committees with jurisdiction over criminal justice often approach substance abuse as a form of deviant behavior, whereas committees with jurisdiction over public health treat substance abuse as a mental or even physical disease. The policies adopted in these venues will likely be radically different—ranging from incarcerating addicts to providing medical care and rehabilitation services.[88]

Organized interests work hard to ensure their pet problems are reviewed in venues favorable to their particular perspective. Because policy monopolies often control certain venues, policy change can necessitate appealing to new, strategically chosen, venues. Venue shopping is the process through which organized interests seek out venues that will support their policy goals.[89] Access to new venues is conditioned by the norms, mores, rules, and customs of the given decision-making entity. Moreover, venues vary in terms of their policymaking capacity, or the "types" of policy outputs they can produce—be it a legal judgment, piece of legislation, or even a regulatory decision. Interest groups must strategically consider all of these factors when selecting venues.[90]

Anticipatory problems also raise important questions about the nature of the venue shopping process. Will the uncertainty of these problems interfere with the selection of a venue? On the other hand, to what extent will organized interest exploit this uncertainty to access new venues? What types of venues will be most receptive to addressing anticipatory problems?

*Policy Formulation*

The streams and punctuated equilibrium theories have been criticized for focusing primarily on predecision activities, like agenda setting, while ignoring the processes of policy formulation and implementation.[91] John offers an evolutionary theory of policy change that extends the streams and punctuated equilibrium theories into these "later" stages.[92] He argues that just as the agenda setting stage is driven by a competition among ideas and definitions, so too are the processes of formulation and implementation a competition of ideas, which he calls "policy memes." Memes discarded during the agenda setting stage "lurk in the background ready to launch a challenge" during the latter stages of the policy cycle.[93]

Evolutionary theory allows scholars to trace policy change across the *entirety* of the policy cycle, as opposed to artificially compartmentalizing each stage or, worse yet, disregarding formulation and implementation entirely. Policy formulation and adoption refer to the process of developing courses of action to deal with a given problem. Solutions are often developed and presented before a problem even enters the policy formulation stage and are packaged as part of larger policy narratives.[94]

Policy formulation involves an extended period of policy design wherein policymakers choose between various instruments for achieving policy goals. The selection of instruments is usually hotly contested, as the outcome of this design process will dictate the types of demands imposed on governmental and nongovernmental actors. Howlett and Ramesh describe four categories of policy instruments.[95] *Organizational instruments* use government, non-profit, or community organizations to provide public goods and services. Reorganization efforts, which aim to improve service delivery, are yet another organizational instrument. Finally, market creation sees organizations create voluntary interactions between consumers and producers. Market creation often involves "property rights auctions," which set a fixed quantity on the right to consume a certain product. This strategy creates a false scarcity and, in turn, creates an artificially engineered market. Those wishing to consume above the allotted amount must purchase these rights from government or other consumers. *Authoritative instruments* require private actors to perform certain functions. Command-and-control regulation demands industry and citizens comply with government regulation or face a penalty. In certain cases, government grants private actors the right to self-regulate. *Treasure instruments* use government resources to encourage or discourage certain actions. Whereas subsidies (e.g., grants, tax incentives, and loans) reward valuable activities, taxes and user charges punish unwanted behavior. Finally, *nodality and information instruments* use government's access to and control over information to alter certain behaviors. Public information campaigns passively provide citizens with information in hopes that this new knowledge will empower them to make better decisions. Suasion is much more calculated in that it not only disseminates information, but also tries to persuade individuals and groups. Governments also generate new and large quantities information in order "to alter the nature of the perceptions held by actors in policy subsystems so as to alter the nature of existing and future policy process."[96]

A fairly extensive arsenal of policy tools exists for regulating risks, ranging from direct regulatory instruments (e.g., mandatory vaccinations, quarantines, zoning regulations) to noncoercive provisions (e.g., public–private partnerships, information sharing, planning initiatives).[97] However, in the midst of crisis, policymakers tend to centralize their response, aggrandizing power in a narrow set of institutions and imposing much more coercive policy demands.[98] In fact, Ripley and Franklin argue subsystems "disappear" in the case of crisis policymaking.[99] How will the policies formulated and adopted to address anticipatory problems differ from reactive policies? What types of policy instruments are best suited for crafting proactive policies? Is the mere threat of crisis enough to induce a centralized response or will policymakers opt for an inclusive policymaking process?

*Policy Implementation*

In his seminal 1977 book, *The Implementation Game: What Happens After a Bill Becomes a Law*, Eugene Bardach conceptualizes the implementation process as a complex political "game" wherein bureaucrats and other administrators are required to navigate a myriad of potential challenges and impediments to the newly enacted law. Challenges can arise from both within government—perhaps even within the very agencies tasked with leading implementation efforts—as well as outside the government. Despite shifting our "analytical lens" to the administrative arena, implementation is far more than a mechanical process carried out by "headless" bureaucrats.[100] Not only does it involve public actors, like bureaucratic agencies or the courts, but also private actors, like industry groups, advocacy organizations, and professional groups.[101] Implementers operate in a potentially treacherous political environment that requires them to engage in their own version of policymaking (often through the drafting and enacting of rules and regulations), interact with a wide array of stakeholders, maneuver to maintain funding, and offer services to diverse client populations.

Policy implementation is rife with potential pitfalls. Legislation is often vague, providing only a basic framework for action. Bureaucrats are expected to devise "rules and regulations to translate general legislative language into detailed prescriptions."[102] In some cases, vague legislation provides an opportunity for the implementing agency to aggrandize power through its own interpretation of a new statute. However, the same situation can also impose huge pressures on implementing agencies and organized interests will try to convince these agencies to embrace an interpretation favoring their goals. Where legislation is not vague, it is often flawed and unworkable, a reflection of legislators' willingness to grapple with complex problems beyond their expertise or in areas where information is limited. In addition, agencies are often linked to certain interest groups and actors, on whom they rely as clients, consumers, and, in some instances, colleagues and collaborators. These mutually symbiotic relationships require the agency to remain highly sensitive to interest group concerns, imposing yet another constraint on the process of implementation. Finally, disunity among agency staff or communication difficulties can derail even the best-intentioned implementation efforts.[103]

Scholars have made a concerted effort to abolish the artificial silos demarcating the various stages of the policy process in lieu of a more encompassing depiction of policy change. This work strives to capture not only policy outcomes and agency performance, but also the ways in which interest groups use the implementation process to reshape power dynamics within existing subsystems.[104] According to John's evolutionary theory, the implementation stage is a useful setting for challenging recently enacted policy because it shifts power to new venues and actors, in this case bureaucrats. What is more, implementation failure can support opposition groups' claims that the adopted policy is inferior to their preferred solutions.[105]

Punctuated equilibrium and streams theories do not neglect the implementation stage completely. Both theories note that implementation produces "policy feedback," as agency officials and policymakers monitor the progress of existing programs. Feedback can include the systematic monitoring of programs, like the mandatory tracking of transit usage or student performance on standardized testing, as well as informal information, like constituent complaints about bureaucratic performance. Feedback expresses a number of messages, including narratives showing that bureaucrats are not conforming to legislative intent, a policy is failing to meet its intended goals, a program is too costly, or a policy is having negative and unintended consequences.[106]

In some instances, this type of information can prompt policymakers to refine or alter the policy, a process known as policy maintenance. In more extreme cases, negative feedback can result in a policy's termination. Of course, feedback can also communicate a "positive" image of an existing policy, further legitimizing an existing monopoly's power and perhaps resulting in program expansion (e.g., more funding or greater responsibility).[107]

Whereas many scholars assume breakdowns during implementation are the product of bureaucratic infighting or a failure to carry out prescribed tasks, Patashnik shows "reforms may crumble not because of anything bureaucrats do, but rather because of the actions (or inactions) of *elected officials themselves.*"[108] As noted before, politicians are often unwilling to make long-term commitments and most major reforms are destined to come under attack during the implementation stage by elected officials hoping to derail the unwanted measures. Reforms are more likely to survive when they cultivate a new clientele, reinforce market forces, generate positive feedback, and realign institutional authority. Successful implementation, in other words, requires wholesale domain reconfiguration. And, even if these important variables converge, getting a reform to "stick" still requires a bit of luck. A policy's "timing" can have enormous implications for its long-term success and seemingly exogenous events (e.g., economic downturns) can quickly overwhelm nascent reform efforts.

Anticipatory problems raise a number of important questions with respect to the implementation process. Policy actions in the area of anticipation tend to consist of three dimensions: 1. Documentation and detection of the emerging threat; 2. Prevention and mitigation of the occurrence of that threat; and 3. Preparedness and adaptation in regard to the consequences of such a threat realized. All three of these activities possess a high degree of ambiguity atypical of reactive policymaking. How is it that this quality interacts with the already existing tendency for policy to gain its true meaning and form through the process of operationalization? How will agencies gauge the adequacy of their implementation efforts in the absence of a clear problem against which they can test their interventions? Moreover, should an anticipatory problem fail to emerge for an extended period of time, what kind of difficulties will arise in terms of maintaining an adequate level of readiness? Can anticipation be an ongoing activity, or is it merely an end achieved through, and comfortably confined to, the implementation process?

## The Road Ahead

This book uses a qualitative case study design to analyze policy change within the pandemic, nanotechnology, and climate change domains. Each domain is examined over a 15- to 20-year period. Although the analysis here primarily relies on "thick description," a number of quantitative metrics of agenda activity are used to demonstrate shifts in policymaker and media attention across time. This book tracks the number of times each problem was mentioned in the *Congressional Record*, in congressional hearings, and in the *New York Times*. It also tracks funding appropriations and legislation, although these measures are typically integrated within my qualitative case analysis. These measures are widely employed in other studies as important indicators of issue attention and agenda activity.[109]

The *Congressional Record* is an official recording of the proceedings of the U.S. Congress and is therefore a good barometer of policymaker concern. Similarly, because much of Congress's work occurs in committee settings, hearing activity usually reflects congressional interest in a particular issue. Data pertaining to *Congressional Record* and hearing activity were derived from the ProQuest Congressional website, which provides a comprehensive record of all policymaking activity within the U.S. Congress.

Although the media do not constitute a formal policymaking institution, scholars have observed a general correlation between media coverage and policymaker attention on an issue.[110] Accordingly, this book records the extent to which each of the following three anticipatory problems was covered in the *New York Times*. Tracking media coverage of an issue, particularly in the *New York Times*, is a common practice among policy scholars.[111] Because the figures presented in this study are purely descriptive, these data cannot be interpreted to imply causation and are, again, only intended to provide a "snapshot" of media coverage trends.

Each of the case studies begins with an outline of the general dimensions of the problem before reviewing the dimensions of the policy conflict, including the problem definition, agenda setting, policy formulation, and implementation processes. While this book primarily focuses on national policymaking institutions, I also consider state and local activity where appropriate and salient to the broader dynamics of policy change. Recall that policy change spans multiple levels and branches of government and that our unit of analysis is the policy subsystem, not a specific government institution. This book closes by considering the distinctive features of anticipatory problems, highlighting their implications for policy theory.

# Notes

1 Leon Fuerth. 2011. "Operationalizing Anticipatory Governance." *Prism* 2(4): 31–46.
2 A full repository of Fuerth's work can be found on The Project on Forward Engagement's website, available at: http://forwardengagement.org.
3 Alvin Toffler. 1970. *Future Shock.* New York: Bantam Books.
4 Aaron Wildavsky. 1988. *Searching for Safety.* New Brunswick: Transaction Books; Karlene Roberts. 2009. "Managing the Unexpected: Six Years of HRO-Literature Reviewed." *Journal of Contingencies and Crisis Management* 17(1): 50–54.
5 Thomas A. Birkland. 1997. *After Disaster: Agenda Setting, Public Policy, and Focusing Events.* Washington, D.C.: Georgetown University Press.
6 Peter J. May. 1991. "Addressing Public Risks: Federal Earthquake Policy Design." *Journal of Policy Analysis and Management* 10(2): 263–285; Peter J. May and Chris Koski. 2013. "Addressing Public Risks: Extreme Events and Critical Infrastructures." *Review of Policy Research* 30(2): 139–159, 139.
7 Kevin B. Smith. 2002. "Typologies, Taxonomies, and the Benefits of Policy Classification." *Policy Studies Journal* 30(2): 379–395; Daniel C. McCool. 1995. "Policy Typologies." In *Public Policy Theories, Models, and Concepts: An Anthology,* ed. Daniel C. McCool. Englewood Cliffs: Prentice Hall: 174–181.
8 Theodore J. Lowi. 1964. "American Business, Public Policy, Case Studies, and Political Theory." *World Politics* 16(4): 687–691; Theodore J. Lowi. 1972. "Four Systems of Policy, Politics, and Choice." *Public Administration Review* 33(4): 298–310; Smith 2002.
9 Lowi 1964, 1972.
10 Ibid.
11 Ibid.
12 George D. Greenberg, Jeffrey A. Miller, Lawrence B. Mohr. 1977. "Developing Public Policy Theory: Perspectives from Empirical Research." *The American Political Science Review* 71(4): 1532–1543; Paul A. Sabatier. 1991. "Toward Better Theories of the Policy Process." *PS: Political Science and Politics* 24(2): 147–156.
13 Smith 2002; Peter J. Steinberger. 1980. "Typologies of Public Policy: Meaning Construction and Their Policy Process." *Social Science Quarterly* 61(1): 185–197; Peter J. May. 1986. "Politics and Policy Analysis." *Political Science Quarterly* 101(1): 109–125.
14 Fred M. Frohock. 1979. *Public Policy: Scope and Logic.* Englewood Cliffs: Prentice Hall; Steven A. Shull. 1983. *Domestic Policy Formation: Presidential-Congressional Partnership?* Westport: Greenwood Press; Randall B. Ripley and Grace A. Franklin. 1991. *Congress, the Bureaucracy, and Public Policy.* Belmont: Wadsworth; Grover Starling. 1988. *Strategies for Policy Making.* Chicago: The Dorsey Press; Raymond Tatalovichand and Byron W. Daynes. 1988. "Conclusion: Social Regulatory Policymaking." In *Social Regulatory Policy: Moral Controversies in America Politics,* eds. Raymond Tatalovich and Byron W. Daynes. Boulder: Westview Press: 210–225; Douglas D. Heckathorn and Steven M. Master. 1990. "The Contractual Architecture of Public Policy: A Critical Reconstruction of Lowi's Typology." *Journal of Politics* 52(4): 1101–1123.
15 Frohock 1979; Shull 1983; Starling 1988; Ripley and Franklin 1998.
16 Anderson 2003.
17 Larry L. Wade and Robert L. Curry. 1970. *A Logic of Public Policy: Aspects of Political Economy.* Belmont: Wadsworth.

18 Brian W. Hogwood. 1987. *From Crisis to Complacency? Shaping Public Policy in Britain*. London: Oxford University Press.
19 Jeremy J. Richardson, Gunnel Gustafsson, and Grant A. Jordan. 1982. "The Concept of Policy Style." In *Policy Styles in Western Europe*, ed. Jeremy Richardson. London: George, Allen, & Unwin: 1–16.
20 Kenneth Dyson. 1982. "West Germany: The Search for a Rationalist Consensus." In *Policy Styles in Western Europe*, ed. Jeremy Richardson. London: George, Allen, & Unwin: 17–46; Johan Olsen, Paul Roness, and Harald Saetren. 1982. In *Policy Styles in Western Europe*, ed. Jeremy Richardson. London: George, Allen, & Unwin; Grant Jordan and Jeremy Richardson. 1987. *British Politics and the Policy Process: An Arena Approach*. London: Unwin Hyman; David Calef and Robert Goble. 2007. "The Allure of Technology: How France and California Promoted Electric and Hybrid Vehicles to Reduce Urban Air Pollution." *Policy Sciences* 40: 1–34; Jonas Hinnfors. 1997. "Still the Politics of Compromise? Agenda Setting Strategy in Sweden." *Scandinavian Political Studies* 20(2): 159–177; Olof Ruin. 1982. "Sweden in the 1970s: Policy-Making Becomes More Difficult." In *Policy Styles in Western Europe*, ed. Jeremy Richardson. London: George, Allen, & Unwin: 141–167; Jan Van Putten. 1982. "Policy Styles in the Netherlands: Negotiation and Conflict." In *Policy Styles in Western Europe*, ed. Jeremy Richardson. London: George, Allen, & Unwin: 168–196.
21 Ronald Manzer. 1984. "Public Policy-Making as Practical Reasoning." *Canadian Journal of Political Science* 17(3): 577–594.
22 Ian Neary. 1992. "Japan." *Power and Policy in Liberal Democracies*, ed. Martin Harrop. Cambridge: Cambridge University Press.
23 Calef and Goble 2007.
24 Colin J. Bennet. 1991. "What Is Policy Convergence and What Causes It?" *British Journal of Political Science* 21(2): 215–233; Elisa Roller and Amanda Sloat. 2002. "The Impact of Europeanisation on Regional Governance: A Study of Catalonia and Scotland." *Public Policy and Administration* 17(2): 68–86; Andrew J. Jordan. 2001. "National Environmental Ministries: Managers or Ciphers of European Environmental Policy?" *Public Administration* 79(3): 643–663; Andrew J. Jordan. 2002. *The Europeanisation of British Environmental Policy: A Departmental Perspective*. London: Palgrave; Brigitte Unger and Frans Van Waarden. 1995. "Introduction: An Interdisciplinary Approach to Convergence." In *Convergence or Diversity? Internationalization and Economic Policy Response*, eds. Frans van Waarden and Brigitte Unger. Aldershot: Avebury: 1–36.
25 Wildavsky 1988.
26 Ibid., 77.
27 Ibid.
28 Ibid.
29 Roberts 2009; Paul R. Schulman. 1993. "The Analysis of High Reliability Organizations: A Comparative Framework." In *New Challenges to Understanding Organizations*, ed. Karlene H. Roberts. London: Macmillan: 33–54; Karl E. Weick and Kathleen M. Sutcliffe. 2007. *Managing the Unexpected: Resilient Performance in an Age of Uncertainty, 2nd Edition*. San Francisco: Jossey Bass.
30 Emery Roe and Paul R. Schulman. 2008. *High Reliability Management: Operating on the Edge*. Stanford: Stanford University Press.
31 Arjen Boin, Louise K. Comfort, and Chris C. Demchak. 2010. "The Rise of Resilience." In *Designing Resilience: Preparing for Extreme Events*, eds. Louise K. Comfort, Arjen Boin, and Chris C. Demchak. Pittsburgh: University of Pittsburgh Press: 1–12, 9.
32 Ibid.

33  Daniel P. Aldrich. 2012. *Building Resilience: Social Capital in Post-Disaster Recovery.* Chicago: University of Chicago Press.
34  Thomas O'Rourke. 2007. "Critical Infrastructure, Interdependencies, and Resilience." *The Bridge* 37(1): 22–30; Douglas Paton and David Johnston. 2001. "Disasters and Communities: Vulnerability, Resilience, and Preparedness." *Disaster Prevention and Management: An International Journal* 10(4): 270–277.
35  Wildavsky 1988.
36  Louise K. Comfort. 1999. *Shared Risk: Complex Systems in Seismic Response.* New York: Pergamon.
37  Wildavsky 1998; Richardson, Gustafsson, and Jordan 1982.
38  Wayne Parsons. 1995. *Public Policy: An Introduction to the Theory and Practice of Policy Analysis.* Cheltenham: Edward Elgar: 188.
39  Wildavsky 1998.
40  Paul R. Schulman. 1980. *Large-Scale Policy Making.* New York: Elsevier.
41  Ibid., 10.
42  Ibid.
43  Ibid., 1.
44  Roger W. Cobb and Charles D. Elder. 1977. *Participation in American Politics: The Dynamics of Agenda Building.* Baltimore: The Johns Hopkins University Press.
45  Ibid., 14.
46  Parsons 1995, 571.
47  Ibid., 571.
48  Frank R. Baumgartner and Bryan D. Jones. 2002. "Positive and Negative Feedback in Politics." In *Policy Dynamics*, eds. Frank R. Baumgartner and Bryan D. Jones. Chicago: University of Chicago Press: 3–28; Frank R. Baumgartner, Bryan D. Jones, and John D. Wilkerson. 2002. "Studying Policy Dynamics." In *Policy Dynamics*, eds. Frank R. Baumgartner and Bryan D. Jones. Chicago: University of Chicago Press: 29–46.
49  Paul Burnstein. 1991. "Policy Domains: Organization, Culture, and Policy Outcomes." *Annual Review of Sociology* 17: 327–350; Peter J. May, Ashley E. Jochim, and Joshua Sapotichne. 2011. "Constructing Homeland Security: An Anemic Policy Regime." *Policy Studies Journal* 39(2): 285–307; Robert S. Wood. 2006. "The Dynamics of Incrementalism: Subsystems, Politics, and Public Lands." *Policy Studies Journal* 34(1): 1–16.
50  Baumgartner and Jones 2002; Baumgartner, Jones, and Wilkerson 2002.
51  Michael Howlett and Benjamin Cashore. 2009. "The Dependent Variable Problem in the Study of Policy Change." *Journal of Comparative Policy Analysis* 11(1): 29–42; Frank R. Baumgartner and Bryan D. Jones. 1993. *Agendas and Instability in American Politics.* Chicago: University of Chicago Press.
52  John W. Kingdon. 2003. *Agendas, Alternatives, and Public Policies, 2nd Ed.* New York: Addison-Wesley Educational Publishers; Baumgartner and Jones 1993.
53  Kingdon 2003.
54  Ibid.
55  Baumgartner and Jones 1993.
56  Ibid.
57  Baumgartner and Jones 1993; Kingdon 2003; Michael Mintrom. 2000. *Policy Entrepreneurs and School Choice.* Washington, D.C.: Georgetown University Press.
58  Elmer E. Schattschneider. 1960. *The Semisovereign People: A Realist's View of Democracy in America.* Hinsdale: Dryden Press.
59  Ibid.

60  Paul 't Hart and Arjen Boin. 2001. "Between Crisis and Normalcy: The Long Shadow of Post-Crisis Politics." In *Managing Crises: Threats, Dilemmas, Opportunities*, eds. Uriel Rosenthal, Arjen Boin, and Louise Comfort. Springfield: Charles C. Thomas: 28–46; See also: Daniel Nohrstedt. 2008. "The Politics of Crisis Policymaking: Chernobyl and Swedish Nuclear Energy Policy." *Policy Studies Journal* 36(2): 257–278.

61  Birkland 1997, 22.

62  Ibid.

63  Birkland 1997; Kingdon 2003.

64  Birkland 1997.

65  Thomas A. Birkland. 2004. "The World Changed Today: Agenda-Setting and Policy Change in the Wake of the September 11 Terrorist Attacks." *Review of Policy Research* 21(2): 179–200; Kingdon 2003; May, Jochim, and Sapotichne 2011.

66  Peter W. Huber. 1986. "The Bhopalization of American Tort Law." *Hazards: Technology and Fairness*. Washington, D.C.: National Academy Press: 89–110, 90.

67  May 1991; May and Koski 2013.

68  Ibid.

69  On the other hand, it can also pose challenges to those desiring sustained, well-funded programs, as evidenced in the vast disparities between preparedness and recovery spending.

70  Kingdon 2003.

71  Bryan D. Jones and Frank R. Baumgartner. 2005. *The Politics of Attention: How Government Prioritizes Problems*. Chicago: The University of Chicago Press.

72  Ibid.

73  Deborah Stone. 2002. *Policy Paradox: The Art of Political Decision Making*. New York: W.W. Norton; Judith Eleanor Innes. 1990. *Knowledge and Public Policy: The Search for Meaningful Indicators*. New Brunswick: Transaction Books; Jones and Baumgartner 2005.

74  Birkland 2006, 7.

75  Ibid., 7.

76  David A. Rochefort and Roger W. Cobb. 1994. "Problem Definition: An Emerging Perspective." In *The Politics of Problem Definition: Shaping the Policy Agenda*, eds. David A. Rochefort and Roger W. Cobb. Kansas: University of Kansas Press: 1–31, 15.

77  Kingdon 2003; Baumgartner and Jones 1993.

78  Rochefort and Cobb 1994; Birkland 1997.

79  Birkland 1997.

80  Ibid.

81  Innes 1990, 24.

82  Stone 2002; Innes 1990.

83  Stone 2002, 193.

84  Annika Brändström, Fredrik Bynander, and Paul 't Hart. 2004. "Governing By Looking Back: Historical Analogy and Crisis Management." *Public Administration* 82(1): 191–219.; Yuen Foong Khong. 1992. *Analogies at War: Korea, Munich, Dien Bien Phu, and the Vietnam Decisions of 1965*. Princeton: Princeton University Press; Richard E. Neustadt and Ernest R. May. 1988. *Thinking in Time: The Uses of History for Decision Makers*. New York: Free Press.

85  Baumgartner and Jones 1993; Sarah B. Pralle. 2006. *Branching Out, Digging In: Environmental Advocacy and Agenda Setting*. Washington, D.C.: Georgetown University Press; Sarah B. Pralle. 2006b. "The 'Mouse That Roared': Agenda Setting in Canadian Pesticide Politics." *The Policy Studies Journal* 34(2): 171–194; Christopher J. Bosso. 1987. *Pesticides and Politics: The Life Cycle of a Public Issue*. Pittsburgh: University of Pittsburgh Press; Christopher

M. Weible. 2008. "Expert-Based Information and Policy Subsystems: A Review and Synthesis." *The Policy Studies Journal* 36(4): 615–635.

86 Pralle 2006; Baumgartner and Jones 1993.

87 Sabatier and Jenkins-Smith 1993; Pralle 2006; Baumgartner and Jones 1993.

88 Pralle 2006.

89 Pralle 2006, 2006b; Baumgartner and Jones 1993.

90 Pralle 2006.

91 Peter John. 1999. "Ideas and Interests; Agendas and Implementation: An Evolutionary Explanation of Policy Change in British Local Government Finance." *British Journal of Politics and International Relations* 1(1): 39–62.

92 Ibid.

93 Ibid., 46.

94 Michael Howlett and M. Ramesh. 2003. *Studying Public Policy: Policy Cycles and Policy Subsystems, 2nd Edition.* London: Oxford University Press; Stella Z. Theodoulou. 1995. "The Contemporary Language of Public Policy: A Starting Point." In *Public Policy: The Essential Readings*, eds. Stella Z. Theodoulou and Matthew A. Cahn. Upper Saddle River: Prentice Hall: 1–9; B. Guy Peters. 2012. *American Public Policy: Promise & Performance, 8th Ed*, Washington, D.C.: CQ Press.

95 Howlett and Ramesh 2003.

96 Ibid., 115.

97 May and Koski 2013; Raymond J. Burby and Peter J. May. 1997. *Making Governments Plan: State Experiments in Managing Land Use.* Baltimore: The Johns Hopkins University Press; David A. Moss. 2002. *When All Else Fails: Government as the Ultimate Risk Manager.* Cambridge: Harvard University Press.

98 Paul 't Hart, Uriel Rosenthal, and Alexander Kouzmin. 1993. "Crisis Decision Making: The Centralization Thesis Revisited." *Administration & Society* 25(1): 12–44.

99 Ripley and Franklin 1991, 183–184.

100 Eugene Bardach. 1977. *The Implementation Game: What Happens After a Bill Becomes a Law.* Cambridge: MIT Press.

101 Peters 2012; Howlett and Ramesh 2003; Malcolm Goggin, Ann Bowman, James Lester, and Laurence O'Toole, Jr. 1990. *Implementation Theory and Practice: Toward a Third Generation.* New York: HarperCollins; James E. Anderson. 2003. *Public Policymaking, 5th Edition.* Boston: Houghton Mifflin Company: 211; Bardach 1977.

102 Eric M. Patashnik and Julian E. Zelizer. 2013. "The Struggle to Remake Politics: Liberal Reform and the Limits of Policy Feedback in the Contemporary American State." *Perspectives on Politics* 11(4): 1071–1087, 1074.

103 Peters 2012.

104 Paul A. Sabatier and Christopher M. Weible. 2007. "The Advocacy Coalition Framework: Innovations and Clarifications." In *Theories of the Policy Process, 2nd Edition*, ed. Paul A. Sabatier, Colorado: Westview Press: 190–220; Michael Hill and Peter Hupe. 2009. *Implementing Public Policy, 2nd Edition.* London: Sage; John 1999.

105 John 1999.

106 Kingdon 2003; Eric M. Patashnik. 2008. *Reforms at Risk: What Happens After Major Policy Changes Are Enacted.* Princeton: Princeton University Press.

107 Baumgartner and Jones 1993.

108 Patashnik 2008, 5.

109 Baumgartner, Jones, and Wilkerson 2002.

110 Maxwell McCombs. 2004. *Setting the Agenda: The Mass Media and Public Opinion.* Cambridge: Polity.

111 Birkland 2007.

# 3   No Small Matter
## Balancing the Risks and Benefits of Nanotechnology

Nanotechnology allows for the manipulation of matter at a subatomic level. The defining feature of nanotechnology is its scale—a single nanometer is one-billionth the size of a meter. At the "nanoscale," scientists are able to alter the very building blocks of everyday materials. Nanotechnology's significance—its promise—is in large part derived from the fact that matter behaves in fundamentally different ways at the nanoscale. Stable in bulk, aluminum exhibits explosive properties at the nanoscale. Gold shows hues of red, blue, and even green. Even zinc oxide, which is normally an opaque substance used to block ultraviolet rays, is transparent at the nanoscale.[1] By working from the very bottom up scientists believe they can construct new materials, harnessing the novel properties found at the nanoscale in ways unimaginable at the atomic level.

The story of nanotechnology epitomizes the balancing act that accompanies any new (let alone potentially revolutionary) technology. On the one hand, it promises enormous social and economic benefits. From improved medical delivery devices to faster communications, more durable materials to longer lasting batteries, the possibilities are endless. American policymakers have a great incentive to leverage the country's robust research and development capacity, establishing the U.S. as a global leader in this emerging field. According to Lux-Research, a Boston-based research firm that studies the likely economic impact of nanotechnology, the global market for nano-products may reach $4.4 trillion by 2018 (in 2012, the market already stood at $731 billion).[2]

On the other hand, history has shown that even the most beneficial technological developments can bring negative consequences for human health, the environment, and society in general. Policymakers, scientists, public health specialists, and environmental activists alike have long expressed concerns about the *potential* implications of nanotechnology. Science has yet to demonstrate definitively any risks associated with this new technology, but warning signs abound. Accordingly, a concerted effort is afoot to anticipate the potential risks associated with nanotechnology prior to its full integration into society.[3]

Nanotechnology's risks represent an anticipatory policy problem in its most extreme form. Many of nanotechnology's risks have not even been discovered, let alone studied and confirmed. However, the simple fact that these risks *could* exist has fueled calls for government to take proactive steps to identify these threats in advance. This particular aspect of the nanotechnology debate makes it different from both the avian influenza and global warming debates. In these latter cases, a relatively known risk (e.g., a deadly pathogen or irreversible climate change) has been identified, although policymakers struggle to determine when, how, or to what extent these hazards will occur. With respect to nanotechnology, it is difficult to even conceptualize what these risks might be, let alone when and where they will emerge first. One could persuasively argue that the nanotechnology debate embodies greater uncertainty. As such, nanotechnology constitutes an optimal starting point for my discussion of anticipatory policymaking.

Moreover, despite its potentially troubling health, environmental, and even social implications, nanotechnology is overwhelmingly regarded as a positive development. To this end, the nanotechnology case explores the inherent tension between promoting a new technology while protecting public health and the environment. And while this chapter primarily focuses on the steps taken to anticipate nanotechnology's health and environmental dangers, the importance of this countervailing narrative simply cannot be dismissed.

## Dimensions of the Problem

Nanotechnology is broadly described as the manipulation and engineering of matter at a subatomic level.[4] Although a universally accepted definition of nanotechnology does not exist, most discussions underscore three key features. First, many equate nanotechnology with "smallness." Nanotechnology often deals with structures of 1–100 nanometers in scale. A single nanometer is one billionth of a meter, smaller than an atom.[5] Second, matter behaves in fundamentally different ways at the "nanoscale," allowing humankind to access an array of once unthinkable and unattainable materials and properties.[6] Third, scientists predict nanotechnology will allow them to control and manipulate the very atoms that make up matter as we know it. Through nanotechnology, scientists can engineer at a subatomic level and "deliberately shape new structures, atom by atom."[7]

Nanotechnology is not a singular technology per se and is used in a wide array of fields, ranging from national defense to health care, transportation and energy to communications. And, while proponents have promised new products and devices in the future, it has largely been used to enhance existing products and technologies.[8] Still, a number of these products promise to be nothing short of revolutionary. Scientists are working on developing nanoparticles that can target cancer cells and deliver medical treatments, like chemotherapy, sparing patients harmful side effects typically associated with having harsh chemicals course through one's entire body.[9] Solar panels fitted with nano-engineered panels are believed to be far more efficient at sunlight conversion. Nanotechnology-enhanced desalinization filters can offer affordable and portable water cleansing systems, increasing access to clean water. Computer scientists are also hopeful that nanotechnology could considerably expand the storage space of traditional memory devices, leading to President Bill Clinton's famous observation that the entire library of Congress may one day be fitted on a memory device the size of a sugar cube.[10] U.S. corporations alone spent $4 billion on nanotechnology in 2014 and the global market for products containing nanotechnology could reach $4.4 trillion by 2018.[11]

Yet, the same features that make nanotechnology so promising are also cause for concern. Nanotechnology's incredible "smallness" poses distinctive risks, the likes of which have not been experienced at a molecular level. John H. Sargent, Jr., Specialist in Science and Technology Policy at the Congressional Research Service, summarizes this "double-edged sword" dilemma:

> On the one hand, some are concerned that nanoscale particles may enter and accumulate in vital organs, such as the lungs and brains, potentially causing harm or death to humans and animals, and that the diffusion of nanoscale particles in the environment might harm ecosystems. On the other hand, some believe that nanotechnology has the potential to deliver important EHS (Environmental, Health, and Safety) benefits such as reducing energy consumption, pollution, and greenhouse gas emissions; remediating environmental damage; curing, managing, or preventing diseases; and offering new safety enhancing materials that are stronger, self-repairing, and able to adapt to provide protection.[12]

Some scientists have even raised concerns about nanotechnology's ability to penetrate the blood–brain barrier structure that protects the brain from harmful substances. When inhaled, nanoparticles have been shown to cross this barrier and perhaps even accumulate in the brain. To be sure, scientists have yet to establish this poses an actual risk to human health.[13] Nonetheless, the ability of nanoparticles to cross the blood–brain barrier is unsettling.[14]

A 2009 study by Trouiller et al. found that nano-titanium dioxide damaged and destroyed DNA in lab animals. The authors concluded "we should be concerned about the potential risk of cancer or genetic disorders especially for people occupationally exposed to high concentrations of [nano-titanium dioxide]."[15] Other studies suggest inhaled nanoparticles produce lung damage alarmingly similar to that of asbestos, the fibrous molecule commonly found in insulation, and can cause cancers of mesothelioma and asbestosis, a chronic inflammatory condition of the lungs.[16] Studies have also shown nanotechnology particulates can kill certain bacteria and microbes critical to sustaining aquatic life, raising fears that residue washed away by rain or even sweat could cause significant environmental damage.[17]

Nanotechnology represents an endeavor of enormous promise that simultaneously suggests extreme caution. Yet, on both accounts, uncertainty remains. As of this writing, few if any of the suspected health and environmental risks have been indisputably confirmed. Similarly, many of the most significant and promising benefits of nanotechnology have yet to come to fruition. In turn, the policymaking debate surrounding nanotechnology is marked by speculation. Below, I investigate the policymaking dynamics associated with anticipating both the desired and feared outcomes.

## Discourse of Conflict

The policy discourse associated with nanotechnology reflects the kind of balance that must be struck when supporting an active public sector role in the promotion of a new technology.[18] Policymakers face considerable pressure, particularly from industry and academia, to ensure America becomes a global leader in nanotechnology research, development, and commercialization. At the same time, however, policymakers are responsible for protecting public health and welfare, and must ensure the technologies they are promoting do not cause harm. Even the perception of harm can cause public backlash, undercutting the adoption of a new technology.[19] However, the proverbial pendulum swinging between benefit maximization and risk minimization tips heavily in favor of the former.

The conceptual origins of nanotechnology stem from a speech delivered to the American Physical Society by Nobel Prize winning theoretical physicist Richard Feynman in 1959. Feynman's speech boldly proclaimed: "[U]ltimately—in the great future—we can arrange the atoms the way we want; the very atoms, all the way down!"[20] While scientists made groundbreaking discoveries throughout the 1960s and 1970s, many observers believe K. Eric Drexler, a student and later a faculty member at the Massachusetts Institute of Technology (MIT), brought nanotechnology to the public's attention.[21] Throughout the 1980s and 1990s, he published a series of books trumpeting nanotechnology's promise. His most famous book, *Engines of Creation*, described what Drexler called "assemblers" or semi-autonomous nanoscale machines capable of self-replicating and producing almost any material imaginable.[22] Written for a general audience, the book cast nanotechnology in an accessible—and positive—light. Drexler envisioned a world where assemblers would drastically improve everyday living conditions, offering new and improved materials and goods.[23] Drexler later testified before a congressional hearing that "molecule-by-molecule control" could become "the basis of a manufacturing technology that is cleaner and more efficient than anything we know today."[24]

If Drexler was largely responsible for communicating an accessible description of nanotechnology to the American public, then Mihail Roco represented the domain's most prominent policy entrepreneur. Roco, a mechanical engineer who began working at the National Science

Foundation (NSF) in the 1990s, was long captivated by the promise of nanotechnology.[25] In 1997, he spearheaded the creation of—and later chaired—the Interagency Working Group on Nanoscience, Engineering, and Technology (IWGN). The IWGN was responsible for reporting directly to the National Science and Technology Council (NSTC), a cabinet-level body established by President Clinton in 1997 to coordinate science and technology policy at the executive level. Over the next several years, the IWGN formulated its vision for a multi-agency nanotechnology R&D effort called the "National Nanotechnology Initiative" (NNI). Roco pitched his idea for the NNI to a handful of the Clinton Administration's top scientific and economic advisors in 1999. The proposal was followed by a personal campaign to educate Congress on the promise of nanotechnology. Roco presented at a number of congressional hearings, distributed brochures describing a future filled with nano-enhanced products, and met with members directly.[26]

His efforts paid off and on January 21, 2000, President Clinton introduced his vision of the federal government's role in facilitating the progress of nanotechnology during a speech delivered before the California Institute of Technology. President Clinton's address echoed the same utopian ideals evoked by Feynman 50 years prior, stating:

> Imagine the possibilities: materials with ten times the strength of steel and only a small fraction of the weight, shrinking all the information housed in the library of Congress into a device the size of a sugar cube, detecting cancerous tumors when they are only a few cells in size. Some of our research goals may take 20 or more years to achieve, but that is precisely why there is an important role for the federal government.[27]

Clinton submitted a FY2001 budget request of nearly $495 million to support the NNI proposal and nanotechnology R&D.[28] In years to follow, Roco continued to represent the most active policy entrepreneur in the nanotechnology debate, focusing most of his effort on securing a permanent statutory authorization for the NNI from Congress. In 2003 *Forbes* magazine recognized him as the first among "Nanotechnology's Power Brokers."[29]

The same optimism associated with the early era of nanotechnology dominated congressional debate surrounding the proposed NNI appropriation. Neal Lane, President Clinton's Science Advisor, stated: "If I were asked for an area of science and engineering that will most likely produce the breakthroughs of tomorrow, I would point to nanoscale science and engineering."[30] Richard Smalley, Rice University scientist and Nobel Prize winner, added: "There is a growing sense in the scientific and technical community that we are about to enter a golden new era."[31] Famed technologist and nanotechnology enthusiast Ray Kurzweil heaped exceptionally high praise on the new technology, calling nanotechnology a "cure-all" solution to many of society's most pressing ills "including pollution, poverty, disease, and aging."[32]

Congress approved $465 million of the $495 million requested to fund the NNI. Of this amount, $150 million would to be dispensed by the NSF in the form of research grants. An additional $90 million was appropriated for basic research to be directed toward scientists and teams of researchers.[33] However, the NNI does not have a centralized funding source. Instead, funding goes to a number of participating agencies. The original recipients included NSF; the Department of Defense (DOD); the Department of Energy (DOE); the Department of Commerce (DOC); the National Institute of Standards and Technology (NIST); the National Aeronautics and Space Administration (NASA); and the National Institutes of Health (NIH). By 2010, the number of participating agencies grew to 25, with 14 receiving specific appropriations to conduct and/or fund nanotechnology R&D.[34] The NNI's annual appropriations have grown considerably, totaling nearly $21 billion since its inception in 2001.[35]

Discourse during this period underscored the temporal dimensions of the new technology, espousing a future vastly improved by nano-technology-enhanced products. Various press releases, as well as the NNI's own website, bore the title "National Nanotechnology Initiative: Leading the Way to the Next Industrial Revolution."[36] Analogizing the NNI, and nanotechnology in general, to the Industrial Revolution allowed the Clinton Administration to communicate a positive and widely understandable narrative that minimized the reality that nanotechnology was in fact a very new technology with a very uncertain feature. Sandler and Kay summarized the strategic importance of this Industrial Revolution analogy writing "to claim that nanotechnology is revolutionary invites a positive socio-ethical evaluation of it, not just a positive scientific or technological one."[37] Nanotechnology, it was argued, promised to improve almost every aspect of human life. So pervasive was this depiction that by 2005 there were "more than a *half million* articles, reports, and web sites that characterize nanotechnology research and development as 'the new,' the 'next,' or 'another' Industrial Revolution."[38]

Somewhat paradoxically, nanotechnology was portrayed as both novel and familiar. It was capable of spawning radically innovative products and materials, but ultimately represented the latest iteration in a long line of life-changing technological advances. More than anything, however, nano-technology was portrayed as an overwhelmingly good thing. Policy actors on both sides of the aisle, not just the Clinton Administration, were enthusiastic. At the 2002 NanoBusiness Conference in New York City, former-House Speaker Newt Gingrich (R-GA) called nanotechnology "the investment with the largest payoff over the next 50 years."[39] He even went so far as to praise the Clinton Administration by calling the NNI "one of the better things the government has done."[40]

For nanotechnology proponents, the only potential "problem" was the federal government's failure to ensure the U.S. would become a global leader in the nanotechnology field. At a 1999 hearing before the U.S. House of Representative's Subcommittee on Basic Research, a subcommittee of the Committee on Science, scores of presenters spoke of the importance of becoming *the* global leader in this emergent technology area. However, this would not be an easy task and many other nations were poised to assume the mantle of leadership. In his opening statement, Representative Nick Smith, Chairman of the Subcommittee on Basic Research, warned: "Unfortunately, while progress has been made, the United States does not dominate nanotechnology. A significant amount of research is currently underway in Europe, especially Japan."[41] Eugene Wong, Assistant Director for Engineering at the National Science Foundation, raised these same concerns: "I think the United States is in the forefront of this new science and technology area, but the other countries, the other developed nations, are not far behind."[42] Ralph C. Merkle, a research scientist at the XEROX Palo Alto Research Center, testified: "Economic progress and military readiness in the twenty-first century will depend fundamentally on maintaining a competitive position in nanotechnology."[43] Other presenters struck the same theme: failure to secure a leading global role in this field could have dire economic and perhaps even military implications. Significant government support was in order.

But, not everyone succumbed to the techno-optimism of the Washington science and technology subsystem. Soon after the NNI's creation, an assortment of advocacy groups voiced concerns about the government's blind faith in this "nano-revolution." Detractors proved equally adept at manipulating the political discourse with colorful language and analogies as a means of undermining nanotechnology. Some took aim at the next Industrial Revolution analogy. Georgia Miller of Friends of the Earth Australia, an international environmental advocacy firm, asked: "What would a 'post-revolutionary' nanotech world look like? Given that past revolutions have resulted in winners, losers, and massive social upheaval is anyone planning to manage this revolution to mitigate its most adverse consequences?"[44] The World Council of Churches, an international "fellowship" of hundreds of churches and clergies, argued that the Industrial Revolution analogy was not an entirely promising comparison because "major new technologies, at least initially, destabilize marginalized peoples while the wealthy anticipate, manipulate and ride the wave's crest."[45]

Weeks before Congress's decision to fund the NNI, Bill Joy, one of the cofounders of Sun Microsystems, published an article in *Wired* magazine describing what has famously come to be known as the "grey goo" scenario. Contrasting the positive depiction of self-replicating machines presented by Drexler, Joy imagined a scenario wherein uncontrolled, self-replicating nano-robots might destroy entire ecosystems, turning the natural world into a mass of grey goo.[46] Michael Crichton's best selling novel *Prey*, the latest in a long line of warnings from this author regarding the reckless pursuit of superscience, offered yet another doomsday scenario and conjured up swarms of nano-robots replicating en masse and hunting humans.[47]

One of the most influential counter-narratives from opponents put forward an analogy between nanotechnology and Genetically Modified Organisms (GMOs). GMOs are organisms that have been altered through genetic engineering techniques, such as recombinant DNA technology. These techniques allow scientists to reconstruct or enhance organisms in order to provide them with, or accentuate, desired traits. The science of genetic modification is especially pervasive in the agriculture industry because it allows scientists to create crops enhanced with beneficial characteristics, including greater resistance to pests, longer shelf life, and higher nutritional value. Ken Roseboro, staff editor at Organic Consumers Association, a nonprofit organization that advocates for, among other things, food safety, sustainability, decreased use of genetic engineering, and "de-industrialization" of the food chain, asserted: "Both [GMOs and nanotechnology] manipulate fundamental levels of nature where the potential for negative unforeseen consequences is great."[48] In other words, opponents held that nanotechnology, like GMOs, violated the sanctity of living things while threatening unforeseeable risks.

The GMO analogy also represented a cautionary tale for policymakers who counted on achieving public support of nanotechnology. In the late 1990s, there was a significant worldwide backlash against GMOs. The backlash was predicated on the moralistic objections raised above, as well as widespread sentiment that governments were introducing a new and potentially transformative technology without considering public concerns. In this light, elected officials were seen as pandering to industry while ignoring the needs of the broader community.[49] In a comment at once analytical and argumentative, Doubleday stated "It is a view shared by many governments and policy analysts that how science policy handles the public dimensions of nanotechnology will be a critical test of whether lessons have been learned from the global controversy over GM foods."[50] The lesson here is quite simple: failure to consider public opinion could result in a consumer revolt.

Most nano-opponents (and even many groups that gave qualified support to the technology) wanted the government to take proactive measures to minimize the uncertainty associated with nanotechnology and, to the extent possible, forecast potential dangers before nano-products became released onto the market. Nanotechnology coordinator in the Environmental Protection Agency's (EPA) Office of Pollution Prevention and Toxics, Jim Atwood, summarized this logic as follows: "There is so much uncertainty about the questions of safety. We can't tell you how safe or unsafe nanomaterials are. There is just too much that we don't yet know."[51] Outside of the most extreme opponents, few argued for an outright ban or moratorium on nanotechnology. Instead, most participants argued that the so-called lessons of the past dictated caution. Patty Lovera, assistant director of Food & Water Watch, a non-governmental organization that focuses on government accountability relating to water and food, argued: "The unsettling track record of other technological breakthroughs—like asbestos, DDT,

PCBs and radiation—should give regulators pause as they consider the growing commercial presence of nanotech products. [A] wait-and-see approach puts consumers and the environment at risk."[52]

The two poles in the debate were set. On the one hand, utopian predictions and a hyper-competitive global R&D market drove a narrative calling for government investment in nanotechnology. On the other hand, opponents believed even the most promising new technologies can harbor incredibly dangerous, indeed lethal, features. Both lines of reasoning were highly anticipatory, conjectural, and indeed marked by extreme uncertainty. Whereas many proponents of nanotechnology urged policymakers to dive headlong into a utopian future, others advocated for government to move with extreme caution. The following section describes policymakers' attempts to satisfy both of these ends.

## Agenda Setting and Policy Formulation

Nanotechnology lingered on the institutional agenda for roughly ten years, between 2000–2010. Two noticeable upticks in attention and policy activity occurred during this period. The first, which began soon after the enactment of the NNI in 2001 and ended in 2004 (spanning the 107th and 108th U.S. Congresses), set out to solidify the U.S. position as a global leader in nanotechnology R&D. Enthusiasts specifically advocated for legislation codifying the federal government's role in this area, specifying the NNI's programmatic responsibilities and funding source. This pro-nanotechnology subsystem ultimately got its wish and in 2003 Congress passed the 21st Century Nanotechnology Research and Development Act. This law amounted to a "policy innovation" in the most classic sense of the term, establishing a new and indeed unprecedented role for the national government in the stewardship of nanotechnology.[53]

The second round of issue attention and policy activity primarily centered on the 110th (2007–2008) and 111th (2009–2010) Congresses. This time, however, increased attention was focused on nanotechnology's potential risks. By and large, groups argued that the federal government needed to close the substantial gap in funding between R&D, which received the lion's share of federal funding, and funding for research on health and safety concerns. Some even advocated for amending parts of the 21st Century Nanotechnology Research and Development Act in order to clarify the responsibilities of regulatory agencies with respect to risk regulation.

Figure 3.1 provides a useful illustration of the above-described pattern of issue attention. Figure 3.1 counts the number of times the terms "nanotechnology" and, searching within these results, "risk" were entered in the *Congressional Record* between 1997 and 2014.[54] Figure 3.1 thus provides a rough measure of issue attention and, more specifically, relative levels of concern. During the initial phase of agenda setting, the number of nanotechnology mentions nearly tripled from 2002 (34 entries) and 2003

(92 entries). What is more, this period saw the biggest "gap" in terms of "risk-to-nanotechnology" mentions. In 2003 the term "risk" was only uttered in 35 out of the 92 nanotechnology entries. This pattern likely reflects the marked "techno-optimism" cultivated by the creation of the NNI and previous executive commitments. Although the number of nano-technology entries remained fairly high between 2004 and 2006 (at least relative to their pre-2000 numbers), a second surge in entries occurred in 2007–2008, the second-round of policy activity. The highest number of "risk" entries was recorded in 2007 (50 out of 66 nanotechnology entries). This at least partially reflects the emergence of a coherent counter-mobilization effort, which worked to stimulate greater discussion of the potential hazards of nanotechnology. However, these numbers dropped precipitously in 2008 and "risk" was only mentioned in 29 out of the 71 entries.

Figure 3.2, which shows the number of congressional hearings on the topic of nanotechnology between the years 1997 and 2014, provides another depiction of nanotechnology's prolonged tenure on the institutional agenda. Once again, a noticeable spike in activity occurred between 2002 (two hear-ings) and 2003 (seven hearings). Hearing activity stabilized in subsequent years, hovering around three hearings per year between 2004 and 2009. In sum, nanotechnology emerged on the institutional agenda in 2003 and remained there until roughly 2009.[55] The following section more closely examines the policymaking process associated with this period.

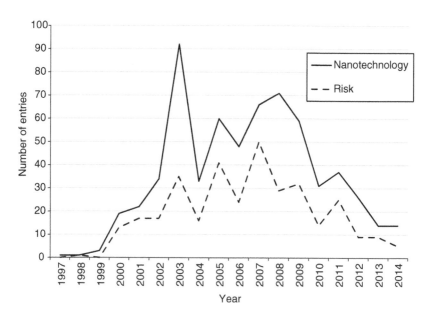

*Figure 3.1*   Number of Nanotechnology Entries in the *Congressional Record*, 1997–2014 (source: Search on ProQuest Congressional for the term "nanotechnology").

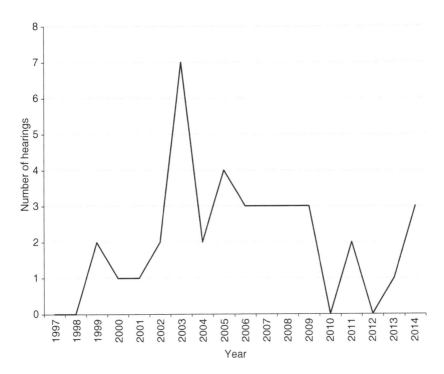

*Figure 3.2* Number of Congressional Hearings on Nanotechnology, 1997–2014 (source: Search on ProQuest Congressional for the term "nano-technology").

## Establishing a Nanotechnology Research Infrastructure

The creation of the NNI in 2001 established an important role for the U.S. national government in stewarding nanotechnology development. Despite this obvious victory, nanotechnology advocates recognized that an organizational infrastructure alone could not bring about the technological revolution envisioned. Instead, a permanent commitment from the federal government was required. Most importantly, they hoped to persuade Congress to develop legislation specifying the U.S.'s commitment to nanotechnology R&D while codifying a lasting funding stream for the NNI.

The surge in agenda activity observed in the wake of the NNI's creation was stimulated by a well-organized and eclectic mix of stakeholders and interests, including representatives from industry, academia, and government bureaucracies, to name a few. No group better exemplified this trend than the NanoBusiness Alliance, an industry association representing corporations, universities, investors, and other stakeholders working on nanotechnology research and manufacturing. The NanoBusiness Alliance was arguably the most active lobbying group and credits itself for stewarding the 21st Century Nanotechnology Research and Development Act through Congress.[56] Mark

Modzelewski, leader and founder of the NanoBusiness Alliance, modeled the group after the Biotechnology Industry Organization, a trade organization representing the biotechnology industry. Modzelewski brandished clear "insider" credentials, having worked for the Clinton Administration. He added a bipartisan flavor to the group by appointing former House Speaker Newt Gingrich as honorary chairman.[57]

The presence of prominent industry groups, like the NanoBusiness Alliance, make the nantochnology case different from many other areas of risk, hazards, and disaster management, which typically lack an organized private interest group structure and an organized public.[58] Moreover, mobilization involved a fairly diverse group of industries and sectors, which sought to frame nanotechnology as something that could enhance their work. Lindquist, Mosher-Howe, and Liu found a plethora of organized interests sought to link nanotechnology as a solution to many existing problems.[59] In other words, nanotechnology was not a problem to be solved, but a problem solver. They write:

> With so many interests seeking money, as a result of the attention and promise for funding for anything nano-related, it also set off attempts to strategically create and control definitions of nanotechnology and of the problems that could be solved by the technology.[60]

McCray echoed this finding, noting that nanotechnology was seen as a "technological fix" and added that: "Topics that politicians care most about—health, the economy, and national security—all looked to benefit from nanotechnology."[61] Their claims of relevancy were again bolstered by the uncertainty surrounding with nanotechnology. Because no one knew what, exactly, a "nano-enhanced" future might look like, stakeholders from an array of sectors could freely propose narratives describing considerable advances in their field—provided the government gave them adequate funding, of course.

Within Congress, nanotechnology R&D was seen as a science and technology issue, and committees with jurisdiction over this particular domain spearheaded most policy activity. The House Committee on Science and Technology and the Senate Committee on Commerce, Science, and Transportation (referred to as the House or Senate "Science Committee" hereafter), both of which have clear jurisdiction over science and technology issues, represented the preeminent policymaking venues throughout the process of agenda setting and policy formulation. The Science Committees convened hearings and led the effort to draft legislation formally authorizing the NNI.

This is not to say that the Science Committees were the *only* venues to consider the nanotechnology issue. Just as nanotechnology accommodated participation from an array of stakeholders and interests, so too did it afford participation from a number of committees. As early as 2004, the Senate Committee on Energy and Natural Resources explicitly considered nanotechnology's capacity to meet future electricity needs.[62] In 2005, the

Senate Foreign Relation's Committee's Subcommittee on European Affairs explored the possibility of U.S.–European Union cooperation on nanotechnology regulation.[63] Similarly, the House Subcommittee on Commerce, Trade, and Consumer Protection explored the capacity of existing statutes to successfully regulate nanotechnology, including the Toxic Substances Control Act (TSCA).[64] And, in 2009, the House Subcommittee on Commerce, Manufacturing, and Trade convened a hearing exploring nanotechnology's ability to drive innovation in the U.S.[65]

Still, most activity occurred within the House and Senate Science Committees, which were highly receptive to advocates' calls for greater federal investment in nanotechnology. In January 2003, Senator Ron Wyden (D-OR) introduced the 21st Century Nanotechnology Research and Development Act (S. 189), which affirmed and expanded the national government's role in nanotechnology R&D. In February 2003, Representative Sherwood Boehlert (R-NY), then Chairman of the House Science Committee, introduced a similar bill in the House, the Nanotechnology Research and Development Act of 2003 (H.R. 766).

Throughout 2003, both the House and Senate Science Committees convened a series of hearings soliciting input on the pending statutes. The tenor of policy discourse was overwhelmingly positive. In the three years since the its creation, the R&D framework established by the NNI was already paying dividends. Optimism was fueled by "indicators and measurements" showing the U.S. was poised to lead the nanotechnology revolution. References to nanotechnology in peer-reviewed scientific journals had more than doubled, from 3,500 in 1997 to more than 10,000 in 2002. Meanwhile the number of patents for nanotechnology products increased from 2,000 in 1995 to 6,425 in 2002.[66]

Still, problem definers were quick to note that these indicators were not cause for complacency. A number of countervailing measures challenged whether the U.S. had a "commanding lead" in this sector. Representative George Allen (R-VA) called on his colleagues in the House and Senate to ensure the U.S. did not fall behind in this vital sector:

> I think that we all ought to recognize that we are not alone in this country being interested in nanotechnology. Indeed, when one will look at the global picture, we are falling a bit behind, insofar as our research and development in nanotechnology, and we're facing some stiff foreign competition in nanotech research from Japan, the European Union, Russia, Korea, and China. Now, this Nation, the United States, has been at the forefront of almost every important transformative technology since the industrial revolution, and we must continue to lead the world in the nanotechnology revolution, in my estimation.
>
> [O]ur role, as elected leaders, should be to creator to foster the conditions precedent for our researchers and innovators to compete and contribute and succeed, both domestically and internationally.[67]

Alan Marty, Executive-in-Residence at JP Morgan Partners, seconded Representative Allen's concerns by noting: "Unlike many past waves of technological development, nanotechnology is not dominated by the United States. In several areas of nanotechnology the U.S. is being outpaced by foreign competition."[68] Once again, scientific publications were leveraged as important indicators. Dr. James Murday, Chief Scientist at the Office of Naval Research, told members of the Senate Committee on Science that the 18,000 articles published on nanotechnology in 2002 are "roughly divided; one-third in the Asian theater, one-third in the U.S., and one-third in Europe. That says, in terms of quantities, we are one-third."[69] Suffice it to say, few policymakers were comfortable with the fact that the U.S. was but one of several key powers on the global nanotechnology stage. Thus, indicators were packaged in a larger narrative expressing that, although the NNI has begun to bear fruit, funding commitments from Congress were needed to secure U.S. supremacy in the area of nanotechnology R&D.[70]

Despite consensus on the need to ensure U.S. dominance, concerns about the inherent uncertainty associated with nanotechnology could not be silenced entirely. Dr. Thomas N. Theis, Director of Physical Science, IBM Research Division, Thomas J. Watson Research Center, urged policymakers "to anticipate that there will be societal implications, not every one of them necessarily good and comfortable, for any rapidly advancing technology. We see this across the board. It is not a particular attribute of nanotechnology, so we need to accompany our research effort with efforts to anticipate and manage those implications."[71] Senator Frank Lautenberg (D-NJ) offered this nervous observation:

> All nanotechnological advances, even the most beneficent, have what are called "externalities." The automobile, for instance, represented an enormous improvement over horse-drawn carriages. But each year, thousands of people are killed in auto accidents and hundreds of thousands more are hurt. Moreover, cars are a leading cause of greenhouse gas emissions.
>
> I'm not suggesting that we would be better off without cars—far from it. My point is that there will be adverse consequences stemming from the development of nanotechnology.
>
> It may not be possible to anticipate all of the unintended consequences of developing nanotechnology, but we should try.... Clearly, the earlier we grapple with the ethical issues and harmful consequences related to nanotechnology, the better off we will be at mitigating them.[72]

Thus, even though nanotechnology was overwhelmingly seen as a "good thing" at this time, risk minimization also represented a goal—albeit a secondary goal—of the U.S. Congress.

The 21st Century Nanotechnology Research and Development Act sailed through Congress, passing both chambers with near unanimous bipartisan support in November 2003.[73] The 21st Century Nanotechnology Research and Development Act represents a profound example of anticipatory policymaking, requiring that federal agencies simultaneously promote a new technology and, to the extent possible, forecast and minimize any potential health, environmental, and social risks. The Act established a National Nanotechnology Program (NNP), while authorizing $3.7 billion for nanotechnology research and development from FY2005 to FY2008. (This funding stream would be up for reauthorization in 2008.) Economic incentives in the form of grants represented the dominant—and indeed logical—policy instrument for incentivizing industry and academia to pursue greater R&D in the area of nanotechnology.

While the Act did not specify the NNP's objectives, it did promulgate 11 "Program Activities," which would be achieved through various agencies and councils. The Act required the NNP to:

1   Develop a fundamental understanding of the manipulation of matter at the nanoscale;
2   Offer grants to researchers working in interdisciplinary teams;
3   Establish a network of technology user centers and facilities;
4   Establish interdisciplinary nanotechnology research centers. This particular provision was significant in that these academic institutions and national research laboratories would drive nanotechnology R&D;
5   Ensure global leadership in nanotechnology;
6   Advance U.S. "productivity and industrial competiveness" through long-term investments in nanotechnology R&D;
7   Accelerate the deployment and application of nanotechnology in the private sector. This Activity, coupled with Activities 5 and 6, was said to mandate a "first-to-the-market" approach to R&D. Merely underwriting R&D would not suffice. Instead Congress demanded that research be quickly translated into commercial products. These Activities clashed with calls for a more precautionary approach to nanotechnology development, as it called for an expedited commercialization process;[74]
8   Encourage interdisciplinary research and ensure "that processes for solicitation and evaluation of proposals under the Program encourage interdisciplinary projects" (P.L. 108–105, Sec. 2.b.8);
9   Provide training programs that promote a culture of interdisciplinary research;
10  Ensure the "ethical, legal, environmental, and other appropriate societal concerns" associated with nanotechnology are considered (P.L. 108–105, Sec. 2.b.10). Activity 10 formally establishes a proactive and anticipatory approach to considering nanotechnology's potential risks and is thus critical to this case;
11  Encourage that nanotechnology be applied to existing processes and technologies.

The Act further specified appropriations for the NSF, DOE, NASA, and EPA. Again, the multiplicity of participating agencies reflected nanotechnology's capacity straddle various policymaking domains. Many sectors, it seemed, had a stake in the new technology.

The Act also utilized organization instruments. The President was responsible for implementing the NNP with the help of a newly formed Nanotechnology Coordination Office and appropriate agencies. Additional program management responsibilities were given to the National Science and Technology Council, a cabinet-level council that advises the White House on science and technology issues and coordinates policy in this area. Otherwise, very little detail was provided regarding administrative responsibilities. Indeed, the Act was relatively vague and brief, amounting to a mere ten pages.

The 21st Century Nanotechnology Research and Development Act highlights the distinctive policymaking dynamics associated with anticipatory policy problems. Because they have yet to occur, anticipatory problems cannot be addressed through a remedial policy design. In the case of nanotechnology, the 21st Century Nanotechnology Research and Development Act amounted to a plan or a strategy for achieving a future policy end, a process that really began with the executive actions of 2000. The 11 activities move the country toward a future end, but, because that end is still marked by incredible uncertainty, offer little in the way of detail. Policymakers may have even seen the 21st Century Nanotechnology Research and Development Act as a work in progress. Indeed, Congress did not even specify a permanent funding stream for NNP operations, instead capping most allocations at five years. While sunset provisions are not entirely uncommon, this action at least implies Congress was cognizant of the fact that information could change in the coming years.

Although R&D was the overriding goal of this Act, Activity 10 established a noticeably anticipatory approach to addressing the potential health, environmental, and social implications of nanotechnology. It mandates that the NNP shall be responsible for: "ensuring that ethical, legal, environmental, and other appropriate societal concerns, including the potential use of nanotechnology in enhancing human intelligence and in developing artificial intelligence which exceeds human capacity are considered" (P.L. 108–105, Sec. 2.b.10). Activity 10 essentially required that nanotechnology's health, environmental, and social risks be considered *during* the R&D process, as opposed to waiting for new technologies to go to market before addressing—or even identifying—their potential dangers.[75]

Beyond Activity 10, other provisions underscore a proactive policymaking approach. For example, the Act requires a number of reports, internal reviews, and even external reviews that are, broadly speaking, designed to update policymakers on the status of the NNP and nanotechnology research in general. With respect to these "oversight" programs, the Act includes language mandating that these reviews describe actions taken to investigate and mitigate nanotechnology's health, environmental, and social risks. Moreover, the Act created a $5 million American Nanotechnology Preparedness Center, which was charged with collecting, disseminating, and coordinating research on nanotechnology's potential risks as well as leading the effort to anticipate potential "responsible development" issues. The Network for Nanotechnology in Society consists of centers at the University of California, Santa Barbara, with participation from Harvard University and the University of South Carolina.[76]

Many observers maintain that Activity 10 represents a revolution in regulatory policy. Fisher and Mahajan stated:

> By prescribing the integration of societal and technical concerns during nanotechnology R&D, the Act could mark a radical shift in [science and technology] policy in so far as it allows consideration of societal concerns to influence technical activities and outcomes.[77]

Indeed, Activity 10 inspired a host of applied models, all of which set out to integrate anticipation into the R&D process. For example, scholars working at the NSF-funded Center for Nanotechnology in Society at Arizona State University tested an "upstream engagement" approach to "real time technology assessment," embedding social scientists and philosophers within laboratory settings in order to facilitate discourse about the likely implications of the new technology.[78] Yet despite these laudable provisions, many observers maintained anticipation and risk reduction were woefully undervalued—and underfunded. Subsequent rounds of policymaking set out to elevate the importance of these concerns.

## Agenda Catalysts

The policymaking process described above was rapid, acute, and relatively conflict free. In many respects, the 21st Century Nanotechnology Research and Development Act represented a form of "distributive policy" in that it provided considerable financial incentives to academia and industry, a far cry from the types of policies considered in subsequent chapters.[79] Nanotechnology rocketed to the top of the institutional agenda in 2003, a year marked by considerable upticks in *Congressional Record* and hearing activity. Most importantly, this activity culminated in an important policy innovation, the 21st Century Nanotechnology Research and Development Act.

Policy change resulted from a number of interrelated factors. Nanotechnology enthusiasts and entrepreneurs, like Mihail Roco, perpetuated a narrative portraying nanotechnology as an important technology worthy of a significant government investment. This narrative was supported by various indicators demonstrating that the U.S. was well-positioned to meet these goals, although competition from abroad abounded. The Science Committees were receptive venues and helped steward the 21st Century Nanotechnology Research and Development Act through Congress. Powerful interest groups, most notably the NanoBusiness Alliance, worked diligently to access these venues and promote policy change. What is more, the Act helped assuage fears, and potential dissent, by integrating mechanisms to identify potential health, environmental, and social hazards in advance. All of the key ingredients for policy change—ideas, institutional support, evidence (e.g., indicators), entrepreneurs—harmoniously converged during this period.[80]

Very few groups mobilized against the 21st Century Nanotechnology Research and Development Act. Low conflict and little visibility are indeed hallmarks of "distributive" policymaking.[81] The 21st Century Nanotechnology Research and Development Act also benefited from the creation of the creation of NNI in 2000. By the time the NanoBusiness Alliance and others requested congressional support, a fully functioning research infrastructure was already established. By funding nanotechnology R&D, the NNI created a political economy for nanotechnology, stoking the interest of the same industry and academic groups that later lobbied for the 21st Century Nanotechnology Research and Development Act. In other words, the NNI helped create a nanotechnology subsystem. The 21st Century Nanotechnology Research and Development Act ultimately crystallized this subsystem's monopoly over the nascent nanotechnology domain.

Other factors converged to support policy change. Figure 3.3, which shows the number of stories published in the *New York Times* containing the word "nanotechnology" between the years 1997 and 2014, documents a growing media interest in nanotechnology after 2001 (32 stories). This

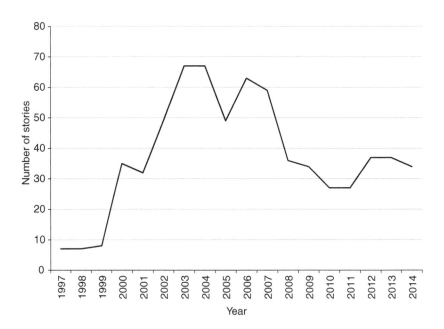

*Figure 3.3* Number of Stories on Nanotechnology in the *New York Times*, 1997–2014 (source: Search on Lexis-Nexis for the term "nanotechnology" in the *New York Times*).

attention conveniently peaked in 2003 (67 stories) and 2004 (67 stories). What is more, Figure 3.3 also reflects the prolonged issue attention described previously, as media coverage rose again in 2006 (63 stories) and 2007 (59 stories), a reflection of the second round of policymaking described later in the chapter.

Opinion polls from this period also indicate a favorable view of nanotechnology, at least among those who were aware of the new technology. A 2002 survey revealed more than 57 percent of respondents felt that "human beings will benefit greatly from nanotechnology."[82] Other surveys, however, showed large segments of the general public were largely unaware of nanotechnology. Roco cites survey data showing that, in 2003, 32 percent of those questioned in U.S. opinion polls responded "don't know" on the topic of nanotechnology. This also helps explain why the agenda setting process was marked by such low visibility. However, "of those who know the majority thinks that nanotechnology will improve the quality of life."[83]

The 21st Century Nanotechnology Research and Development Act constitutes a watershed moment—a policy innovation—in the nanotechnology domain. A new subsystem solidified its monopoly over this emerging area, dictating the dominant narrative and controlling the institutions associated with this domain. However, before this pro-nanotechnology subsystem could comfortably settle into a period of outright stasis, a second round of policymaking was needed to address growing concerns about the potential dangers associated with nanotechnology.[84] In particular, a counter-mobilization movement argued that the proverbial pendulum between risk minimization and R&D had swung too far in favor of the latter, and that Congress had to balance this divide.

### Asserting Risk: Policy Maintenance and the 110th and 111th Congresses

Policymaker attention to nanotechnology spiked for a second time in the 110th Congress (2007–2008) and 111th Congress (2009–2010). Perhaps more accurately, the 21st Century Nanotechnology Research and Development Act, and nanotechnology in general, never fully faded from Congress's institutional memory. Throughout the 2005–2006 Congress, the Senate and House Science committees held a series of hearings on nanotechnology. Often times, these hearings were more or less a formality in that they set out to review the progress of the research infrastructure developed by the 21st Century Nanotechnology Research and Development Act.[85]

However, this period also saw the emergence of a concerted effort to reduce nanotechnology's risks. In 2005 the Woodrow Wilson International Center for Scholars and the Pew Charitable Trusts partnered to form the Project on Emerging Nanotechnologies (PEN), which worked to minimize nanotechnology's risks.[86] PEN's leaders, most notably David Rejeski, the group's Director, and Andrew Maynard, the group's Chief Science Advisor, were two of the most important policy entrepreneurs during this period. Both individuals were unrelenting in their quest for stronger risk regulation, regularly appearing before the House and Senate Science Committees. Maynard was so respected that his personal blog, "2020 Science", became an important venue for information exchange in its own right. Members of the nanotechnology community frequented his page, interacting with Maynard through the "comments" section and even offering "guest" postings themselves.[87]

Moreover, a fairly sophisticated risk assessment infrastructure was in place by 2005 that produced a number of studies demonstrating the potential dangers of nanotechnology. These studies came to represent important indicators of nanotechnology's risks. Figure 3.4 illustrates this trend and tracks the number of peer-reviewed publications on nanotechnology's health, environmental, and safety risks between the years 1997 and 2013. The number of studies on nanotechnology's risks skyrocketed over this period, from a mere 98 publications in 1997 to 1,432 in 2013. Particularly robust growth was reported between the years 2005 (179 publications) and 2010 (928 publications).[88]

As early as 2005, *Washington Post* reporter Rick Weiss observed:

Amid growing evidence that some of the tiniest materials ever engineered pose potentially big environmental, health, and safety risks, momentum is building in Congress, environmental circles and in industry itself to beef up federal oversight of the new materials, which are already showing up in dozens of consumer products.[89]

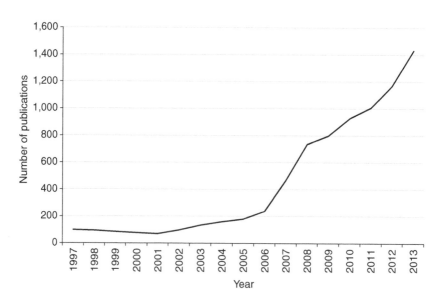

*Figure 3.4* Number of Peer-Reviewed Publications on Nanotechnology's Health and Environmental Risks, 1997–2013 (source: International Council on Nanotechnology's "Nano-EHS Database Analysis Tool").

By this time, it was abundantly clear that the very same characteristics that made nanotechnology so promising could also pose its greatest risks. That nanotechnology possesses the capacity to permeate biological systems otherwise inaccessible to normal sized particles was one of the most notable findings of this period. Although some felt this characteristic represented a unique opportunity to deliver drugs (e.g., cancer treatment),[90] others were petrified by the thought of nanoparticles accumulating in individuals' brains. Scientists also confirmed nanoparticles' ability to permeate the placenta of pregnant women as well as human lung tissue[91] and, in 2004, environmental scientist Dr. Eva Oberdörster found exposing fish to nanoparticles for only 48 hours resulted in severe brain damage.[92]

But no finding did more to galvanize opponents than the discovery that nanotechnology could produce asbestos-like damages in the lungs. In 2004, NASA researchers found that when carbon nanotubes reached the lungs they were more toxic than both carbon black and quartz. A year later, researchers at University of Texas, El Paso demonstrated the cell toxicity of carbon nanotubes was identical to that of chrysotile asbestos.[93]

Unlike the "blood–brain" barrier discovery, which was a first in the world of environmental and biological science, the asbestos finding allowed problem definers to analogize nanotechnology to a highly negative experience from the past. The asbestos finding sparked concern within insurance and legal industries, as well as among worker protection groups, that nanotechnology would result in a deluge of product liability cases similar to the "asbestos litigation crisis" of the 1990s.[94] Supported by a growing body of medical evidence that asbestos inhalation causes lung cancer and mesothelioma, a wave of litigation beginning in the 1970s and cresting in the 1990s charged the asbestos industry with knowingly suppressing knowledge of the hazardous effects of its products. In a 2005 report published in *Nanotechnology Law & Business*, Attorney John Monica and colleagues underscored the relevance for nanotechnology:

> Some critics have already made the asbestos analogy when discussing carbon nanotubes, which are currently one of the most commercially promising and widely researched nanomaterials. The asbestos litigation crisis provides a useful case study to explain the importance of keeping products liability law in mind from the beginning of a product's life cycle to the end.[95]

Beyond these indicators, the various evaluative reports and reviews mandated by the 21st Century Nanotechnology Research and Development Act began to produce evidence and feedback implying executive agencies were struggling to develop an adequate regulatory regime for the new technology. In December 2008, the NRC concluded that the 18 government agencies charged with assessing chemical safety failed to prove nanoparticles are not dangerous. Their report stated that nanotechnology risk research needed to be "proactive—identifying possible risks and ways to mitigate risks before the technology has a widespread commercial presence."[96] As noted in Chapter 2, this type of negative feedback can be an important catalyst for policy change *in-and-of-itself*, let alone in conjunction with troublesome indicators of risk.[97]

Many observers felt the NNP had failed to make good on its promise to consider health and environmental risks. "The government agencies responsible for protecting the public from the adverse effects of these technologies seem worn and tattered," wrote former EPA assistant administrator Clarence Davies in a PEN commissioned study.[98] Without conclusive scientific proof as to the dangers of nanotechnology the government agencies responsible for protecting human health and the environment simply could not adopt nano-specific regulations. Keith Blakely, Chief Executive Office NanoDynamics, Inc., a private company that uses nanotechnology to, among other things, produce "clean" energy, argued that reducing this uncertainty was important not only for protecting human health but for ensuring a prosperous business environment as well:

> In looking at the Nanotech space from an EHS perspective it is clear that we must drastically reduce uncertainty surrounding environmental, health, and safety issues of nanomaterials. This is important not only for the safety of the public but also for the success of nanotech industries that depend on consumers not harboring unfounded or ill-informed fears that will keep them from buying nanotech products.[99]

Of course, by limiting certain agency expenditure lines, the 21st Century Nanotechnology Research and Development Act set the stage for a second burst in policy activity. But what made the period between roughly 2007 and 2010 so distinctive was that Congress not only entertained substantive funding changes but also alterations to the actual content of the 21st Century Nanotechnology Research and Development Act in the face of new knowledge of the technology's risk. Some groups called for an outright ban on all nanotechnology products. Friends of the Earth advocated for a "moratorium on the further commercial release of sunscreens, cosmetics and personal care products that contain engineered nanomaterials, and the withdrawal of such products currently on the market, until adequate, peer-reviewed safety studies have been completed."[100] The ETC Group, an advocacy organization that addresses socioeconomic and ecological issues with new technologies, also called for a moratorium, indicating that:

> Given the concerns raised over nanoparticle contamination in living organisms, Heads of State ... should declare an immediate moratorium on commercial production of new nanomaterials and launch a transparent global process for evaluating the socio-economic, health and environmental implications of the technology.[101]

Of course, such an extreme proposal did not go unopposed. Under Secretary of Commerce for Technology Phillip J. Bond responded in this way:

> Those who would have us stop in our tracks argue that it is the only ethical choice. I disagree. In fact, I believe a halt, or even a slowdown would be the most unethical of choices.... Given the promise of nanotechnology, how can our attempt to harness its power at the earliest opportunity—to alleviate so many of our earthly ills—be anything other than ethical? Conversely, how can a choice not to attempt to harness its power be anything other than unethical?[102]

Many groups, including PEN, opted for a moderated strategy, arguing the problem was not that nanotechnology imposed risks, but that government—despite the provisions in the 21st Century Nanotechnology Research and Development Act—simply had not done *enough* to anticipate and prepare for these risks. Although they also drew on the various studies demonstrating nanotechnology's risk, PEN and others argued the amount of federal money spent on health and environmental concerns was disproportionately lower than the amount spent on R&D—less than 1 percent according to some estimates. What is more, they added, the fact that larger numbers of products containing nanotechnology were entering the marketplace magnified the importance of solid risk research.[103]

Against this backdrop, the House Committee on Science held a hearing on April 16, 2008, to solicit input on a proposal to amend the 21st Century Nanotechnology Research and Development Act. Few presenters more succinctly summarized the concerns of the counter-mobilization movement than PEN science advisor Andrew Maynard. During one portion of his comments, fittingly titled "The Need for Foresight," Maynard stated:

> Moving towards the nanotechnology future without a clear understanding of the possible risks, and how to manage them, is like driving blindfolded. The more we are able to see where the bends in the road occur, the better we will be able to navigate around them to realize safe, sustainable and successful nanotechnology applications. But to see and navigate the bends requires the foresight provided by strategic science.[104]

Maynard estimated that an additional $20 million to $100 million per year should be invested in risk research over and above current funding levels.[105] In a separate analysis, the Environmental Defense Fund, a non-profit environmental advocacy group, asked for $100 million or more in funding to study environmental, health, and safety risks.[106]

Some policymakers scoffed at the notion that the U.S. was not doing enough to ensure health and safety. White House Office of Science and Technology Policy (OSTP) director John Marburger stated that the U.S.

"leads the world not only in spending for nanotechnology development, but also, by an even larger margin, in its investment in research to understand the potential health and safety issues."[107] Yet, PEN's (and others') concerns about nanotechnology safety found traction in Congress. In his opening statement at a 2008 committee hearing, Chairman of the House Committee on Science Bart Gordon (D-TN) highlighted the importance of using sound science to guide policymaking activity:

> Although the NNI has from its beginning realized the need to include activities for increasing understanding of the environmental and safety aspects of nanotechnology, it has been slow to put in place a well-designed, adequately-funded, and effectively executed research program.
>
> The environmental and safety component of NNI must be improved by quickly developing and implementing a strategic research plan that specifies near-term and long-term goals, sets milestones and timeframes for meeting near-term goals, clarifies agencies' roles in implementing the plan, and allocates sufficient resources to accomplish the goals.
>
> This is the essential first step for the development of nanotechnology to ensure that sound science guides the formulation of regulatory rules and requirements. It will reduce the current uncertainty that inhibits commercial development of nanotechnology and will provide a sound basis for future rule making.[108]

The counter-mobilization groups' efforts paid off and policymakers dramatically increased funding for safety research. PEN estimated that only about $13 million was spent on risk research in 2006.[109] Funding for environmental, health, and safety research jumped to $74.5 million in FY2009 and then again to $91.6 million in FY2010. The Obama Administration's FY2011 budget proposed to increase funding for health and safety research again, this time to the tune of $116.9 million.[110] Proponents of greater federal recognition of nanotechnology's risks were elated, especially Maynard who joyously proclaimed in his blog "The unthinkable has happened!"[111]

However, some policymakers had grander visions for the future of the 21st Century Nanotechnology Research and Development Act. A series of fairly comprehensive amendments were introduced in the House and Senate, which set out to clarify many Act's provisions relating to risk. In fact, the House version, entitled the National Nanotechnology Initiative Amendments Act of 2008 (H.R. 5940), offered more than 32 pages of amendments alone, a significant addition to a law that was only ten pages in length. Among H.R. 5940's various changes, one provision mandated the National Nanotechnology Coordination Office to maintain a public database of all NNI-funded environmental, health, and safety research. The database was intended to serve as a centralized source of information for regulators looking to make sense of, and interpret, the disparate studies relating to nanotechnology's safety. The amendments also required the National Nanotechnology Advisory Panel establish a subpanel to assess whether the NNI was sufficiently considering health, environmental, and social concerns. Further administrative restructuring required the associate director of the White House Office of Science and Technology Policy serve as Coordinator for the Societal Dimensions of Nanotechnology, a position that would be responsible for establishing annual research plans for federal nanotechnology environmental, health, and safety activities; monitoring all environmental, health, and safety programs; and encouraging public–private partnerships to support environmental, health, and safety research.

The Senate amendments bill (S. 3274), like the House version, also established a separate subpanel on environmental, health, and safety concerns, and created a Coordinator for Societal Dimensions of Nanotechnology. S. 3274 also proposed a number of measures not directly integrated into the House bill. For example, it called on the National Nanotechnology Program to solicit and draw on the advice of industry groups when determining which nanoparticles might pose risks. This cooperative regulatory approach was, again, an outgrowth of nanotechnology's enormous uncertainty. Logic dictated industry groups would be the first to identify many of nanotechnology's risks and thus would be well-positioned to help reduce scientific uncertainty, assuming they could be forthright about these risks.

On June 5, 2008, H.R. 5940 passed the House by an overwhelming majority (407–6). Yet securing the legislation's passage in the Senate proved difficult. Despite momentum for stronger oversight of environmental, health, and safety issues, the Senate Committee on Science did not report its bill to the floor before the end of the 2007–2008 Congress. An identical bill was introduced in the 111th Congress (2009–2010), once again passing House but stalling in the Senate Science Committee. No comprehensive reauthorization legislation was introduced in the 112th (2011–2012) or 113th Congresses (2013–2014).

A handful of other legislative proposals have attempted to address the environmental, health, and safety concerns associated with nanotechnology. For example, the Nanotechnology Safety Act of 2010, introduced by Senator Mark Pryor (D-AR) in January of 2010, amends the Federal Food, Drug, and Cosmetic Act to require that the Secretary of HHS establish a program within the FDA specifically dedicated to scientific investigation of the potential risks of nanoscale materials and, in particular, the implications of their interaction with biological systems. The bill was referred to the Senate Committee on Health, Education, Labor, and Pensions, but has yet to find its way to the Senate floor.

Despite these failed statutory efforts, the second round of nanotechnology policymaking proved a success for proponents of more robust risk regulation. In response to the proposed 2011 funding increases, Maynard blogged:

We're not out of the woods yet on ensuring we have the information needed to develop and use new nanotechnology-based materials and products safely. But it looks like the U.S. is making progress. And that's good news for anyone hoping to see the emergence of strong nanotechnology-based solutions to a whole host of challenges.[112]

Evidence suggests investments have been sustained. Environmental, health, and safety investments have risen from under 3 percent of all NNI funding in 2005 to over 7 percent in the proposed 2015 budget, totally close to $900 million over this ten-year period.[113]

But for the agencies responsible for protecting human health and the environment, greater funding did not eliminate the extraordinary difficulties associated with regulating nanotechnology. In the years following the 2003 enactment of the 21st Century Nanotechnology Research and Development Act, a host of federal regulatory bodies struggled to determine how best to cope with the potential dangers of nanotechnology without imposing unnecessary barriers to innovation and basic research. This struggle is examined in greater detail below. But first, let us examine the various factors that converged to induce this second round of policy-making activity.

*Agenda Catalysts*

The second round of policy activity followed a much different trajectory than the first. Whereas the first round was marked by abrupt policy change, the second round of activity and attention spanned nearly four years—two Congresses—as policymakers gradually developed policies to address nanotechnology's risks. And, unlike the first round, this second iteration of activity did not end in policy innovation, although the considerable increase in funding for risk research and regulation can certainly be characterized as a form of policy maintenance.[114]

What factors catalyzed this uptick in issue attention? For one, it was partially induced by the 21st Century Nanotechnology Research and Development Act itself, which capped important expenditure lines. But more importantly policymaking resulted from a much more sophisticated counter-mobilization effort, the likes of which did not exist when the 21st Century Nanotechnology Research and Development Act was enacted. PEN, which is home to some of the most strident policy entrepreneurs in Rejeski and Maynard, was founded in 2005—two years after the Act was signed. Moreover, just as a plethora of different industry and academic groups lobbied on behalf of increased R&D money, so too did a variety of environmental, public health, and social justice groups mobilize for tighter risk regulation.

Still, proposals advocating for an outright ban on nanotechnology or even stringent regulations were never truly entertained, as pro-nanotechnology groups controlled—dominated—important policymaking venues and narratives. Instead, counter-mobilization efforts succeeded at the margins by arguing within the dominant frames and policymaking norms perpetuated by the NanoBusiness Alliance and other pro-nanotechnology groups. Nothing was more important to this strategy than the marshalling of peer-reviewed publications and other forms of scientific evidence as problem indicators, the same "types" of measures used to demonstrate the need for R&D funding during the first round of policymaking. In this instance, however, the articles tabulated focused on nanotechnology's dangers as well as the deficiencies in the current regulatory environment, not "positive" advances in nanotechnology R&D. Counts and measures of peer-reviewed publications were coupled with data showing a gulf between R&D and risk research funding. Taken together, this data helped justify calls for increased spending on health, environmental, and safety research. In other words, PEN conformed to the cultural norms associated with the nanotechnology domain, which, since its inception, has put a tremendous amount of stock in scientific and data-driven decision making. One could also speculate PEN's affiliation with the Woodrow Wilson International Center for Scholars, a prominent think-tank in Washington, D.C., only bolstered their legitimacy.

Of course, a more cynical reading of this second round of policymaking could argue that the health, environmental, and safety research provisions included in the 21st Century Nanotechnology Research and Development Act were primarily an outgrowth of industry's desire to squelch dissent and stave off the type of public backlash observed in the GMO and biotech debates. The NNI allocated only a fraction of money toward risk research and the pro-nanotechnology subsystem essentially ignored calls for tighter regulations and balked at the notion that the government was not doing enough to ensure safety. Indeed, policy change during this period was, at best, minimal and did little to tarnish Washington's overwhelmingly positive opinion of nanotechnology. In other words, one could posit that industry "co-opted" calls for a precautionary approach by proactively integrating risk regulation mechanisms in the NNI's authorizing statute, an important strategy of agenda denial.[115] This strategy was effective in that it undercut claims that the government was ignorant to these dangers. This subsystem continued to object to any proposal or narrative casting nanotechnology in a negative light during the implementation stage.

Feedback also played an important role in promoting policy change.[116] Bureaucrats and stakeholders relayed evidence of implementation difficulties back to Congress, as hamstrung regulatory agencies struggled to grapple with the enormous complexities associated with this new technology. Policy feedback, which was partially stimulated by the reporting requirements established by the 21st Century Nanotechnology Research and Development Act, was obviously not possible in the first round of policymaking given that, prior to 2001, the national government had little experience with nanotechnology R&D or risk regulation.

Finally, as shown in Figure 3.3, there was an uptick in the number of stories published in the *New York Times* 2006 (63 stories) and 2007 (59 stories), which likely helped stimulate policymaker interest. Public opinion polls also indicated growing recognition of the new technology among the American public. For example, a 2009 poll conducted by Hart Research Associates revealed: "One in three (31%) adults has heard a lot (5%) or some (22%) about nanotechnology."[117] With this policy framework in place, the stage was set for the federal bureaucracy to usher in the next Industrial Revolution. The next section reviews the challenges associated with this momentous task.

## Policy Implementation

The 21st Nanotechnology Research and Development Act created three somewhat conflicting implementation goals. First, and certainly foremost, the Act sought to promote nanotechnology R&D. To this end, implementing agencies have been very successful. By 2012, more than 1,200 U.S. companies, universities, government laboratories, and other organizations were involved in nanotechnology R&D and marketization.[118] This number constitutes a nearly 50 percent increase from 2010.[119]

The second goal, which is closely related to R&D, called for rapid marketization of nanotechnology products. Scores of products containing nanotechnology were already reaching the marketplace by 2005. Market forecasters BCC Research anticipate total worldwide revenues for nanotechnology, which stood at roughly $11.7 billion in 2009, will increase to more than $26 billion by 2015. Nanomaterials are expected to see the biggest bump, increasing from $9 billion in 2009 to $19.6 billion in 2015.[120]

However, the third goal has proven much more difficult to achieve. The Act mandated that a variety of agencies take steps to protect against the potential health and environmental risks associated with nanotechnology. This charge was twofold. First, it required funding of research on what these "risks" might be in the first place. Although proponents of more robust risk research claimed victory during the second round of policymaking, funding for environmental, health, and safety research only constitutes a fraction of the NNI's budget—less than 7 percent.[121] Second, it required that the information gathered through this research be translated into action, meaning the agencies responsible for protecting human health and the environment would take proactive measures to ensure consumers are safe. The second charge could, in part, be met by adopting rules and regulations to guide private business and consumer action.

The demand that policymakers anticipate the health and environmental implications of nanotechnology clashed with the other two goals. Regulatory policy requires adequate understanding of the source of concern. But gathering this information requires time and resources, and many believe nanotechnology's potential risks are far too great to allow an extended delay. Moreover, the 21st Century Nanotechnology Research and Development Act's first-to-the-market approach heightened the pressure to come up with regulations quickly and in the absence of solid information.[122] In fact, the World Economic Forum went so far as to argue that, given these regulatory difficulties, nanotechnology constituted one of the most significant technological risks facing the planet.[123]

Policymakers, bureaucrats, and interest groups were divided on whether the risks posed by nanotechnology were truly unique or similar to those posed by existing chemicals. Moments after the President signed the 21st Century Nanotechnology Research and Development Act in 2003, John Marburger, Policy Director at the Office of Science and Technology Policy, stated:

> The risks from nanotechnology do not differ substantially from those of other technology hazards, such as toxicity of new chemicals or new biological materials, or environmental impacts. I believe many of these concerns can be addressed with existing regulatory mechanisms.[124]

Sean Murdock, former executive director of the NanoBusiness Alliance, agreed with Marburger: "The apparatus for effective nanotechnology regulation is largely in place through various statutes and agencies."[125]

In the eyes of many, existing regulatory statutes could not possibly suffice. How can a technology so revolutionary be expected to conform to existing rules? According to Davies, it cannot:

> Nanotechnology is difficult to address using existing regulations. There are a number of existing laws—notably the Toxic Substances Control Act; the Occupational Safety and Health Act; the Food Drug and Cosmetic Act; and the major environmental laws (Clean Air Act, Clean Water Act, and Resource Conservation and Recovery Act)—that provide some legal basis for reviewing and regulating [nanotechnology] materials. However, all of these laws either suffer major shortcomings of legal authority, or from a gross lack of resources, or both. They provide a very weak basis for identifying and protecting the public from potential risk, especially as nanotechnologies become more complex in structure and function and the applications become more diverse.[126]

Steffen Foss Hansen of the Technical University of Denmark argued that existing statutes and regulations could be applied, but they would need to be amended considerably. Foss Hansen estimated that "almost every aspect of the various laws and legislations along the life-cycle of products containing nanomaterials needs to be adapted so that nanomaterials are covered within the scope of the laws, definitions are applicable, etc."[127]

A diversity of agencies, ranging from NASA to the EPA, Food and Drug Administration (FDA) to DOD, National Institute for Occupational Health (NIOSH) to DOE, grappled with nanotechnology regulation at one time or another. While it is far beyond the scope of this study to examine each and every agency's implementation experience, the remainder of this section will illustrate some of the above-noted complexities by reviewing developments at the EPA, FDA, and NIOSH. All three of these agencies have been at the forefront of the debate surrounding how best to regulate nanotechnology's risk and thus constitute ideal examples. The EPA case is particularly instructive, as nanotechnology triggered a movement to amend one of its most significant regulatory statutes, the Toxic Substances Control Act (TSCA).

## Environmental Protection Agency (EPA)

The EPA was one of the earliest federal agencies to call for more information on the relative uncertainties surrounding nanotechnology and the environment.[128] The self-proclaimed leader in research on the environmental and health effects of nanotechnology, the EPA supports investigations of nanoparticle toxicology, exposure to humans, and transport, to name a few activities. Generally, this research is funded through grants and other funding opportunities awarded on a competitive basis. The EPA's presumed authority to regulate nanomaterials stems from a number of existing statutes, including the Clean Air Act (CAA), Clean Water Act (CWA), Federal Insecticide, Fungicide, and Rodenticide Act (FIFRA), and TSCA. Together, these statutes amount to a broad writ of authority to protect against toxins that may occur in air, water, land in general, or insecticides and other pest control chemicals.[129]

Most of the agency's efforts have focused on the applicability of TCSA and FIFRA. TSCA provides the EPA with authority to maintain an inventory of all chemicals and to regulate all "new" chemicals. But nanoparticles have strained this regulatory structure, and the EPA has struggled to determine whether nanoparticles constitute "new" chemicals under TSCA. Despite their novel properties, nanoparticles are generally used to enhance existing chemical formulas, many of which are already regulated by TSCA.

The regulatory dilemmas engendered by this question are manifold. One of the most basic problems relates to the applicability of premanufacture information notices (PMNs). TSCA's PMN provision helps the EPA identify potential chemical risks prior to their commercialization. Producers of a new chemical are required to provide a PMN, after which the EPA has 90 days to approve the chemical's manufacture, require more information from the chemical's producer, or restrict the chemical's use. Because it is undetermined if nanoparticles should be viewed as new chemicals, regulators and producers alike were perplexed for some time as to whether they were required to adhere to PMN requirements.[130]

Given this confusion, the EPA began soliciting advice in 2005 through a series of public meetings. The agency established a pilot program in 2008 called the Nanoscale Material Stewardship Program (NMSP), which encouraged companies to voluntarily submit information on any nanoscale materials they produce. The EPA hoped the NSMP would allow them to gather data on nanotechnology manufacturing. A total of 29 companies reported data on 123 nanomaterials.[131] Despite relatively small participation numbers, the EPA concluded the program was a success in that it allowed them to advance their knowledge of nanoscale engineering.[132]

Soon after launch of the NMSP, the EPA returned to the question of whether or not nanomaterials constituted new chemicals subject to review under TSCA. In 2008, the EPA indicated it would be developing a "Significant New Use Rule" (SNUR) pursuant to section 5(a)(2) of TSCA, which allowed the agency to require persons and companies who manufacture, import, or process new nanoscale materials that integrate materials listed in the current TSCA inventory to submit notice to the EPA at least 90 days prior to any development activities. The EPA also proposed an information gathering rule that would require individuals processing nanomaterials to submit information on production volume, methods of manufacturing and processing, exposure and release information, and available health and safety data. Finally, the EPA proposed a rule to require testing for certain nanoscale materials already in commerce. The agency was particularly concerned about carbon nanotubes, characterizing them as different from conventional carbon compounds and likely subject to regulation as a new chemical.[133]

As predicted by John's evolutionary theory of policy change, the conflict surrounding the EPA's proposed rule was largely an extension of the debate that took place during the agenda setting and policy formulation stages, although the questions raised entailed much more technical complexity.[134] While the pro-nanotechnology policy monopoly encountered little in the way of significant opposition from Congress, the shift to new venues prompted industry groups to adopt a defensive and guarded position, as agency officials cast a more critical eye on the technology than their legislative counterparts. Industry balked at the proposed SNUR—despite the fact that it was relatively innocuous—and filed an array of legal injunctions arguing that such rules violated industry's right to maintain confidential trade secrets.[135] In fact, Dr. Richard Dennison of the Environmental Defense Fund held that the *most significant* opposition to the proposed rule came from within the Obama Adminstration, as executive branch officials argued the regulations would "stigmatize" nanotechnology products and undermine marketization efforts. Indeed, the pro-nanotechnology subsystem permeated even the highest echelons of the executive branch, it seemed.[136] In the face of this opposition, the EPA redacted their proposals and submitted a revised set of rules to the Office of Management and Budget (OMB). The revised SNUR would require companies currently working with nanotechnology to report basic information to the EPA. This was, in many respects, a watered-down version of the original rule, which would have allowed the EPA to conduct a full safety review of any company hoping to commercialize nanoproducts.[137]

The EPA has also attempted to regulate nanotechnology through FIFRA, which allows the agency to regulate products containing pesticides and other anti-microbial agents.[138] Similar to the TSCA debate, questions relating to nanotechnology's "newness" have complicated FIFRA regulation.

Nanosilver, for example, has been added to many existing pesticides, which are already registered and approved through FIFRA. Still, in 2010, the EPA opted for a precautionary approach and decided that the presence of nanoscale materials would be considered "new," regardless of whether the parent substance was already registered with the EPA.[139]

Similar to the proposed TSCA SNUR, the 2010 declaration was met with extreme skepticism by industry, which charged the regulations were arbitrary—there was little evidence confirming nanosilver truly posed novel dangers—and that "the new policy will unquestionably stigmatize the use of nanomaterials as commentators will equate nanomaterials with 'adverse effect reports'."[140] In practice, the EPA was less stringent in its application of the rule. A major test case came in December 2011, after the EPA issued a conditional registration to a company, HeiQ Material AB, making anti-microbial powders containing nanosilver. The conditional registration rule allows the EPA to register a pesticide containing an active ingredient not currently registered for a specified period on the assumption that, at the end of that period, the company producing the chemical will provide data on the potential dangers of the product.[141]

The National Resources Defense Council filed a lawsuit in the United States Court of Appeals for the Ninth Circuit challenging the EPA's decision. First, NRDC charged EPA's application of its risk calculation methodology was flawed in that it was based on the characteristics of toddlers, not infants, and therefore put consumers at risk. Second, NRDC argued the EPA failed to consider short- and intermediate-term aggregate oral and dermal exposure for substances coated with antimicrobial powders, yet another deviation from standard protocols. The Court partially vacated the EPA's decision on the grounds that the failure to consider the short- and intermediate-aggregate oral and dermal effects violated the agency's technical rules.[142] Although the EPA has yet to explicitly respond to the ruling, David L. Wallace and Justin Schenk, attorneys at the product liability firm Herbert Smith Freehills LLP, believe the ruling will force the EPA to adopt a more stringent regulatory approach.[143]

The EPA continues to face pressure to expand its regulation of pesticides through FIFRA. Environmental advocacy groups Beyond Pesticides and the Center for Food Safety (CFS) filed a second lawsuit against the EPA in December 2014, which charged the agency with failing to regulate novel nanomaterials in pesticides. The groups estimate that more than 400 unregulated products are currently on the market. In a separate public statement, plaintiffs cited the HeiQ ruling, noting that court actions are seemingly the only thing that can stir EPA action. The case has yet to be adjudicated.[144]

As of this writing, the EPA has been in a regulatory "holding pattern," at least with respect to TSCA. As outlined below, Congress has made various attempts to revise the law since 2010, although none have been enacted. These changes would likely be sweeping, effectively negating the EPA's approach to not only nanotechnology regulation but an array of products. The next section reviews these efforts in greater detail.

## Nanotechnology: A Catalyst for TSCA Reform?

A number of policymakers and organized interests have called for more wholesale reform of TSCA, which has essentially remained unchanged for 30 years. While nanotechnology no doubt magnified these concerns, many observers believe the law has long stymied the EPA. With respect to chemical safety, the current statute places the burden of proof on the EPA, thus favoring industry. In other words, the EPA must prove a chemical poses a health threat before it can act to regulate the chemical, a requirement that clashes with the proactive regulatory approach outlined in the 21st Century Nanotechnology Research and Development Act. In fact, the EPA must prove danger before it can even request additional information.[145]

In Summer 2010, legislation to amend TSCA was introduced in the House, the Toxic Chemicals Safety Act (H.R. 5820), and, in the Senate, the Safe Chemicals Act of 2010 (S. 3209).[146] Both proposals sought to strengthen the EPA's regulatory power by expanding the information sharing responsibilities of the chemical industry. Among the most notable provisions were increased reporting requirements, a "right to know" provision that would allow the public to access information on manufacturing procedures, and a requirement that vulnerable populations be explicitly considered in risk-assessment programs.[147]

Bill sponsors described these bills as an opportunity to fix the EPA's regulatory inadequacies. "America's system for regulating industrial chemicals is broken," stated Senator Frank Lautenberg (D-N.J.), sponsor of S. 3209, in a public statement on April 15, 2010.[148] He added:

> Parents are afraid because hundreds of untested chemicals are found in their children's bodies. The EPA does not have the tools to act on dangerous chemicals, and the chemical industry has asked for stronger laws so that their customers are safe.[149]

Representative Henry Waxman (D-CA), co-sponsor of H.R. 5820, commented:

> Reform of the Toxic Substances Control Act is long-overdue. Under [H.R. 5820], all chemicals will be reviewed for safety, dangerous chemicals will be restricted or eliminated, and new, safer chemicals will be developed more rapidly to move our economy toward a sustainable future.[150]

Environmental and public health advocates applauded the legislation. Speaking at a conference on the day when Lautenberg's bill was announced, the EPA Administrator Lisa Jackson stated that:

> I'm really thrilled to know today, as we all sit here, in Congress for the first time we're going to see the introduction of a modern TSCA act, a brand new environmental law to deal with chemicals that are finding their way into our bodies, into our environment.[151]

Andy Igrejas, Director of Safer Chemicals, Healthy Families, a coalition of 250 environmental and public health groups, echoed Jackson's sentiments in stating that H.R. 5820 "will reduce chronic disease in this country, a burden that scientists have increasingly linked to toxic chemicals found in our homes and places of work."[152]

Industry groups countered that the legislation would create an unworkable business environment. Cal Dooley, President and CEO of the American Chemistry Council, remarked: "[W]e are concerned that the bill's proposed decision-making standard may be legally and technically impossible to meet. The proposed changes to the new chemicals program could hamper innovation in new products, processes and technologies."[153] The Society of Chemical Manufacturers & Affiliates, an association representing chemical manufacturers, voiced similar concerns indicating that H.R. 5280 would stifle innovation by imposing excessive regulatory burdens on chemical manufacturers.[154]

None of the 2010 laws were enacted. More recent proposals have set out to appeal to environmentalists and industry alike. In 2013, Senator Lautenberg and Louisiana Republican David Vitter co-sponsored the Chemical Safety Improvement Act (S. 1009). The Act required testing for chemicals on the market as well as new chemicals. It also gave the EPA authority to phase out harmful chemicals. However, Chair of the Environment and Public Works Committee Barbara Boxer (D-California) refused to refer the bill out of committee. Boxer expressed concerns that the proposal would preempt state legislation, including her home state of California's Proposition 65, which has been touted as one of the most stringent regulatory statutes in country.[155]

In the absence of a unifying national standard, states and localities have created their own nanotechnology regulations. Berkeley, California initiated most of this subnational activity in 2006 when it enacted an ordinance requiring all handlers of nanomaterials to submit toxicology reports.[156] Other states and localities have adopted or considered similar regulations, including Cambridge, Massachusetts as well as the states of Massachusetts, Maine, Wisconsin, Pennsylvania, South Carolina, and Washington.[157]

In September 2014 Senators Mark Udall (D-CO) and Vitter released a redraft of the Chemical Safety Improvement Act (S. 1009). Key provisions included maintenance of the current TSCA safety standard of a reasonable risk of injury; aggressive deadlines for establishing EPA policies, procedures, guidance documents, and actions; a requirement that the EPA identify ten high priority substances, which would be addressed immediately; explicit specification of the differences between a final agency action versus actions that would be subject to judicial review; and maintenance of the existing TSCA language regarding preemptions for state chemical statutes.[158] The American Chemistry Council as well as other chemical trade associations expressed support for the bill.[159] Boxer released a "counterproposal," on which significantly altered the Udall/

Vitter proposal. The Boxer bill included much more stringent safety standards and risk provisions; extended the number of high priority substances to 15; emphasized deadlines (like the Udall/Vitter bill); introduced a new fee structure to fund EPA actions; and eliminated any TSCA preemption of state laws.[160]

The 113th Congress could not resolve the stark differences between these proposals. However, many observers believe TSCA will be reformed soon. The Republican takeover of the Senate means Senator Inhofe (R-OK), a long time support of TSCA modernization, will replace Senator Boxer (D-CA), the most fervent opponent of the Chemical Safety Improvement Act, as chair of the Committee on Environment and Public Works. And in the House, John Shimkus (R-IL), Chairman of the House Subcommittee on Environment and Energy, introduced a TSCA reform proposal in 2014 and is a steadfast proponent of reform.[161] While nanotechnology is clearly not the sole impetus for this movement, its novel characteristics have only highlighted TSCA's flaws.[162]

### National Institute for Occupational Safety and Health (NIOSH)

Other agencies have encountered similar difficulties. NIOSH, the worker safety research arm of the Centers for Disease Control and Prevention (CDC), has scrambled to develop ways to protect workers handling nanotechnology. The agency worries that workers, including scientists handling nanotechnology products in laboratories, are likely to be the first group exposed to the new technology's risks. In turn, it established the NIOSH Nanotechnology Research Center in 2004 to coordinate the agency's research objectives.[163]

However, because NIOSH lacks regulatory authority (it can only make recommendations to OSHA), its ability to intervene directly on behalf of workers is severely limited. At best, NIOSH can study the risks associated with nanotechnology; it does not have the authority to adopt binding regulations. Nonetheless, the agency has pursued a comprehensive research agenda, publishing more than 500 peer-reviewed studies on the health effects of nano-exposure between 2004 and 2012.[164]

NIOSH established recommendations for the safe handling of carbon nanotubes and nanofibers in 2010, suggesting a reduction in worker exposure to an airborne concentration of no more than 1 microgram per cubic meter (the lowest possible number). The recommendations also established guidelines for the safe handling of nanoparticles, strategic engineering processes to control occupational exposure, and health surveillance of workers exposed to nanoparticles.[165]

In 2011, it issued an additional guidance document, focusing specifically on nano-titanium dioxide. This document sent shockwaves through the scientific and occupational hazard communities because it constituted the first time the agency distinguished between the toxicological effects of inhaling very small, nanosized particles of titanium dioxide versus larger particles of titanium dioxide. NIOSH held that size and surface area should constitute primary criteria for determining a substance's toxicity, an important methodological assumption given that much of the controversy over nanotechnology has centered on the toxicological implications of scale. The document concluded ultrafine, smaller particles are particularly worrisome and therefore deserving of an exposure limit of .3 milligrams per cubic meter—100 nanometers in size. Richard Denison, a senior scientist at the Environmental Defense Fund, judged that NIOSH's approach was "more clearly stated and defended and justified here than I've seen before."[166]

In 2014, NIOSH released its much anticipated 2014–2016 nanotechnology research and guidance strategic development plan, which provides a comprehensive overview of nanoparticple production and usage in commerce, offers resources for safe handling of nanoparticles, describes analytical and sampling methods, describes mechanisms for evaluating the effectiveness of workplace controls, and summarizes published research on nanotechnology's risks. NIOSH has signaled it will next turn its attention to nanosilver, a relatively high volume commercial product that received a great deal of regulatory attention from the EPA. Field and risk assessments are expected to be completed by 2016.[167] Thus, OSHA and NIOSH have largely limited their role to information sharing and education. As of this writing, OSHA has not adopted any of NIOSH's recommendations by drafting concrete regulations.[168]

### Food and Drug Administration (FDA)

Few regulators seem to have drawn as much ire as the FDA, which has the authority to regulate cosmetics, medical devices, food, drugs, and biological products. Products in these areas are anticipated to readily incorporate nanotechnology and many have already done so. Thus, the FDA will have to regulate a variety of "combination products," such as drug-devices and device-biological products, which integrate nanotechnology with current products and systems.[169]

Because the FDA originally assumed nano-ingredients were the "bioequivalent" of bulk products many feared the agency would not proactively regulate nano-enhanced products. In a PEN-sponsored review of the FDA's ability to regulate nanotechnology, Michael Taylor concluded that the FDA has absolutely no capacity to complete a "pre-market review" of nanoproducts, meaning it cannot review the safety of nanoproducts before they have been commercialized and released onto the marketplace. For Taylor, the lack of pre-market review coupled with nanotechnology's uncertainty "is an important issue in the nanotechnology cosmetic arena because of genuine uncertainty about the actual composition and properties of ingredients that are claimed on cosmetic product labels to be produced through or otherwise incorporate the benefits of nanotechnology."[170]

Although the FDA has yet to enact any binding regulations, it formed a Nanotechnology Task Force in 2006. The Nanotechnology Task Force was broadly tasked with assessing the extent to which existing rules and guidelines can sufficiently govern nanoproducts. In 2007, the Nanotechnology Task Force released a report acknowledging that the characteristics relevant to product safety may change at the nanoscale. However, the report was careful to add that more testing was required before any regulatory decisions could be rendered. The Nanotechnology Task Force also called for agency guidelines specifying disclosure and reporting requirements for private companies, and suggested that the FDA improve its data assessment program.[171]

After several years of public commentary, the FDA announced in summer 2011 that it would have an "open dialogue" about nanotechnology regulation. In May 2011, it published a series of draft guidelines in the federal register that, according to FDA Commissioner Margaret A. Hamburg, M.D., served as a "starting point" for the nanotechnology regulation discussion. Hamburg went on:

> Our goal is to regulate these products using the best possible science. Understanding nanotechnology remains a top priority within the agency's regulatory science initiative and, in doing so, we will be prepared to usher science, public health, and FDA into a new, more innovative era.[172]

In June 2014, the FDA finalized its information gathering process and issued a series of documents outlining its intended regulatory approach for nanoproducts. The FDA expressed a commitment to a thorough but non-stigmatizing regulatory approach, noting that it "does not categorically judge all products containing nanomaterials or otherwise involving the application of nanotechnology as intrinsically benign or harmful."[173] The agency specified that it intends to regulate nanotechnology products using existing statutes and, therefore, pre-market review would continue for new drugs, food additives, color additives, animal products, human devices, and certain dietary products. However, the onus of submitting data remains on industry.[174] How this approach will work in practice remains to be seen. Thus, much like NIOSH, the FDA's role has largely been limited to information gathering and sharing.

## Conclusion

The nanotechnology case represents an example of anticipatory policymaking under conditions of *extreme* uncertainty. All three cases examined in this book will feature highly uncertain problems. However, the nanotechnology case is distinctive because, in many ways, policymakers, scientists, and the general public alike have no real idea what its associated "problems" are in the first place, although scientific evidence suggest potentially dangerous interactive effects. The nanotechnology case is also made distinctive by the fact that it likely represents the purest example of anticipatory policymaking, not only in this study, but maybe—just maybe—over the last decade or so. Policymakers made a calculated attempt to construct an anticipatory governance regime, although the jury is still out on whether this framework will be able to forestall the emergence of hazardous health and environmental effects.

The uncertainty associated with nanotechnology, coupled with policymakers' desire to act with prescience, impacted the policymaking process considerably. Aside from the obvious fact that the 21st Century Nanotechnology Research and Development Act amounted to an aggressively anticipatory policy design, there remain a number of important findings relating to anticipatory problems revealed in this study. For example, the nanotechnology debate illustrated the importance of analogy as a strategic rhetorical device for limiting uncertainty—or at least the perception of uncertainty. Congressional policymaking was fairly protracted, spanning ten years and consisting of two bursts in activity (2001–2004 and 2007–2010). Indicators demonstrating both the promise and later the dangers of nanotechnology succeeded in capturing policymaker attention. Moreover, well-funded interest groups and academic specialists controlled the agenda setting process, which largely occurred outside the watchful eye of the general public. Policy design sought to encourage information gathering and R&D, and agencies struggled to determine whether new regulatory regimes are needed or if—and when—existing statutes are sufficient. Interestingly, nanotechnology has also played an important role in stimulating institutional attention to another problem, the deficiencies of TSCA. As demonstrated in subsequent chapters, many of these findings apply to my other cases as well.

The findings revealed in this chapter are preliminary in so far as the nanotechnology debate continues to unfold as of this writing. While it is impossible to predict how unresolved conflicts will unfold, one thing is certain, and that is that time will vastly improve understandings of nanotechnology. How will better information, improved knowledge, and greater clarity impact the current policymaking and governance regime? How will policymakers respond if the risks they so highly anticipate do in fact arrive? What if they never emerge? In the absence of federal regulations, is a "bottom-up" nanotechnology regulatory structure possible? The answers to these questions will prove most fruitful to our understanding of the nanotechnology debate and anticipatory policymaking in general. For now, however, the best we can do is wait and see.

## Notes

1 Sean T. O'Donnell and Jacqueline A. Isaacs. 2010. "A World of Its Own? Nanotechnology's Promise—and Challenges." In *Governing Uncertainty: Environmental Regulation in the Age of Uncertainty*, ed. Christopher J. Bosso. Washington, D.C.: Resources for the Future: 12–27.

2  Lux Research. 2014. "Nanotechnology Update: Corporations Up Their Spending as Revenues for Nano-enabled Products Increase." State of the Market Report. February 17. Available at https://portal.luxresearchinc.com/research/report_excerpt/16215 (accessed March 27, 2015).
3  Robert E. McGinn. 1990. *Science, Technology, and Society*. New Jersey: Prentice Hall; Charles More. 2000. *Understanding the Industrial Revolution*. London: Routledge.
4  National Science Foundation (NSF). 2012. "Dr. Mihail C. Roco." Available at www.nsf.gov/eng/staff/mroco.jsp (accessed March 27, 2015).
5  O'Donnell and Isaacs 2010.
6  Ibid.
7  Patrick W. McCray. 2005. "Will Small be Beautiful? Making Policies for Our Nanotechnology Future." *History and Technology* 21(2): 177–203, 178.
8  John F. Sargent. 2011. *Nanotechnology: A Policy Primer*. Congressional Research Service. January 19. Washington, D.C.: Library of Congress. Available at https://fas.org/sgp/crs/misc/RL34511.pdf (accessed March 27, 2015).
9  Sargent 2011.
10 Sargent 2011; National Science and Technology Council. 2000. *National Nanotechnology Initiative: The Initiative and Its Implementation Plan*. Committee on Technology, Subcommittee on Nanoscale Science, Engineering and Technology. July. Available at www.nsf.gov/crssprgm/nano/reports/nni2.pdf (accessed March 27, 2015).
11 Lux Research 2014.
12 John F. Sargent. 2011b. *Nanotechnology and Environmental, Health, and Safety: Issues for Consideration*. Congressional Research Service. January 20. Washington, D.C.: Library of Congress: A1. Available at www.fas.org/sgp/crs/misc/RL34614.pdf (accessed March 27, 2015).
13 Sargent 2011b.
14 "Nanotechnology Risks: How Buckyballs Hurt Cells." 2008. *Science Daily* May 27. Available at www.sciencedaily.com/releases/2008/05/080527091910.htm (accessed March 27, 2015).
15 Benedicte Trouiller, Ramune Reliene, Aya Westbrook, Parrisa Solaimani, and Robert H. Schiestl. 2009. "Titanium Dioxide Nanoparticles Induce DNA Damage and Genetic Instability *In vivo* in Mice." *Molecular Biology, Pathology, and Genetics* 69(22): 8784–8789, 8788.
16 Paul Stimers. 2008. "The Implications of Recent Nanomaterials Toxicity Studies for the Nanotech Community." *Nanotechnology Law & Business* 5(3): 313–318.
17 Organic Consumers Association. 2009. "Studies Show Nanoparticles Used in Sunscreens and Makeup can Harm the Environment." March 26. Available at www.organicconsumers.org/scientific/studies-show-nanoparticles-used-sunscreens-and-makeup-can-harm-environment (accessed March 27, 2015).
18 Erik Fisher and Roop L. Mahajan. 2006. "Contradictory Intent? U.S. Federal Legislation on Integrating Societal Concerns Into Nanotechnology Research and Development." *Science and Public Policy* 33(1): 5–16.
19 Ronald Sandler and W.D. Kay. 2006. "The National Nanotechnology Initiative and the Social Good." *Journal of Law, Medicine, & Ethics* 34(4): 675–681.
20 Geoffrey Hunt and Michael D. Mehta. 2006. *Nanotechnology: Risk, Ethics, and Law*. London: Routledge: 2.
21 Rudy Baum. 2003. "Nanotechnology: Drexler and Smalley Make the Case For and Against 'Molecular Assemblers.'" *Chemical & Engineering News* 81(48): 37–42; McCray 2005; Colin Milburn. 2008. *Nanovision: Engineering the Future*. Durham: Duke University Press.
22 Eric Drexler. 1987. *Engines of Creation: The Coming Era of Nanotechnology*. Norwell: Anchor Press.

23 Ironically, Drexler's utopian story was flipped on its proverbial head years later and used to communicate a horrifying portrayal of nanotechnology—a world overrun by aggressive and unstoppable "nanobots." See: Center for Responsible Nanotechnology. 2005. "Nanobots Not Needed." Briefing Document, March 2. Available at www.crnano.org/BD-Nanobots.htm (accessed March 27, 2015).

24 Glenn Harlan Reynolds. 2003. "Nanotechnology and Regulatory Policy: Three Futures." *Harvard Journal of Law & Technology* 17(1): 179–209, 183.

25 McCray 2005; Milburn 2008; David M. Berube. 2005. *Nano-Hype: The Truth Behind the Nanotechnology Buzz*. Amherst, N.Y.: Prometheus Books.

26 Ibid.

27 National Science and Technology Council 2000, 11.

28 Ibid.

29 National Science Foundation 2012.

30 Mihail C. Roco. 2011. "The Long View of Nanotechnology Development: The National Nanotechnology Initiative in 10 Years." In *Nanotechnology Research Directions for Societal Needs in 2020: Retrospective and Outlook*, eds. Mihail C. Roco, Chad A. Mirkin, and Mark C. Hersam. New York: Springer: 1–28.

31 Milburn 2008, 9.

32 Fisher and Majahan 2006, 6.

33 McCray 2005.

34 John F. Sargent. 2011c *The National Nanotechnology Initiative: Overview, Reauthorization, and Appropriations Issues*. Congressional Research Service. March 25. Washington, D.C.: Library of Congress: 4. Available at: http://assets.opencrs.com/rpts/RL34401_20110325.pdf (accessed March 27, 2015)/.

35 National Nanotechnology Initiative. 2014. "NNI Supplement to the President's 2015 Budget." March 25. Available at www.nano.gov/node/1128 (accessed March 27, 2015).

36 White House. 2001. "National Nanotechnology Initiative: Leading the Way to the Next Industrial Revolution." Office of the Press Secretary, January 21. Available at http://clinton4.nara.gov/WH/New/html/20000121_4.html (accessed March 27, 2015).

37 Sandler and Kay 2006, 675.

38 W.D. Kay and Christopher J. Bosso. 2003. "A Nanotech Velvet Revolution? Issues for Science Inquiry." *STEP Ahead* 3(2): 2–4, 2.

39 Patrick Di Justo. 2002. "Newt Gingrich Gets Small." *Wired*, May 20. Available at www.wired.com/science/discoveries/news/2002/05/52673 (accessed March 27, 2015).

40 Ibid.

41 U.S. Congress. House. 1999. Subcommittee on Basic Research. "Nanotechnology: The State of Nano-Science and Its Prospects for the Next Decade." June 22. 106th Cong., 1st sess. Washington, D.C.: Government Printing Office: 1.

42 Ibid., 11.

43 Ibid., 10.

44 Georgia Miller. 2008. "Contemplating the Implications of a Nanotechnology 'Revolution.'" In *The Yearbook of Nanotechnology in Society: Presenting Futures*, eds. Erik Fisher, Cynthia Selin, and Jameson M. Wetmore. New York: Springer: 215–225, 216.

45 World Council of Churches. 2005. "A Tiny Primer on Nano-scale Technologies and the Little Bang Theory." January 15. Available at www.oikoumene.org/en/resources/documents/wcc-programmes/justice-diakonia-and-responsibility-for-creation/science-technology-ethics/nano-scale-technologies.html (accessed March 27, 2015).

46 McCray 2005.

47 Michael Crichton. 2008. *Prey*. New York: Harper.
48 Ken Roseboro. 2006. "From GMO to Nano: A Familiar Debate Over a New Technology." Organic Consumers Association, September. Available at: www.organicconsumers.org/articles/article_1946.cfm (accessed March 27, 2015).
49 Ronald Sandler. 2006. "The GMO-Nanotech (Dis)Analogy?" *Bulletin of Science, Technology & Society* 26(1): 57–62.
50 Robert Doubleday. 2007. "Risk, Public Engagement, and Reflexivity: Alternative Framings of the Public Dimensions of Nanotechnology." *Health, Risk & Society* 9(2): 211–227, 212.
51 Andrew Schneider. 2010. "Amid Nanotech's Dazzling Promise, Health Risks Grow." *AOL News*, March 24. Available at www.aolnews.com/2010/03/24/amid-nanotechs-dazzling-promise-health-risks-grow/ (accessed March 27, 2015).
52 Schneider 2010.
53 Frank R. Baumgartner and Bryan D. Jones. 1993. *Agendas and Instability in American Politics*. Chicago: University of Chicago Press.
54 To clarify, all searches were conducted using ProQuest Congressional, a comprehensive database of government material and publications. The initial search gathered all *Congressional Record* entries of the term "nanotechnology." I then searched *within* these results for the term "risk."
55 Committee hearing data was also derived using ProQuest Congressional. My initial search included the terms "nanotechnology." Hearing data was retrieved by searching all fields *except* full text. This approach helps minimize the collection of hearings that only make a passing reference to nanotechnology. Hearings were then manually reviewed in order to ensure nanotechnology was, in fact, the focal topic.
56 The NanoBusiness Alliance. 2012. "Mission." Available at www.nanobca.org/about-2/ (accessed March 27, 2015).
57 Adam Keiper. 2003. "The Nanotechnology Revolution." *The New Atlantis* Summer: 17–34.
58 Peter J. May and Chris Koski. 2013. "Addressing Public Risks: Extreme Events and Critical Infrastructures." *Review of Policy Research* 30(2): 139–159, 139.
59 Eric Lindquist, Katrina N. Mosher-Howe, and Xinsheng Liu. 2010. "Nanotechnology ... What is it Good For? (Absolutely Everything): A Problem Definition Approach." *Review of Policy Research* 27(3): 255–271.
60 Ibid., 255.
61 McCray 2005, 183.
62 U.S. Congress. Senate. 2004. "Electricity Generation." Committee on Energy and Natural Resources. April 27. 108th Cong., 2nd sess. Washington, D.C.: Government Printing Office.
63 U.S. Congress. Senate. 2005. "U.S.–E.U. Regulatory Cooperation on Emerging Technologies." Subcommittee on European Affairs. May 11. 109th Cong., 1st sess. Washington, D.C.: Government Printing Office.
64 U.S. Congress. House. 2009. "Revisiting the Toxic Substances Control Act of 1976." Subcommittee on Commerce, Trade, and Consumer Protection. February 26. 111th Cong., 1st sess. Washington, D.C.: Government Printing Office.
65 U.S. Congress. House. 2014. "Nanotechnology: Understanding How Small Solutions Drive Big Innovation." Subcommittee on Commerce, Manufacturing, and Trade. July 29. 113th Cong., 1st sess. Washington, D.C.: Government Printing Office.
66 Fisher and Majahan 2006, 8.
67 U.S. Congress. Senate. 2003b. "S. 189, 21st Century Nanotechnology Research and Development Act." Committee on Commerce, Science, and Transportation. May 1. 108th Congress, 1st sess. Washington, D.C.: Government Printing Office: 1.

68  U.S. Congress. Senate. 2003. "Nanotechnology." Subcommittee on Science, Technology, and Space. September 17. 107th Cong., 2nd sess. Washington, D.C.: Government Printing Office: 60.
69  U.S. Congress. Senate. 2003b, 28.
70  Judith Eleanor Innes. 1990. *Knowledge and Public Policy: The Search for Meaningful Indicators.* New Brunswick, NJ: Transaction Books.
71  U.S. Congress. House. 2003. "H.R. 766, Nanotechnology Research and Development Act of 2003." Committee on Science. March 19. 108th Cong., 1st sess. Washington, D.C.: Government Printing Office: 31.
72  U.S. Congress. Senate. 2003b, 72.
73  Robert Cresanti. 2007. "Technology Administration Speech." National Institute of Standards and Technology, April 3. Available at www.nist.gov/tpo/publications/speechtransroundtablenanotech.cfm (accessed March 27, 2015).
74  Fisher and Mahajan 2006.
75  David H. Guston and Daniel Sarewitz. 2002. "Real-Time Technology Assessment." *Technology in Society* 23(4): 93–109.
76  Sargent 2011b, 2011c.
77  Fisher and Mahajan 2006, 5.
78  Guston and Sarewitz 2002.
79  Theodore J. Lowi. 1964. "American Business, Public Policy, Case Studies, and Political Theory." *World Politics* 16(4): 687–691; Theodore J. Lowi. 1972. "Four Systems of Policy, Politics, and Choice." *Public Administration Review* 33(4): 298–310.
80  Baumgartner and Jones 1993; John W. Kingdon. 2003. *Agendas, Alternatives, and Public Policies, 2nd Ed.* New York: Addison-Wesley Educational Publishers.
81  Lowi 1964, 1972.
82  William Sims Bainbridge. 2002. "Public Attitudes Towards Nanotechnology." *Journal of Nanoparticle Research* 4(6): 561–570, 563.
83  Mihail C. Roco. 2003. "Broader Societal Issues of Nanotechnology." *Journal of Nanoparticle Research* 5: 181–189, 187.
84  Baumgartner and Jones 1993.
85  U.S. Congress. House. 2006. "Research on Environmental and Safety Impacts of Nanotechnology: What are the Federal Agencies Doing?" Committee on Science. September 21. 109th Cong., 2nd sess. Washington, D.C.: Government Printing Office; U.S. Congress. House. 2005. Subcommittee on Research. "National Nanotechnology Initiative: Review and Outlook." May 18. 109th Cong., 1st sess. Washington, D.C.: Government Printing Office.
86  Project on Emerging Nanotechnologies 2012. "Mission." Available at www.nanotechproject.org/about/mission/ (accessed March 27, 2015).
87  See: http://2020science.org.
88  This data comes from International Council on Nanotechnology's "Nano-EHS Database Analysis Tool", which provides a comprehensive collection of all peer-reviewed articles on nanotechnology's health, environmental and safety risks. See: http://icon.rice.edu/report.cfm.
89  Rick Weiss. 2005. "Nanotechnology Regulation Needed, Critics Say." *Washington Post*, December 5, A08.
90  Gabriel A. Silva. 2007. "Nanotechnology Approaches for Drug and Small Molecule Delivery Across the Blood Brain Barrier." *Surgical Neurology* 67: 113–116.
91  Asgeir Helland. 2004. *Nanoparticles: A Closer Look at the Risks to Human Health and the Environment.* Master's Thesis in Environment Management and Policy, International Institute for Industrial Environmental Economics. October.
92  Eva Oberdörster. 2004. "Manufactured Nanomaterials (Fullerines, C60) Induce Oxidative Stress in the Brain of Juvenile Largemouth Bass." *Environmental Health Perspectives* 122(10): 1058–1062.

93 Alan Dang. 2006. "Nanotech the New Asbestos: Carbon Nanotube Toxicity." *Daily Tech*, May 5. Available at www.dailytech.com/Nanotech+the+New +Asbestos+Carbon+Nanotube+Toxicity/article2132.htm (accessed March 27, 2015).
94 Keay Davidson. 2005. "Big Troubles May Lurk in Super-tiny Tech/ Nanotechnology Experts Say Legal, Ethical Issues Loom." *SFGate.com*, October 31. Available at http://articles.sfgate.com/2005–10–31/news/17396870 _1_foresight-nanotech-institute-nanotechnology-industry-nanomaterials (accessed March 27, 2015).
95 John C. Monica, Jr., Patrick T. Lewis, and John C. Monica. 2006. "Preparing for Future Health Litigation: The Application of Products Liability Law to Nanotechnology." *Nanotechnology Law & Business* 3(1): 54–63, 55–56.
96 National Research Council. 2009. *Review of the Federal Strategy for Nanotechnology-Related Environmental, Health, and Safety Research*. Washington, D.C.: The National Academies Press: 5.
97 Kingdon 2003; Eric M. Patashnik. 2008. *Reforms at Risk: What Happens After Major Policy Changes Are Enacted*. Princeton: Princeton University Press.
98 J. Clarence Davies. 2009. *Oversight of the Next Generation of Nanotechnology*. Woodrow Wilson International Center for Scholars, Project on Emerging Technologies. Available at www.nanotechproject.org/process/assets/files/7316/ pen-18.pdf (accessed March 27, 2015) 7.
99 U.S. Congress. House. 2005. "Environmental and Safety Impacts of Nanotechnology: What Research is Needed?" Committee on Science. November 17. 109th Cong., 1s sess. Washington, D.C.: Government Printing Office: 130.
100 Sargent 2011b, 15.
101 Ibid.
102 Ibid., 16.
103 David Rejeski. 2008. "Gearing Up for the Reauthorization of the Nanotechnology R&D Act." *Nanotechnology Now*, May 16. Available at www. nanotech-now.com/columns/?article=195 (accessed March 27, 2015).
104 U.S. Congress. House. 2008. "The National Nanotechnology Initiative Amendments Act of 2008." Committee on Science. April 16. 111th Cong., 1st sess. Washington, D.C.: Government Printing Office: 41.
105 Andrew Maynard. 2010. "US Government Kicks Nanotechnology Safety Research Up a Gear." *2020 Science*, February 18. Available at: http://2020science. org/2010/02/18/us-government-kicks-nanotechnology-safety-research-up-a-gear/ (accessed March 27, 2015).
106 Sargent 2011b, 14.
107 Sargent 2011b, 12.
108 U.S. Congress 2008, 10.
109 Rejeski 2008.
110 Sargent 2011c, 20.
111 Andrew Maynard. 2009. "Nanotechnology Safety Research Funding Up." *2020 Science*, May 21. Available at http://2020science.org/2009/05/21/ nanotechnology-safety-research-funding-on-the-up/ (accessed March 27, 2015).
112 Andrew Maynard, 2010. "US Government Kicks Nanotechnology Safety Research Up a Gear." *2020 Science*, February 18. Available at http://2020science. org/2010/02/18/us-government-kicks-nanotechnology-safety-research-up-a-gear/ (accessed March 27, 2015).
113 National Nanotechnology Initiative 2014.
114 Wayne Parsons. 1995. *Public Policy: An Introduction to the Theory and Practice of Policy Analysis*. Cheltenham: Edward Elgar: 571.
115 Roger W. Cobb and Marc Howard Ross. 1997. "Agenda Setting and the Denial of Agenda Access: Key Concepts." In *Cultural Strategies of Agenda*

*Denial: Avoidance, Attack, and Redefinition*, eds. Roger W. Cobb and Marc Howard Ross. Kansas: University of Kansas Press: 3–24.

116 Kingdon 2003; Baumgartner and Jones 1993.

117 Hart Research Associates. 2009. *Nanotechnology, Synthetic Biology, & Public Opinion: A Report of Finding*. Conducted on behalf of Project on Emerging Nanotechnologies, The Woodrow Wilson International Center for Scholars. September 22: Available at www.nanotechproject.org/process/assets/files/8286/nano_synbio.pdf (accessed March 27, 2015).

118 Project on Emerging Nanotechnologies. 2012b. "US Nanometro Map." Available at www.nanotechproject.org/inventories/map/ (accessed March 27, 2015).

119 Ibid.

120 BCC Research. 2010. "Nanotechnology: A Realistic Market Assessment." July. Available at www.bccresearch.com/market-research/nanotechnology/nanotechnology-realistic-market-assessment-nan031d.html (accessed March 27, 2015).

121 Sargent 2011, 2011c.

122 J. Clarence Davies. 2008. *Nanotechnology Oversight: An Agenda for the New Administration*. Woodrow Wilson International Center for Scholars. Project on Emerging Technologies. July. Available at www.nanotechproject.org/process/assets/files/6709/pen13.pdf (accessed March 27, 2015).

123 World Economic Forum. 2008. *Global Risks 2008: A Global Risk Network Report*. January. Available at https://members.weforum.org/pdf/globalrisk/report2008.pdf. (accessed on March 27, 2015): 51.

124 Natalie J. Mikhail. 2003. "Bush Approves Billions for Nanotechnology Research." *The Badger Herald*, December 4. Available at http://badgerherald.com/news/2003/12/04/bush_approves_billio.php (accessed March 27, 2015).

125 Sargent 2011b, 25.

126 Ibid.

127 Michael Berger. 2007. "Regulating Nanotechnology: Incremental Approach or New Regulatory Framework?" *Nanowerk*, June 5. Available at www.nanowerk.com/spotlight/spotid=2027.php (accessed March 27, 2015).

128 Sargent 2011b.

129 Ibid.

130 American Bar Association. 2006. "Regulation of Nanoscale Materials under the Toxic Substances Control Act." Section of Environment, Energy, and Resources. June. Available at www.americanbar.org/content/dam/aba/migrated/environ/nanotech/pdf/TSCA.authcheckdam.pdf (accessed March 27, 2015).

131 Vahbiz Karanjia. 2011. "United States: Nanosteps Toward Regulating Nanotechnology, Part 2." Minority Corporate Counsel Association, July 19. Available at www.mondaq.com/unitedstates/x/139436/Environmental+Law/Nanosteps+Towards+Regulating+Nanotechnology+Part+II (accessed March 27, 2015).

132 Lynn L. Bergeson. 2009. "EPA Publishes NMSP Interim Report." Nano and Other Emerging Chemical Technologies Blog. Bergeson & Campbell, PC, January 14. Available at http://nanotech.lawbc.com/2009/01/articles/united-states/federal/epa-publishes-nmsp-interim-report/ (accessed March 27, 2015).

133 Karanjia 2011; Matthew M. Hoffman. 2008. "EPA Takes First-Ever Regulatory Actions Aimed at Potential Nanomaterial Risks." *Goodwin Proctor Alert*, November 6. Available at www.goodwinprocter.com/Publications/Newsletters/ClientAlert/2008/1106_EPA-Takes-First-Ever-Regulatory-Actions-Aimed-at-Potential-Nanomaterial-Risks.aspx?device=print (accessed March 27, 2015).

134 Peter John. 1999. "Ideas and Interests; Agendas and Implementation: An Evolutionary Explanation of Policy Change in British Local Government Finance." *British Journal of Politics and International Relations* 1(1): 39–62.

135 Andrew Schneider. 2010b. "Obsession With Nanotech Growth Stymies Regulators." *AOL News*, March 24. www.aolnews.com/2010/03/24/obsession-with-nanotech-growth-stymies-regulators/ (accessed March 27, 2015).

136 Richard Dennison. 2014. "A Hint of Movement in the Super Slo-Mo that is Nanoregulation at EPA under TSCA." Environmental Defense Fund Blog, October 8. Available at http://blogs.edf.org/health/2014/10/08/a-hint-of-movement-in-the-super-slo-mo-that-is-nanoregulation-at-epa-under-tsca/ (accessed March 27, 2015).

137 Ibid.; Lynn Bergeson. 2014. "EPA Withdraws Direct Final SNUR for Functionalized Carbon Nanotubes." Nano and Other Emerging Chemical Technologies Blog. Bergeson & Campbell, PC, December 22. Available at http://nanotech.lawbc.com/2014/12/articles/united-states/federal/epa-withdraws-direct-final-snur-for-functionalized-carbon-nanotubes-generic/ (accessed March 27, 2015).

138 Karanjia 2011.

139 David L. Wallace and Justin A. Schenk. 2014. "EPA Targets Nanotechnology: Hi-Ho, Nanosilver Away?" *Nanotechnology Law & Business* 207 (Fall 2014): 207–218.

140 Rosalind Volpe. 2010. "Industry Comments on EPA OPP Proposed Nano-pesticide Policy." Silver Nanotechnology Working Group. Presentation to the Office on Management and Budget. The White House. Available at www.whitehouse.gov/sites/default/files/omb/assets/oira_2070/2070_08192010–3.pdf (accessed March 27, 2015): 14.

141 Wallace and Schenck 2014.

142 *National Resources Defense Council* v. *EPA*. 2013. United States Court of Appeals for the Ninth Circuit. No. 12–70268. Filed November 7. Available at http://cdn.ca9.uscourts.gov/datastore/opinions/2013/11/07/12–70268.pdf. (accessed March 27, 2015); Wallace and Schenk 2014.

143 Wallace and Schenk 2014.

144 Beyond Pesticides. 2014. "Lawsuit Challenges EPA's Failure to Regulate Nanomaterial Pesticides." *Daily News Blog.* Available at www.beyondpesticides.org/dailynewsblog/?p=14685 (accessed March 27, 2015).

145 Schneider 2010b; Sara Goodman. 2010. "Sen. Lautenberg Introduces Chemicals Reform Bill, Saying Current Regulation 'Is Broken'" *New York Times*, April 15. www.nytimes.com/gwire/2010/04/15/15greenwire-sen-lautenberg-introduces-chemicals-reform-bil-25266.html (accessed March 27, 2015).

146 Goodman 2010.

147 Daniel Rosenberg. 2010. "It's Official: Strong TSCA Reform Bill Debuts in the House." *Switchboard*, July 23. Available at http://switchboard.nrdc.org/blogs/drosenberg/its_official_strong_tsca_refor.html. (accessed March 27, 2015); Daniel Rosenberg. 2011. "Decades of Delay: TSCA Turns 35—Chemical Industry Still Stifles Protection from Toxic Chemical." *Switchboard*, October 18. Available at http://switchboard.nrdc.org/blogs/drosenberg/decades_of_delay_tsca_turns_35.html (accessed March 27, 2015).

148 "Senator Lautenberg Introduces 'Safe Chemicals Act.'" 2010. Public Statement. Available at www.facebook.com/notes/healthy-child-healthy-world/senator-lautenberg-introduces-safe-chemicals-act/380285517798 (March 27, 2015).

149 Ibid.

150 "Chairman Rush, Waxman Release H.R. 5820, The Toxic Chemicals Safety Act." 2010. Committee on Energy and Commerce, July. Available at http://democrats.energycommerce.house.gov/index.php?q=news/chairmen-rush-waxman-release-hr-5820-the-toxic-chemicals-safety-act (accessed March 27, 2015).

151 Goodman 2010.

152 Cheryl Hogue. 2010. "Mixed Receptions for Chemical Bill." *Chemical & Engineering News* 88(31): 1.
153 Goodman 2010.
154 Christine Sanchez. 2010. "New TSCA Reform Bill Threatens Domestic Chemical Manufacturing and Innovation." Society of Chemical Manufactures & Affiliates, July 23. Available at www.socma.com/pressroom/index.cfm?subSec =3&sub=71&articleID=2528 (accessed March 27, 2015).
155 Jason Plautz. 2013. "Hearing Lays Path for TSCA Reform, but Boxer Role Unclear." *E&E Daily*, August 1. Available at www.eenews.net/stories/ 1059985441 (accessed March 27, 2015); Lynn L. Bergeson. 2014. "TSCA Reform a Viable Contender For Serious Legislative Attention Next Year." JD Supra Business Advisor, December 4. Available at www.jdsupra.com/legalnews/ tsca-reform-a-viable-contender-for-serio-04794/ (accessed March 27, 2015).
156 Suellen Keiner. 2008. *Room at the Bottom? Potential State and Local Strategies for Managing the Risks and Benefits of Nanotechnology*. Woodrow Wilson International Center for Scholars, Project on Emerging Nanotechnologies, March. Available at www.nanotechproject.org/process/assets/files/6112/ pen11_keiner.pdf (accessed March 27, 2015); "Berkeley to be first city to regulate nanotechnology." 2006. *New York Times*. November 12. Available at www.nytimes.com/2006/12/12/technology/12iht-nano. 3870331.html?_r=0 (accessed March 27, 2015).
157 John DiLoreto. 2011. "State Regulation of Nanotechnology: All Politics are Local." Global Chem Conference, March 22, Baltimore, MD. Available at: www.socma.com/assets/File/socma1/PDFfiles/gcrc/2011/JD-GlobalChem-Presentation-State-Regulation-of-Nanotechnology-3–22–11.pdf (accessed March 27, 2015).
158 Bergeson 2014.
159 Ibid.
160 Bergeson 2014; Dennison 2014.
161 Bergeson 2014.
162 John DiLoreto. 2009. "Nanotechnology a Driver in TSCA Reform Push." *NanoReg News*, February 26. Available at www.nanoregnews.com/article. php?id=299 (accessed March 27, 2015).
163 Centers for Disease Control and Prevention. 2011. "Nanotechnology." Available at www.cdc.gov/niosh/topics/nanotech/ (accessed March 27, 2015).
164 National Institute for Occupational Safety and Health. 2013. *Protecting the Nanotechnology Workforce: NIOSH Nanotechnology Research and Guidance Strategic Plan, 2013–2016.* Department of Health and Human Services, Center for Disease Control and Prevention. Available at www.cdc.gov/niosh/ docs/2014–106/pdfs/2014–106.pdf (accessed March 27, 2015).
165 Nura Sadeghpour. 2013. "NIOSH Recommends New Level of Exposure for Nanomaterials." National Institute for Occupational Safety and Health, April 24. Available at www.cdc.gov/niosh/updates/upd-04–24–13.html (accessed March 27, 2015).
166 Gwyneth Shaw. 2011. "NIOSH: Nano-Titanium Dioxide 'A Potential Occupational Carcinogen.' " *The New Haven Independent*, April 27. Available at www.merid.org/Content/News_Services/Nanotechnology_and_Development_ News/Articles/2011/Apr/27/A_NIOSH.aspx. (accessed March 27, 2015).
167 James G. Votaw. 2014. "NIOSH Releases Nanomaterial Safety Research Plans." *Environmental Leader*, May 22. Available at www.environmentalleader. com/2014/05/22/niosh-releases-nanomaterial-safety-research-plans/ (accessed March 27, 2015).
168 National Institutes of Health, Office of Research Services, Division of Occupational Health and Safety. 2014. "Nanotechnology Safety and Health Program." October. Available at www.ors.od.nih.gov/sr/dohs/Documents/

Nanotechnology%20Safety%20and%20Health%20Program.pdf (accessed March 27, 2015).
169 Sargent 2011b.
170 Michael Taylor. 2006. *Regulating the Products of Nanotechnology: Does the FDA Have the Tools It Needs?* Woodrow Wilson Center for International Scholars, Project on Emerging Technologies. October. Available at: www.nanotechproject.org/file_download/files/PEN5_FDA.pdf. (accessed March 27, 2015). 29.
171 Food and Drug Administration. 2007. *Nanotechnology Task Force Report 2007.* July. 27. Available at www.fda.gov/ScienceResearch/SpecialTopics/Nanotechnology/UCM2006659.htm (accessed March 27, 2015).
172 Food and Drug Administration. 2011. "FDA Opens Dialogue on 'Nano' Regulation." Consumer Health Information, June. Available at http://nanotechnology.wmwikis.net/file/view/Nanotech_0611.pdf. (accessed March 27, 2015): 5.
173 Food and Drug Administration. 2015. "FDA's Approach to Regulation of Nanotechnology Products." March 13. Available at www.fda.gov/ScienceResearch/SpecialTopics/Nanotechnology/ucm301114.htm (accessed March 27, 2015).
174 Ibid.

# 4 An Ounce of Prevention
## Bird Flu, Swine Flu, and the Politics of Pandemic

H5N1 avian influenza, commonly known as "avian flu" or "bird flu," is a strain of influenza virus originating in birds that can be transmitted to humans through close proximity. H5N1 was first detected in bird populations in China's Guangdong Province in 1996. The following year, the virus reemerged in Hong Kong, resulting in the deaths of thousands of birds and infecting 18 humans, six of whom died. The virus lay dormant for roughly a dozen years, before reemerging in 2003. Since then, it is known to have infected 718 individuals, killing 413 of them, all in Asia, the Middle East, and Eastern Europe. No cases have been reported in the U.S.[1]

Concerns about H5N1 influenza are twofold. First, unlike the seasonal flu, which cycles through human populations annually and typically has a similar genetic composition regardless of yearly strain, H5N1 influenza is foreign to the human immune system. This biological novelty—humankind's lack of natural immunity—makes H5N1 a very lethal virus, killing more than 50 percent of those infected. Second, scientists fear H5N1 will mutate into an airborne strain, capable of rapid human-to-human transmission. As of this writing, bird flu can only be contracted through direct contact with an infected host, such as a sick animal or human, or through direct contact with an infected host's excrement. However, scientists believe it is very probable H5N1 will develop the capacity to transmit itself from one host to another through a cough or sneeze. Such a mutation would likely spark a global H5N1 pandemic, or an infectious disease epidemic rapidly spreading through human populations across the globe.

Scientists do not know if, when, or to what extent an outbreak might occur. Yet, based on knowledge acquired from recent cases and past pandemics scientists can estimate, with varying degrees of probability, what an H5N1 pandemic might look like. The symptoms and health effects of the virus, its pathology, even the number of people who would likely be sickened and killed can all be approximated, although such approximations vary widely.

Despite the fact that not a single case has occurred on American soil, the President and Congress have allocated billions of dollars and enacted substantive policy change in order to prepare the nation for the possibility of pandemic. What factors catalyzed this dangerous but highly unpredictable—and by no means certain—threat onto the national agenda? Moreover, as described later, a different influenza strain known as the H1N1 influenza or "swine influenza" did, in fact, lead to a pandemic in 2009. How did the policymaking dynamics associated with the swine flu differ from those associated with bird flu?[2]

## Dimensions of the Problem

Avian influenza viruses (or influenza "A viruses") typically develop in the intestines of wild birds. Influenza A viruses are highly contagious and can readily spread from wild to domestic bird species, including chickens, ducks, and turkeys, through direct contact with an infected wild bird or through contaminated materials. Domesticated birds typically have no natural immunity to the virus, with consequently high incidence of sickness. Highly pathogenic strains or rapidly spreading viruses can have a mortality rate of 90–100 percent.[3]

The majority of human influenza A infections are contracted through direct contact with a sickened animal or their secretions. Human-to-human infections are rare and also necessitate direct contact in most cases. Depending on the subtype and strain of the virus, avian-borne viruses have been shown to cause "typical" flu-like symptoms, such as fever, cough, sore throat, muscle aches, and eye infections. In other instances, avian influenza viruses can cause serious and even life threatening conditions, such as pneumonia and severe acute respiratory distress.[4] H5N1 avian influenza is one of the deadliest influenza strains in decades. The virus often settles in the lungs of its victims, triggering a flood of immune cells to the lung tissue. This barrage of cells, a "cytokine storm," can suffocate victims, reducing their lungs to a sodden mass of dead tissue.[5]

While sporadic infection with avian-borne influenza virus is disconcerting, the major concern surrounding influenza A viruses stems from their ability to trigger a global pandemic. A pandemic is a substantial number of human cases of a disease that exceeds normal expectations spreading through human populations across a large geographic area. Pandemic viruses are infectious, meaning they allow for rapid and direct human-to-human transmission. Many deadly diseases are not considered pandemics. For example, cancer, despite its enormous human toll, is not considered a pandemic because it is not infectious.[6]

Airborne transmissibility or the ability to spread via a cough or sneeze is often a precursor for pandemic. Many viruses acquire these traits over time and can evolve through a variety of evolutionary processes. One of the most common scenarios sees a single host (e.g., a bird, pig, or even a human) simultaneously infected with a non-airborne strain of novel influenza and an airborne strain of seasonal influenza. Within the infected host, the two viruses can swap genetic characteristics, creating a new airborne strain.[7] Logically, each and every case provides a new opportunity to evolve.[8]

Prolonged cases of avian influenza infection—even in animal populations—tend to raise concerns within the global public health community. And within the policymaking arena, animal and especially human cases represent important problem indicators, as they point to the possibility of pandemic.

Unlike seasonal flu, humans are very rarely exposed to novel avian-borne influenza viruses and lack the antibodies necessary to combat these types of viruses. As a result, pandemic influenza strains are often highly lethal. For example, the H2N2 virus or Asian flu, which circulated among human populations from 1957 to 1968, resulted in upward of two million deaths worldwide. Nor is influenza A the only "type" of pandemic causing virus. In fact, H1N1 influenza subtypes are responsible for some of the deadliest influenza outbreaks in modern history. H1N1, which can originate in birds and pigs, was responsible for the 1918 Spanish flu pandemic, which killed 50 to 100 million people worldwide. The 1918 outbreak is widely considered the deadliest outbreak in modern history.[9] An H1N1 virus also caused the 2009 swine flu pandemic, which spread to 214 countries and resulted in close to 18,138 deaths.[10]

Combating novel influenza is difficult. Antiviral drugs, which are administered as a prophylactic to mitigate the symptoms of seasonal influenza, have yielded mixed results combating H5N1. The majority of experts agree vaccination is the most effective treatment. In a *New England Journal of Medicine* editorial, Dr. Gregory Poland writes:

> Safe and effective vaccines are likely to be the single most important public health tool for decreasing the morbidity, mortality, and economic effects of pandemic influenza—particularly in view of the reported resistance of influenza A (H5N1) to antiviral agents.[11]

Unfortunately, because they are in a constant state of flux, influenza A viruses often defy a single vaccine formula. As the virus evolves, previous vaccines tend to lose their efficacy, requiring scientists to develop new solutions. Seasonal influenza, by contrast, is more consistent in its genetic composition and circulates with greater frequency, allowing scientists to quickly draw on a variety of virus samples.[12]

H5N1 avian influenza was first reported in China's Guangdong Province in 1996. The Guangdong case was isolated to a single farm goose. The virus reappeared in 1997 and ravaged poultry populations on farms and live poultry markets in Hong Kong, killing thousands of birds. (Many of these birds were killed via mandatory culling practices, which were instituted by the Chinese government in an attempt to halt the virus's spread.) More concerning, the 1997 outbreak hospitalized 18 people, most of whom had direct contact with birds in the open-air poultry markets of Hong Kong. Six of the 18 hospitalizations ended in death. Many of the victims died as a result of acute respiratory distress syndrome (ARDS), a condition characterized by extreme difficulty breathing.[13]

International and U.S. public health agencies, such as the World Health Organization (WHO) and Centers for Disease Control and Prevention (CDC), saw the Hong Kong outbreak as a dark foreshadowing of things to come. The CDC dispatched teams of scientists to Hong Kong to draw samples of the virus, which were later subjected to scientific testing at their headquarters in Atlanta, Georgia.[14]

Fortunately, the Hong Kong outbreak did not spark the global health crisis feared by many. From 1998 through 2002, not a single case of H5N1 avian influenza was reported. Dormancy did not last, however, and, in February 2003, H5N1 reemerged, infecting a family in Hong Kong. By the end of 2003, H5N1 infected and killed individuals in China, Thailand, and Vietnam. In 2004, 46 avian influenza cases were reported, of which 32 resulted in death. Although the 2003 and 2004 infections were contracted through direct contact with infected poultry, public health authorities suspected at least one case of human-to-human transmission in Thailand. Many feared this human-to-human transmission represented yet another step toward pandemic. Public health experts warned the longer H5N1 circulated, the more likely it was to mutate into a pandemic strain.[15] Guénaël Rodier, director of the WHO Department of Communicable Disease Surveillance and Response, stated:

> There is no evidence of a big crisis. But there are enough elements to say there may be something going on.... We have enough data to be concerned. At the same time we don't have enough data to be sure.[16]

Influenza experts Robert Webster and Elena Govorkova argued the growing number of H5N1 cases was cause for concern:

> There is no question that there will be another influenza pandemic someday. We simply don't know when it will occur or whether the H5N1 avian influenza virus will cause it. But given the number of cases of H5N1 influenza that have occurred in humans to date (251 as of late September 2006) and the rate of death of more than 50%, it would be prudent to develop robust plans for dealing with such a pandemic.[17]

This prolonged evolutionary period was a rare, but welcome, opportunity to act with anticipation. Dr. Anthony Fauci, Director of the National Institute of Allergy and Infectious Disease, argued:

> Previous influenza pandemics have arrived with little or no warning, but the current widespread circulation of H5N1 viruses among avian populations and their potential for increased transmission to humans and other mammalian species may afford us an unprecedented opportunity to prepare for the next pandemic threat.[18]

He added: "Rather than react in panic, however, we need to determine what can be done now with the knowledge and resources currently available to prevent or minimize the impact of a potential pandemic."[19] Fauci, like many others in the public health community, felt H5N1 demanded an anticipatory policy approach.

The growing number of cases and deaths abroad, coupled with a heightened sense of alarm within the scientific community, crystallized a coherent policy debate. On the one hand, scores of global and domestic public health specialists advocated for sweeping preparedness policies, capable of readying the U.S.—and the world—for an H5N1 pandemic. On the other hand, some observers warned against overreaction. These individuals and groups held it was far too early to determine if H5N1 constituted a viable pandemic threat and, therefore, policymakers should restrain from enacting costly and potentially invasive public health interventions. The following section reviews this debate in greater detail.

## Discourse of Conflict

At least initially, avian influenza was primarily seen as a bureaucratic issue. Public health institutions, like the WHO and CDC, worked diligently to construct epidemiological accounts or descriptions of the disease, tracking its transmission, assessing the factors that correlate with patterns of illness, and counting the number of cases and deaths.[20] The public health domain's emphasis on data-driven decision making gave these measures heightened importance within bureaucratic and, as we will see later, legislative settings. Indeed, the avian influenza case epitomizes the pattern of "indicator lock" described by Jones and Baumgartner, as policymakers fixated on cases and deaths as the most important symbols of the pandemic problem.[21]

Representatives from the CDC travelled to Hong Kong in 1997, gathered samples of the virus, and then shipped those samples back to headquarters in Atlanta for analysis.[22] The 1997 Hong Kong incident barely caused a stir outside of the public health policy community, let alone in the halls of Congress or the White House. The CDC's dominance over this potential problem was made apparent in a 1998 article appearing in *Time* magazine, which referred to the CDC as the "Ghostbusters" of the epidemic and noted that the agency "grabbed most of the headlines" during this period.[23]

The number of H5N1 cases increased to 98 (43 of whom died) in 2005. In 2006, the virus's most lethal year to date, H5N1 infected 115 individuals in nine countries (Azerbaijan, Cambodia, China, Djibouti, Egypt, Indonesia, Iraq, Thailand, and Turkey). Seventy of those cases resulted in death.[24] Increases in the number of cases and deaths expanded issue attention, fueling a divisive problem definition conflict. Uncertainty was—and remained for some time—the defining feature of the avian influenza problem. As one public service announcement by the Minnesota Department of Health explains:

Public health experts are concerned that the H5N1 virus could change (mutate) into a form that is easily spread from one person to another. We don't know for sure whether that will happen—or when it might happen. But if it does, the result could be a global influenza pandemic.[25]

A similar announcement by the City of Long Beach, California's Department of Health and Human Services asked: "Will H5N1 cause the next flu pandemic? It is unknown at this time."[26]

Some scientists attempted to sidestep the problem's inherent uncertainty by arguing it was not a question of "if" the next pandemic would occur, but "when." Infectious disease experts Richard Webby and Robert Webster noted: "Influenza experts agree that another influenza pandemic is inevitable and may be imminent."[27] And while scientists could not conclusively determine whether H5N1 would evolve into a pandemic strain, the virus's high lethality rate, which hovered around 59 percent in the years following the Hong Kong outbreak, only magnified their concern.[28] In describing pandemic influenza as a cyclical process, these individuals minimized the perception that policy action might be in vain. Even if this incredibly threatening disease did not evolve into a pandemic strain, another virus surely would, in due time, emerge, as an influenza pandemic was long overdue.

Discourse also underscored the virus's novelty. Humankind has no immunities to H5N1, heightening the virus's lethality. Lawrence Altman, a columnist for the *New York Times*, wrote "Why should so few cases cause such drastic measures locally and apprehension globally? The main reason is [H5N1's] novelty to humans."[29] Even so, another observer commented, somewhat cynically, "novelty alone doesn't make for a pandemic ... and, so far, novelty is all the 'bird virus' has going for it."[30] Such skepticism aside, novelty matters. The scholarship on problem definition has aptly demonstrated that novel problems often capture the attention of the public and policymakers, who are attracted to new and unique issues.[31]

Analogy constituted arguably the most important rhetorical device of this period. Analogy allowed problem definers to extrapolate a future pandemic's human and economic toll. Proponents of immediate government intervention compared the avian influenza to the 1918 Spanish flu pandemic. Mike Davis, author of *The Monster at Our Door: The Global Threat of Avian Influenza*, called the Spanish flu "a template for the public-health community's worst fears about the imminent threat of avian influenza."[32] Dr. Michael Osterholm, director of the University of Minnesota Center for Infectious Disease and Policy, felt a virus similar to the 1918 flu could kill at least 70 million people worldwide.[33] Even the Congressional Budgeting Office (CBO) used the analogy to offer a slightly less extreme estimate of H5N1's human toll. They predicted a pandemic resembling the 1918 flu could infect 90 million Americans, killing two million of them.[34]

Other comparisons were drawn between the 1918 Spanish flu and the H5N1 strain of avian influenza as well. Dr. William Schaffner, chairman of preventive medicine at Vanderbilt University Medical Center in Nashville, Tennessee, argued the two strains were similar with respect to the characteristics of their victims. Schaffner noted that both viruses tend to infect young people between the ages 20 and 35. Most influenza viruses, by contrast, disproportionately impact the elderly and immune compromised groups.[35] Because young people normally have a stronger immune system, they are more susceptible to a cytokine storm or a flooding of the lungs with cells.[36] *U.S. News & World Report*'s Bernadine Healy painted a horrific picture of the avian influenza's potentially devastating impact on the young in her article "The Young People's Plague." Wrote Healy: "The victims tend to overreact to the alien virus, triggering a massive immune response called a cytokine storm, turning healthy lungs into a sodden mass of dying tissues congested with blood, toxic fluid, and rampaging inflammatory cells."[37] She went on to describe hospital corridors lined with dead young people. This frightening scenario communicated a narrative that identified a very sympathetic problem population, as few observers could accept losing a generation of young men and women to pandemic influenza.[38]

Others argued analogies to the 1918 flu were misconstrued. In the aptly entitled "Asian 'Bird Flu' Isn't Similar to 1918," Dr. Fred Levit wrote, "If appropriate measures were instituted, an epidemic would be more limited and much less dangerous than that of 1918."[39] Some even contended that avian influenza was more analogous to the 1976 swine flu outbreak. Following the outbreak of a rare strain of swine influenza among U.S. army recruits in Fort Dix, New Jersey, public health experts convinced President Gerald Ford to order a nationwide $135 million immunization program. Yet the feared global swine flu pandemic never materialized. The swine flu case is seen as one of the greatest public health miscalculations of the twentieth century, both because of its cost and because the vaccine administered itself resulted in the deaths of at least 25 Americans.[40]

Altman and others argued the "specter" of the swine flu debacle haunted "federal health officials" as they tried to cope with uncertainty surrounding H5N1.[41] Marc Siegel, an associate professor at the New York University School of Medicine, called the H5N1 a "pandemic of fear," noting:

> The swine flu fiasco of 1976 is an example of the damage that can be done by fear of a mutated virus that never quite lives up to expectations. About 1,000 cases of ascending paralysis occurred from a rushed vaccine given to more than 40 million people in response to a feared pandemic that never arrived.[42]

Some skeptics advocated a tempered approach. Dr. Jeremy Farrar of the Hospital for Tropical Diseases in Ho Chi Minh City conceded that H5N1 was a very alarming virus and that it might even be prudent to begin preparedness planning, but concluded the likelihood of the disease mutating into a pandemic strain was still very much unknown. Farrar stated:

> I think you have to say we really don't know the odds of pandemic, and people are not comfortable with that. It could fizzle out and kill 98 people—one more than the number dead today. Or it could be something like 200 million. It's terrifying if it happens, but it is very, very unlikely, I think—and it is difficult to balance those facts.[43]

The discourse surrounding avian influenza was marked by considerable uncertainty. This uncertainty no doubt contributed to the rampant use of analogy, which painted a very dire depiction of the dangers of *both* government inaction, and, for skeptics, government overreaction. Those demanding government action eventually won the day and, in 2005, the avian flu reached the agendas of both the U.S. Congress and the President.

## Agenda Setting and Policy Formulation

Similar to the nanotechnology case, issue attention in the pandemic influenza case spiked on two distinct but interrelated occasions. The first flurry of activity occupied roughly a four-year period, between 2003 and 2007. President George W. Bush initiated most of this activity in February 2003 when he requested funding from Congress to support the development of a pandemic vaccine. Years later, in 2005, the President launched a *National Strategy for Pandemic Influenza*, which set out to coordinate the national government's influenza planning.

A number of important national and even state laws were also enacted during this period, reshaping power dynamics within the pandemic policy domain. The Pandemic and All-Hazards Preparedness Act (PAHPA), which was signed into law in 2006, reorganized the national government's public health preparedness infrastructure while bolstering the capacity of governments at all levels to respond to emergency events. In addition, the Public Readiness and Emergency Preparedness Act (PREPA), which was enacted in 2005, shielded vaccine manufacturers from tort liability during public health emergencies. Finally, 26 states adopted parts of the Model State Emergency Health Powers Act (MSEHPA) between 2003 and 2007. MSEHPA, which was developed by the Center for the Law and Public's Health, a collaborative partnership between Georgetown University and Johns Hopkins University, provided states with a menu of policy options to help them better control disease outbreaks and respond to bioterrorism.

After a lull in policymaking activity in 2008, pandemic influenza burst back onto the institutional agenda in 2009. Whereas the first round of policymaking encompassed a number of years, this iteration was brief, occupying a period of less than 12 months. This time the concern was not avian influenza, but a new strain, H1N1 swine influenza. Unlike H5N1, swine flu afforded little time for anticipatory action. In the summer of 2009, H1N1 quickly spread through Mexico and the Southwestern U.S. The outbreak evolved into a full-blown global pandemic in a matter of weeks, infecting thousands of people worldwide. This second round of policymaking set out to reinforce the laws, strategies, and plans put in place years prior. Congress immediately allocated billions of dollars to support the national response effort. Soon after, President Obama issued a national emergency declaration.

Figure 4.1, which shows the number of times "pandemic influenza," "avian influenza," and "swine influenza" were entered in the *Congressional Record*, is demonstrative of the two stages of policy activity described above.[44] A significant uptick in entries occurred between 2004 (35 entries) and 2005 (252 entries). The number of *Congressional Record* entries remained relatively high in 2006 (195 entries) and 2007 (130 entries), before dropping dramatically in 2008 (41 entries). During the 2009 swine flu outbreak, *Congressional Record* entries surged to 219, but plummeted in subsequent years.

Figure 4.2, which shows the number of hearings held on the topic pandemic influenza, also reflects this trend.[45] The number of hearings significantly increased from 2004 (two hearings) to 2005 (13 hearings) and 2006 (13 hearings). Hearing activity dropped considerably in 2007 (five hearings) and 2008 (two hearings), before peaking during the 2009 swine flu outbreak (22 hearings). The following section examines these two distinct phases of policy activity in greater detail.

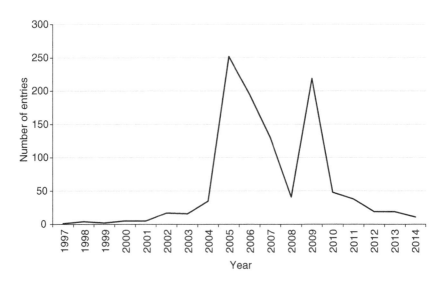

*Figure 4.1* Number of Pandemic Influenza Entries in the *Congressional Record*, 1997–2014 (source: Search on ProQuest Congressional for the terms "pandemic influenza," "swine influenza," "avian influenza," and others. Please see endnote for complete search string).

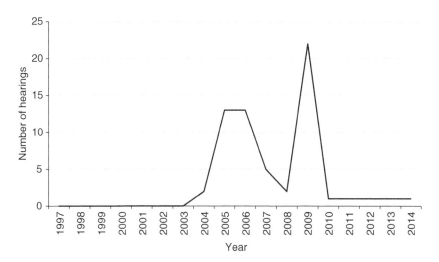

*Figure 4.2* Number of Congressional Hearings on Pandemic Influenza, 1997–2014 (source: Search on ProQuest Congressional for the terms "pandemic influenza," "swine influenza," "avian influenza," and others. Please see endnote for complete search string).

### 2003–2008: Agenda Setting and H5N1 Avian Influenza

Avian flu's ascent onto the institutional agenda was primarily driven by an accumulation of cases and death abroad, the primary indicators of the pandemic issue. Congressional attention, which began in 2003 before peaking in 2006 and 2007 (see Figures 4.1 and 4.2), largely mirrors H5N1's transmission in human populations. After a period of dormancy between 1998 and 2002, the virus reemerged in 2003. The number of cases and deaths increased incrementally between 2003 and 2005 before spiking in 2006 (115 cases, 79 deaths)—the same year most agenda activity occurred (see Figure 4.3).[46]

Evidence from this period indicates sensitivity to the growing number of H5N1 incidences. In 2005, *CNN News* reported: "Amid a growing number of cases of bird flu around the world, there is increasing global concern that the virus may mutate to a human-to-human strain and eventually lead to a pandemic."[47] At a 2005 congressional hearing, Andrew Pavia of the Infectious Diseases Society of America, a medical association of infectious disease doctors and scientists, expressed concern with the growing number of avian influenza cases:

At least 97 confirmed human cases of H5N1 infections have been documented by the World Health Organization (WHO) since January 2004 with 53 deaths. A recent WHO consultants meeting found evidence of further mutation and a suggestion that person-to-person transmission

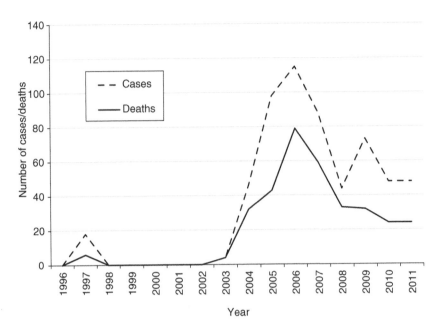

*Figure 4.3* Avian Influenza Cases and Deaths 1996–2011 (source: WHO. "Cumulative number of confirmed human cases of avian influenza A(H5N1) reported to WHO." May 2, 2012. Available at www.who.int/influenza/human_animal_interface/H5N1_cumulative_table_archives/en/index.html (accessed August 22, 2013).

might be occurring in Northern Vietnam. Should the virus become readily transmissible from human to human, the disease could easily spread beyond Asia's borders and initiate a global pandemic.[48]

Public health researchers Christina Mills, James Robins, Carl Bergstrom, and Marc Lipsitch added that:

> Over 160 human cases, about half of them fatal, have occurred, from Indonesia to Turkey. These trends suggest an increasing risk that the virus may acquire the ability to transmit efficiently from human to human, equipping it to cause a new pandemic.[49]

The Bush Administration initiated the agenda setting process, requesting $100 million from Congress in February 2003 to help prepare a pandemic influenza vaccine. This appropriation was to be made directly to the Department of Health and Human Services (HHS), a logical destination given its jurisdiction over health-related issues. Mobilizing Congressional support was difficult, and the legislative branch only granted half of the Administration's request, allocating $50 million to HHS.[50]

In February 2004, the Bush Administration requested another $100 million to help strengthen vaccine production. This time, Congress accommodated the entire request, awarding $100 million to HHS. A number of factors converged to heighten policymaker anxiety about pandemic influenza. For one, the virus reemerged in late 2003–2004 after several years of dormancy. Second, the U.S. lost more than half of its annual flu vaccine in the summer of 2004 after one of the two companies that manufactured the vaccine for the U.S., Chiron, was forced to close its manufacturing plant in Liverpool, England because of a failed inspection. The Chiron crisis came at a particularly inopportune time as 2004 was an election year and many Americans were unable to vaccinate. Various members of Congress indicated that the Chiron incident, which revealed severe inadequacies in the nation's vaccine production capacity, influenced their decision to fund the 2004 avian influenza request.[51]

Leadership changes in the executive branch also played an important role in elevating policymaker interest in avian influenza. In 2004, former-Utah Governor Michael Leavitt replaced Tommy Thompson as Secretary of HHS. According to Stewart Simonson, former Assistant Secretary of HHS, Leavitt became deeply concerned with pandemic influenza after reading John Barry's 2005 bestseller *The Great Influenza: The Story of the Deadliest Pandemic in History*, which detailed the enormous human, social, and economic toll of the 1918 Spanish flu pandemic. In fact, he distributed highlighted copies of the book to members of Congress serving on committees with jurisdiction over HHS.[52] Leavitt's concerns were magnified by the fact that there were 98 human cases of avian influenza in 2005, more than double the 2004 total of 46 cases.

With Leavitt at the helm of HHS and a growing number of cases and deaths abroad, the push for preparedness began in earnest in 2005. Executive branch agencies and departments released a series of pandemic influenza "plans," including the *National Strategy for Pandemic Influenza*, HHS's pandemic preparedness plan, the Department of Defense's (DOD) *Pandemic Influenza Preparation and Response Planning Guidance*, and an informal—unpublished—plan by the Department of Veterans Affairs (DVA).[53] Planning quickly became an integral part of the executive branch's approach to managing this emerging problem.

While all of these plans established short- and long-term goals, a severe pandemic outbreak would trigger the *National Response Plan* (*NRP*). Enacted by the Department of Homeland Security (DHS) in 2004, the *NRP*, which was renamed the National Response Framework in 2008, specified coordination among local, state, and national governments during national emergencies. While the *NRP* required that state and local governments initiate most disaster planning and guide emergency response efforts, it also specified that the response to a "large scale" crisis would be coordinated through DHS, although HHS maintained a number of powers during public health crises.[54]

The *NRP* was actually part of a much broader emergency management framework to ready the nation for "all-hazards."[55] Initiated in the wake of the September 11, 2001 terrorist attacks, the all-hazards approach to emergency and disaster management construed terrorism as but one of a variety of naturally occurring and man-made threats to the U.S. Other threats included, but were not limited to, extreme climate events, tornados, earthquakes, asteroids, and disease outbreaks. Hurricane Katrina only bolstered policymaker commitment to preparing for all-hazards, as it affirmed the tragic consequences of failing to adequately prepare for and respond to disaster.[56]

HHS planning can be traced to a "draft plan" circulated in August of 2004. It established a number of specific priorities including: (1) developing and testing antiviral drugs; (2) increasing the speed of pandemic vaccine production; (3) detecting new influenza subtypes and antiviral resistance; and (4) developing research protocols for determining the effectiveness of vaccines during a pandemic.[57] Critics charged the plan was too vague and delegated too many critical activities to the states. The American Biological Safety Association (ABSA), an organization of biological safety practitioners, argued the plan lacked solid risk communication protocols and noted that:

There will be a critical need to provide accurate technical information to health care professionals in the medical community. There will also be a need to provide accurate information that can be readily understood by the public at large during a pandemic.[58]

The President's *National Strategy for Pandemic Influenza* (*National Strategy* hereafter), released in November 2005, was arguably the most important plan of this period. The *National Strategy* established three "pillars" of implementation activities—(1) Preparedness and Communication; (2) Surveillance and Detection; and (3) Response and Containment. In conjunction with the release of the *National Strategy*, the Bush Administration requested $7.1 billion in emergency funding, which would be dispersed among the HHS, DHS, Department of Agriculture (DOA), DOD, Department of Interior (DOI), Department of State (DOS), and Department of Veterans Affairs (DVA). The lion's share of this funding ($6.24 billion) would be devoted to HHS.[59]

Vaccine development was the centerpiece of the Bush plan and the vast majority ($4.7 billion) of HHS's appropriation was allocated to this end. The Bush Administration hoped to produce 300 million courses of the vaccine within six months of an outbreak.[60] Vaccine production was projected to occur in two phases. In the first phase, roughly 80 million doses (the first 20 million would go to first responders) would be developed via the "egg-based" vaccine production method.[61] However, the Administration worried the egg-based method, which has a high rate of vaccine contamination and relies on a large quantity of eggs, was too slow and would not produce enough vaccine in time for the pandemic.[62] The remaining doses would therefore come from "cell based" vaccine production, an alternative method that grows the virus in cells. While the cell method could potentially cut production time by several weeks, it relied on noticeably underdeveloped technologies. The Bush Administration promised to invest in cell-production, although some estimated it would take as many as five years before scientists finally perfected this new approach.[63]

The *National Strategy* also requested $259 million to strengthen the CDC's global surveillance efforts. Monitoring and tracking changes in the disease would obviously be critical to ensuring the country could rapidly respond to and manage a pandemic outbreak. Another $644 million was requested to support federal, state, and local preparedness efforts.[64] States were anticipated to face a number of strains, including overtaxed medical facilities and worker absenteeism. Moreover, state health agencies were required to develop their own preparedness plans, outlining their emergency communication networks, quarantine protocols, and vaccine distribution strategies.[65]

Because they address problems that have yet to occur, anticipatory policies inevitably seek to strategize, plan, and prepare. The *National Strategy* achieved all of these ends. In pitching his plan, the President himself relied on the rhetoric of anticipation, comparing an H5N1 pandemic to a forest fire—a problem that, while potentially devastating, can be mitigated through proactive measures. "A pandemic is a lot like a forest fire," President Bush proclaimed. "If caught early, it might be extinguished with limited damage; if allowed to smolder undetected, it can grow to an inferno that spreads quickly beyond our ability to control it."[66] In another statement, President Bush evoked similar sentiments: "There is no pandemic flu in our country, or in the world, at this time. But if we wait for a pandemic to appear, it will be too late to prepare."[67]

The *National Strategy* gave the President a pragmatic and politically advantageous approach for coping with uncertainty. Government could not stop a pandemic, but it could certainly prepare for one by obtaining a vaccine, purchasing antivirals, developing plans, and surveying the disease's spread. Sarah Lister, a Specialist in Public Health and Epidemiology at the Congressional Research Service, called the development of a vaccine the "cornerstone" of the Bush plan and the "primary measure for influenza prevention."[68] To inoculate someone in advance of a pandemic is to avert illness through anticipation, nothing more, nothing less. Surveillance represents an equally proactive policymaking approach. Similar to nanotechnology policymaking, which readily utilized nodality and treasure instruments to incentivize the production of information documenting nanotechnology's risks, identifying and interpreting indicators of the pandemic problem was critical to helping policymakers determine an appropriate policy response.

In December 2005, a month after the President introduced the *National Strategy*, Congress appropriated $3.8 billion of the requested $7.1 billion.

The appropriation was said to represent a "first installment" and an additional $1.8 billion was allocated before the end of FY2006.[69] Reaction to the President's plan was mixed. Jim Greenwood, president and CEO of the Biotechnology Industry Organization (BIO), applauded "the President for his leadership"[70] Conversely, Ruth Berkelman, professor of public health at Emory University, praised the plan for being "well reasoned" but indicated that the U.S. still had a "long way to go" before it is fully prepared for pandemic influenza.[71]

But many Democrats worried the strategy was too little, too late.[72] Senator Edward Kennedy (D-MA) took aim at what he believed was an overreliance on a flu vaccine: "We need to strengthen the capacity of hospitals and health care facilities to respond and react to a pandemic. Stockpiles alone aren't enough without the capacity to make use of them."[73] Respected medical journalist Jane Parry shared Kennedy's concerns noting: "Influenza vaccines are no magic bullet. Work on a pandemic vaccine continues in several countries, but the true efficacy of a vaccine would become apparent only when used."[74]

Congress overwhelmingly framed avian influenza as a public health issue. From FY2004 to FY2009, Congress appropriated roughly $7 billion for avian influenza-related programs. Of this, 86 percent (approximately $6 billion) was allocated to HHS agencies, like the CDC and National Institutes of Health (NIH).[75] Congressional committees with jurisdiction over health issues dominated the policy debate and the vast majority of laws introduced during the 108th, 109th, and 110th Congresses were referred to the House Committee on Commerce and Energy's Subcommittee on Health (Subcommittee on Health hereafter) or the Senate Committee on Health, Education, Labor, and Pensions (Senate HELP committee hereafter).

Both of these legislative venues supported the idea that avian influenza required a proactive public health approach. On May 26, 2005, the Subcommittee on Health held a hearing entitled *The Threat of and Planning for Pandemic Flu*, which solicited advice from various of public health experts on how best to prepare and plan, including Dr. Julie Gerberding, Director of the CDC; Dr. Bruce Gellin, Director of the National Vaccine Program Office; and Dr. Anthony Fauci, Director of the National Institute of Allergy and Infectious Diseases. The policies discussed during this hearing incorporated many of the same goals outlined in the *National Strategy*, including the development of vaccines, stockpiling of antiviral drugs, and cultivation of state and local preparedness measures.

Much of the legislation introduced during this period put a special emphasis on acquiring a vaccine. The Influenza Preparedness and Prevention Act of 2005 (H.R. 4245), which was referred to the Subcommittee on Health, sought to incentivize greater vaccine production by loosening the liability restrictions on drug manufacturers. The proposed law also required HHS to purchase any excess vaccine doses that were not used during a pandemic. Similarly, The Influenza Vaccine Security Act of 2005 (S. 1828) set out to stimulate influenza vaccine production by relieving drug manufacturers of liability for personal harms during a pandemic. The Seasonal Influenza and Pandemic Preparation Act of 2005 (S. 2112) provided for a free pandemic and seasonal influenza vaccine program. A slew of other statutes similarly sought to strengthen the nation's capacity to produce an effective and readily available pandemic influenza vaccine.

Although avian influenza was overwhelmingly depicted as a classic public health problem, certain dimensions of the issue spanned other domains as well. While the vast majority of the appropriations made between FY2004 and FY2009 went to HHS, nearly $10 billion went to other executive branch departments and agencies including the DHS, DVA, DOI, DOA, Farm Credit Administration (FCA), and the U.S. Agency for International Development (USAID). Nor were the HELP committee and the Subcommittee on Health *the only venues* to hold hearings on avian influenza. The House Committee on Homeland Security, House Committee on Government Reform, Senate Committee on Agriculture, as well as a host of others, investigated the policy implications of avian influenza.

Avian flu's capacity to cut across various policy venues and jurisdictions accommodated at least two influential sub-narratives. The first cast avian flu as the latest in a long line of global health problems. A number of policy entrepreneurs hoped to convince policymakers to reconfigure the national public health infrastructure in order to more aggressively combat health threats on an international stage, as opposed to waiting for them to reach our shores. Proponents of this "global health" perspective argued increased global interconnectedness facilitated greater disease transmission from foreign countries—most notably, developing countries with poor health infrastructures—to the U.S.[76] Laurie Garrett, a Senior Fellow for Global Health at the Council of Foreign Relations, was especially active during this period, appearing before Congress on several occasions. In a 2005 article in *Foreign Affairs*, a widely distributed journal on foreign diplomacy and international affairs, Garrett wrote:

> The majority of the world's governments not only lack sufficient funds to respond to a superflu; they also have no health infrastructure to handle the burdens of disease, social disruption, and panic. The international community would look to the United States, Canada, Japan, and Europe for answers, vaccines, cures, cash, and hope. How these wealthy governments responded, and how radically death rates differed along worldwide fault lines of poverty, would resonate for years thereafter.[77]

A number of members of Congress offered legislation integrating these concerns about the global dimensions of pandemic influenza. Senator Barack Obama (D-IL) and Representative Nita Lowey (D-NY) offered the Attacking Viral Influenza Across Nations Act of 2005 (S. 969), which sought to expand U.S. disease surveillance capabilities by requiring the Secretary of HHS to coordinate all public and private surveillance efforts. It also mandated that the Secretary of HHS submit a proposal to the Director of WHO regarding the development of a Pandemic Fund that would be available to all countries impacted by the avian influenza. The Global Network for Avian Influenza Act (S. 1912/H.R. 4476), which was sponsored, in the Senate, by Joseph Lieberman (I-CT) and, in the House, by Representative Rosa DeLauro (D-CT), also included global health provisions, most notably the establishment of a global network for monitoring and surveying H5N1 in migratory birds.

The second sub-narrative had even greater traction in Congress and portrayed avian influenza as tantamount to bioterrorism. Preparing for influenza, it argued, was an extension of the U.S.'s broader homeland security agenda, which set out to ready for "all-hazards." Advancing this stance, the *Washington Times* wrote: "[T]he avian flu is just one of many potential biological problems—whether natural or man-made—that we must grapple with."[78] No law was more symbolic of this securitization of public health than the Project BioShield Act of 2004 (P.L. 108–276). Signed by President Bush in July 2004, Project BioShield amended the Public Health Service Act and launched a ten-year, $6 billion counter-bioterrorism program focusing specifically on preventing and combating the use of biological, chemical, radiological, and nuclear agents. Project BioShield also removed certain vaccine-testing requirements, allowing for quicker availability of important drugs.

The legacy of Project BioShield loomed large as Congress debated avian influenza. The Project BioShield II Act (S. 975), introduced by Senator Joseph Lieberman (D-CT) in the 109th Congress, expedited the drug development process for all public health emergencies, not just bioterror attacks. Project BioShield Act II, as well as two related policy proposals (the National Biodefense and Pandemic Preparedness Act and the Biodefense and Pandemic Preparedness Act), sought to couch avian influenza within the larger bioterror policy domain, as opposed to framing it as a "classic" public health issue. However, Project BioShield II ultimately died in committee.

The growing number of cases and deaths abroad coupled with the Bush Administration's bioterror agenda (and perhaps even the release of Barry's *The Great Influenza*) elevated pandemic influenza to the top of the institutional agenda. However, concern with avian influenza increased precipitously after Hurricane Katrina battered the Gulf Coast on August 29, 2005. Christine Gorman of *Time* noted:

> For the past two years, scientists, public-health officials and even a high ranking government official or two have warned about the potential danger of a deadly worldwide outbreak, or pandemic, of avian flu. But it took a couple of furies named Katrina and Rita to really bring home how much can go wrong if you don't plan for major emergencies.[79]

Katrina drove home the enormous dangers of being unprepared for disaster, resulting in widespread skepticism of government's ability to manage large-scale crises. A *Washington Post-ABC News* Poll revealed President Bush's public approval rating dipped to 42 percent in the weeks following Katrina, the lowest of his presidency up to that point in time.[80] Even HHS Secretary Leavitt called Hurricane Katrina a "wake-up call" for government officials hesitant to act on the avian flu.[81]

Members of Congress framed Katrina as an example of policy failure and harshly criticized the Bush Administration's H5N1 preparedness plan. Senator Ted Kennedy (D-MA) declared:

> We need to act, because the administration has failed to prepare adequately for pandemic flu. The danger of a major hurricane hitting the Gulf Coast was ignored until it was too late. We must not make the same mistake with pandemic flu. Other nations have taken effective steps to prepare, and America cannot afford to continue to lag behind.[82]

Representative Bill Pascrell, Jr. (D-NJ) shared Senator Kennedy's concerns: "the pandemic flu scenario is affording us much more time to prepare, but as of today it appears the nation is poised to repeat a grave error by not heeding the lessons learned from Katrina."[83]

To some, Hurricane Katrina reinforced the need for stronger public health measures. To others, it demonstrated the importance of an all-hazards approach to preparedness. The Pandemic and All-Hazards Preparedness Act (PAHPA) (S. 3678), which was enacted in the winter of 2006, appeased both camps. PAHPA amends the Public Health Service Act and adds four new titles. The first title elevates the Secretary of HHS to lead the federal public health and medical responses under Activity 8 of the NRP, which was dedicated to public health and medical preparedness and emergency response. Title one also created a new Assistant Secretary for Preparedness and Response (ASPR) to coordinate the national government's public health preparedness and emergency response. The second title establishes specific guidelines for medical providers (e.g., hospitals, universities, and laboratories) hoping to secure federal funding for public health emergencies and mandates that HHS establish benchmarks for determining public health preparedness. It established a nationwide "real-time" public health situational awareness and information sharing system that could be used by national, state, and local governments. It also required HHS to track the distribution of influenza vaccines. Title three expands the federal government's oversight of emergency volunteerism by establishing a system for verifying the credentials and licensure of potential volunteers. In addition, title three expands the Epidemic Intelligence Program, a post-graduate program run by the CDC that trains health professionals interested in epidemiology. Finally title four requires the Secretary of HHS to develop a strategic plan on biodefense and infectious disease R&D. Equally important, it establishes within HHS the Biomedical Advanced Research and Development Authority (BARDA), which is tasked with promoting the development of new medicines and vaccines to combat biological, chemical, radiological, nuclear, and other health threats.

A so-called "milestone" act for public health preparedness, PAHPA encountered very little opposition within Congress.[84] While most agreed PAHPA strengthened the nation's public health infrastructure, a number of individuals warned certain provisions amounted to federal overreach. James Hodge, Lawrence Gostin, and Jon Vernick of the Center for Law & the Public's Health argued PAHPA expanded federal power, potentially paving the way for increased national government meddling in state public health activities.[85] In many respects, their concerns echoed the broader debate over the value of "recentralizing" emergency management in the national government through the post-9/11 all-hazards homeland security regime. In the eyes of many, shifting *any* power away from states and localities, which are typically the first "levels" of government to respond to domestic crises, would undermine the country's ability to manage disasters and other catastrophic events.[86]

A similar, but much more controversial law, was also enacted in 2005. The Public Readiness and Emergency Preparedness Act (PREPA) (P.L. 109–148) authorizes the Secretary of HHS to shield vaccine manufacturers from tort liability for damages arising from disease countermeasures, including a pandemic influenza vaccine. Drug industry representatives, who lobbied heavily for PREPA, argued that, without liability exemptions, drug manufacturers would not produce vaccines during a pandemic. A number of Democrats countered by characterizing PREPA as a handout to vaccine manufacturers that ultimately compromised public safety. In a joint press release, Senators Edward Kennedy (D-MA), Tom Harkin (D-IA), and Chris Dodd (D-CT) stated: "Without a real compensation program, the liability protection in the defense bill provides a Christmas present to the drug industry and bag of coal to everyday Americans."[87] The Association of Trial Lawyers of America supported the Democrats in their opposition to the bill, arguing the indemnification of pharmaceutical companies was a gross violation of consumer rights.[88]

This period also saw a number of states increase their public health powers through the Model State Emergency Health Powers Act (MSEHPA). At the request of the CDC, MSEHPA was written by the Center for Law and the Public's Health, a collaboration between researchers at Georgetown University and Johns Hopkins University. The Act is intended to serve as a "model" and states can pick and choose provisions to meet their specific needs. MSEHPA is perhaps best known for *broadly* expanding state power during public health emergencies, allowing states to seize and destroy contaminated private property, enforce mandatory quarantines, and impose mandatory vaccinations.[89]

Despite widespread agreement that many state public health laws were antiquated and in need of revision, MSEHPA encountered criticism, particularly from civil liberties groups. The American Civil Liberties Union (ACLU) lambasted the Act for letting:

a governor declare a state of emergency unilaterally and without judicial oversight, fails to provide modern due process procedures for quarantine and other emergency powers, it lacks adequate compensation for seizure of assets, and contains no checks on the power to order forced treatment and vaccination.[90]

Sue Blevins, President of the Institute of Health Freedom, added that MSEHPA would "eliminate our freedom to choose our medical care and health treatment and potentially eliminate a broader range of our basic civil liberties."[91]

As of 2007, 33 states introduced 133 bills and resolutions drawing from the various provisions of MSEHPA. Forty-eight of these bills and resolutions were enacted.[92] Thus, like their colleagues in the 109th and 110th Congresses, many states found the threat of pandemic influenza too great to ignore. In a period of roughly three years, the pandemic domain experienced substantive policy change at both the federal and state levels of government. The following section considers the various factors that ushered in this dynamic period of change.

*Agenda Catalysts*

From 2004 to 2007 the national government and a number of state governments passed important public health legislation, restructuring federal agencies, loosening liability requirements on vaccine manufactures, and, in some respects, altering the very dynamics of national and state power. At the same time, the executive branch created a number of important strategies and plans, which set out to prepare the country for a looming pandemic. In sum, the pandemic policy domain unequivocally experienced substantial policy change and, most importantly for this study, this change *preceded* the onset of an actual emergency event.

Untangling the actual political implications of these various changes is complex. One thing that is abundantly clear is that this period saw a marked expansion of government power. From PAHPA to MSEHPA, the *National Strategy* to PREPA, laws and plans provided national and state governments with vastly expanded powers to prepare for and respond to all public health crises, including avian influenza. Indeed, even PREPA, despite being criticized as a "kickback" for pharmaceutical companies, gave HHS the extraordinary authority to suspend tort liability.

This broad expansion of government power, particularly at the national level, is in part a testament to the proliferation of the all-hazards approach to risk governance, which successfully infiltrated the public health domain and equated disease management to homeland security. The encroachment of this so-called "boundary spanning regime" into the public health arena began shortly after 2001 anthrax attacks and seemingly continued with every subsequent crisis.[93] Hurricane Katrina reaffirmed the need to be ready for anything, including a potential influenza pandemic. Defining avian influenza as a threat to national security not only legitimized an expansion of government power over private individuals, but also supported the federal government's control over certain aspects of state public health activities, as evidenced in the provisions in PAHPA mandating that the federal government oversee emergency volunteer programs. These laws elevated the importance of those factions within the federal government working to reframe disease as an extension of the broader bioterrorism and homeland security agenda. Indeed, the President himself even floated the possibility of imposing a military quarantine in the face of an outbreak, an idea that many believed ran counter to public health best practices.[94]

While these changes certainly tightened the bond between homeland security and public health, they did not, on balance, *diminish* the role of the public health community. If anything, this marriage strengthened the power of public health officials, providing them increased authority within the broader national security regime. Whether they actually wanted this authority is obviously fodder for a different discussion. For example, PAHPA *elevated* the importance of HHS within the homeland security bureaucracy and practically gave the department unfettered control over National Response Framework's Activity 8, which guided the nation's response to health emergencies. Nor is it accurate to deem this a period of unbridled federal expansion. As noted throughout, the threat of avian influenza jumpstarted state adoption of the MSEHPA, a policy model that vastly expanded the emergency powers of state governors and other public health officials.

In sum, the policy change experienced during this period was so dynamic—so multifaceted—that almost everyone seemed to benefit. Homeland security gurus succeeded in linking pandemic influenza to national security, while public health experts enjoyed an elevated role in the all-hazards regime as well as expanded emergency powers. The federal government continued its emergency management power grab, but states responded by expanding their authority in their own territories. Public health preparedness was strengthened, but private vaccine manufacturers received a bevy of new incentives and kickbacks. Perhaps the only "losers" were civil liberties groups, which could do little to stem the tide of expanded government power on the eve of crisis.

In some, but certainly not all, respects, the pandemic policy domain is very similar to public risk and disaster domains described in Chapter 2.[95] In these domains, government actors and individuals with a highly specialized skill-set tend to dominate the policymaking process. These domains are often characterized as "lacking a public," meaning they are devoid of widespread issue expansion and mobilization by private actors. While this characterization was accurate for *most* of the influenza debate, certain pieces of legislation triggered fairly robust issue expansion. A host of private actors weighed in on the PREPA debate, for example, including vaccine manufacturers and lawyers. What is more, proposed quarantine measures proved equally divisive, as evidenced in the civil liberty community's response to the MSEHPA. Thus, the pandemic domain seems to represent a proverbial "middle ground" between more classic disaster domains and the other anticipatory domains examined in this book, which engendered fairly widespread and diverse mobilization patterns.

What drove the avian influenza onto the institutional agenda? The 1918 Spanish flu analogy coupled with claims of novelty, severity, and the identification of a sympathetic problem population (young adults) heightened concerns with avian influenza. Most importantly, the dominant policy narrative from this period cast avian influenza as an inevitable problem—a question of "when" not "if." Policymakers resigned themselves to the reality that a pandemic would occur, although they could not predict when and to what extent it would happen. Secretary of HHS Michael Leavitt expressed this sentiment concisely in a 2006 *Department of Health and Human Services Pandemic Planning Update* writing: "We are in a race, a race against time and complacency."[96] A failure to prepare, in other words, would portend crisis.

Issue attention was indicator driven and upticks in policymaker attention closely coincided with increases in H5N1 cases and deaths. However, Hurricane Katrina provided the final "nudge" for policy change. While hurricanes and influenza viruses share few similarities on the surface, Hurricane Katrina represented such a watershed example of government failure that its implications—the "lessons" it revealed—reverberated across a variety of risk and disaster domains. Avian flu, in many respects, was low hanging fruit, as previous executive actions coupled with accumulating indicators paved the way for policy action. The prospect of being "caught off guard" by yet another disaster—regardless of its source—was politically unacceptable in the aftermath of Katrina. Influenza planning, it seemed, allowed policymakers to redeem themselves for the follies of Katrina.

Media attention paralleled the number of avian influenza cases and deaths, and supported the momentum toward policy change. Figure 4.4 shows the number of *New York Times* stories published on avian influenza. Coverage grew from 2004 (155 stories) to 2005 (317 stories) before peaking in 2006 (367 stories). A second surge in media attention was recorded in 2009 (678 stories), the year of the swine flu outbreak. There was also an uptick in public concern between October of 2005 and March of 2006. Ho et al. found that 60 percent of the public was "very or somewhat concerned" about an avian flu pandemic.[97] Similarly, 60 percent of the public surveyed between October of 2005 and April of 2005 said they "felt it was very or somewhat likely that a variant of the avian flu virus would strike the United States."[98] Having reviewed the agenda setting process surrounding the avian influenza, we now turn our attention to a very different pandemic threat: the H1N1 swine flu.

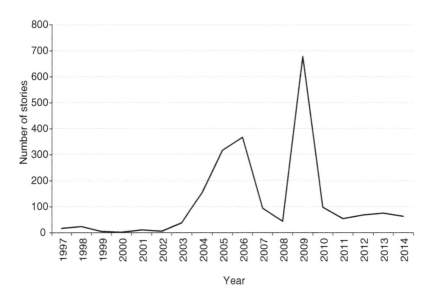

*Figure 4.4* Number of Stories on Pandemic Influenza in the *New York Times*, 1997–2014 (source: Search on Lexis-Nexis for the terms "pandemic influenza," "swine influenza," "avian influenza," and others in the *New York Times*. Please see endnote for complete search string).

## 2009: Agenda Setting and the H1N1 Swine Influenza

The H1N1 virus or swine flu first appeared in Veracruz, Mexico, in April of 2009. Swine flu is believed to have originated in pigs, although H1N1 is technically a combination of human, swine, and bird influenza viruses. The symptoms associated with the virus are milder (e.g., cough, fatigue, diarrhea, muscle aches) than those associated with avian flu, and only a small percentage of victims suffer respiratory distress.[99]

Figure 4.5, which shows the number of new H1N1 human cases between April 2009 and November 2009, indicates swine influenza emerged much more rapidly than avian influenza. There were nearly 75,000 H1N1 cases in August alone. November saw 180,821 cases, bringing the cumulative total to 622,482 cases worldwide. In fact, the number of cases became so great the WHO determined it was futile to continue counting after November 2009. But swine flu disappeared just as quickly as it emerged and, in July 2010, a little over a year after the disease first appeared, only 95 cases were reported worldwide. Thus, the proverbial "ramp-up" time for H1N1 influenza was much more constrained than in the H5N1 case.

On June 11, 2009, less than two months after the first cases were discovered in Mexico, WHO declared H1N1 a global pandemic.[100] This is not to say preparedness and anticipation were absent. Even after this declaration,

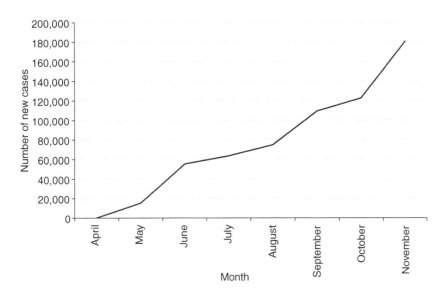

*Figure 4.5* Number of New Swine Influenza Cases by Month, 2009 (source: WHO. "Situation updates – Pandemic (H1N1) 2009." Available at www.who. int/csr/disease/swineflu/updates/en/index.html (accessed May 22, 2013).

the Obama Administration indicated planning and preparedness were needed for the fall flu season because influenza cases typically spike during the autumn and winter months.[101] While many media outlets were quick to deem the outbreak a "crisis," the Obama Administration offered a far more tempered definition of H1N1.[102] In a 2009 speech the President indicated: "This is, obviously, cause for concern and requires a heightened state of alert. But it's not cause for alarm."[103] Because swine influenza did, in fact, constitute a tangible emergency, the President did not need to "sell" the problem to the public. The number of domestic and international cases spoke for themselves. Instead, the President faced a critical test of his crisis management skills, requiring that he strike a balance between instilling fear and encouraging complacency.

Still, in some respects, the swine flu discourse paralleled that of the avian flu case. For example, analogy was also used to describe H1N1. In spring of 2009, public health experts and the media contemplated whether the swine flu was similar to the 1918 Spanish influenza. *Fox News* reported that researchers at the University of Wisconsin found "The new H1N1 influenza virus bears a disturbing resemblance to the virus strain that caused the 1918 flu pandemic, with greater ability to infect the lungs than common seasonal flu viruses."[104]

In the face of severe vaccine shortages in the fall of 2009, a number of pundits held that swine flu constituted President Obama's Hurricane Katrina. Conservative political pundit Rush Limbaugh argued: "This ought to be Obama's Katrina. [The vaccine shortage] is no different than FEMA supposedly screwing up and not getting down there in time to New Orleans."[105] The *Washington Times* published an editorial entitled "Will swine flu be Obama's Katrina?" where author Martin Schram wrote: "They are racing against the uncertainties of nature and science—desperately hoping the autumn return of the H1N1 virus, known as swine flu, will not become the Obama presidency's Katrina."[106] Analogies were once again marshaled as an important strategic tool.

Issue attention skyrocketed in 2009. The number of hearings soared to 22, eclipsing the 2004 and 2005 highs of 13 hearings. The number of the *Congressional Record* entries reached 219 in 2009, the highest total since 2005 (see Figures 4.1 and 4.2). However, unlike the avian influenza case, attention was not sustained and by 2010 pandemic influenza all but faded from the institutional agenda.

Other important distinctions can be identified. Consider, for example, the actual appropriations process. In FY2010, Congress appropriated more than $8 billion in response to the H1N1 outbreak. In July, the Obama administration earmarked $1.8 billion to develop an H1N1 vaccine. However, whereas the H5N1 policy process saw a small chunk of money go to a variety of federal agencies other than HHS, *all* appropriations during this period went to HHS and the vast majority of funding was dedicated to tried-and-true public health interventions, like vaccine development. In other words, policymaking was much more centralized during this time of crisis.

Moreover, many policies introduced during this period proposed a reactive, emergency response approach. The Emergency Influenza Containment Act (H.R. 3991) and the Pandemic Protection for Workers, Families, and Businesses Act (H.R. 4092) established paid sick programs for workers sickened by virus. The Public Health Emergency Response Act of 2009 (H.R. 2231) presented a series of provisions that would be activated by the onset of a pandemic event, delegating authority to the Secretary of HHS to streamline the provision of nationwide healthcare services during a public health emergency. The School Protection Act of 2009 (H.R. 3798) created a pandemic and bioterror response training program for elementary and secondary school nurses. In all instances, key provisions were triggered by the onset pandemic—not in anticipation of one. However, none of these laws were enacted.

Congressional hearings reflected the fact that, as early as the spring of 2009, many believed a pandemic was inevitable. The Senate Committee on Homeland Security and Government Affairs held a hearing on April 29, 2009, entitled *Swine Flu: Coordinating the Federal Response*, which set out to investigate the severity of the Mexico outbreak and examine the government's capacity to respond. Less than a month into the initial outbreak, policymakers demanded an aggressive federal response to H1N1. Senator Susan Collins (R-ME) indicated:

The American people have the right to expect that the Federal Government is doing everything possible to combat this potential pandemic, and to date, I would agree with the Chairman that it appears that our Federal officials have taken this threat very seriously and responded very effectively.[107]

As the number of H1N1 cases exploded across the country in the fall of 2009, Congress shifted its attention from crafting new laws to monitoring the ongoing response to H1N1. The legislature was especially concerned with the Obama Administration's vaccine development program, a topic described in greater detail in the implementation section. By September of 2009, there was growing evidence of a widespread pandemic vaccine shortage. A number of hearings devoted considerable attention to determining the cause of this shortage and, of course, assigning blame.[108]

The Obama Administration did not devise new strategies or plans, and largely relied on the preparedness infrastructure put in place by the Bush Administration.[109] Conversely, one of the President's most significant actions was a swine flu emergency declaration on October 24, 2009. The Administration empowered the Secretary of HHS to voluntarily waive federal laws protecting patient privacy. The Secretary could also allow hospitals to create "satellite facilities" to accommodate a potential surge in swine flu patients. Satellite facilities might include a local armory, tent, another hospital, or other suitable locations. Finally, the declaration allowed for a more rapid distribution of emergency medicines, including antiviral drugs.[110]

Administration officials sought to cast the emergency declaration as an anticipatory measure. "This is not a response to any new developments. It is an important tool in our kit going forward," stated White House spokesman Reid Cherlin.[111] Another official indicated: "It's important to note that this is a proactive measure—not a response to a new development."[112] Despite these claims, the emergency declaration cannot be construed as anticipatory policy. By the time the Obama Administration made its emergency declaration, thousands of Americans were infected with the virus and nearly 46 states reported widespread influenza activity. By October 2009, the virus was estimated to have killed at least 1,000 Americans and hospitalized more than 20,000.[113]

While many feared the declaration centralized too much control in the hands of the President and would amount to a broad restriction on civil liberties, the American Civil Liberties Union (ACLU) concluded the "Obama Administration has so far acted appropriately in response to the outbreak and has helped to establish and maintain a sense of calm throughout the country."[114] They added that the President's approach was far less invasive than the military enforced quarantines proposed by the Bush Administration. In this respect, the movement to militarize disease appeared to hold far less sway within the Obama Administration. The ACLU did, however, cite a number of instances of state overreaction, including New York's rule that all health officials be vaccinated against swine influenza.[115]

## Agenda Catalysts

The second round of pandemic influenza policymaking, which was initiated in response to the 2009 outbreak of swine influenza, differed considerably from the first. Most obviously, this activity did not lead to major policy change, but instead focused on overseeing, implementing, and funding the policies established between 2004 and 2006. Nor did swine influenza "linger" on the agenda for an extended period of time. Indeed, swine influenza quickly receded from the institutional agenda and there has been a precipitous drop in issue attention in the wake of the 2009 outbreak.[116] Even *New York Times* coverage jumped from 2008 (44 stories) to 2009 (678 stories), only to plummet in 2010 (98 stories) (see Figure 4.4).

The second round of activity was not emblematic of anticipatory policymaking, but more a classic example of crisis management. Cases and deaths remained important agenda setting catalysts in the swine flu case, yet, unlike the avian flu case, these indicators did not reveal themselves slowly and over an extended period of time. Instead they aggregated rapidly, bringing the number of swine flu cases and deaths well into the thousands less than four months after the initial outbreak. In the case of swine influenza, one could argue the *cascading accumulation* of indicators prompted policymakers to adopt a strategy of crisis management and emergency response, as opposed to pursuing proactive and preventative measures. To be sure, a great deal of authority was vested in the executive branch, although Congress and the states remained active participants in the response effort. Whereas Congress took to overseeing and monitoring (and funding) the Administration's response, states played a critical role in managing the outbreak, tending to patients and distributing vaccines. This finding echoes Boin and 't Hart's conclusion that a uniform and centralized crisis management framework cannot exist in large industrialized states.[117] And while analogical reasoning remained an important rhetorical device, there was near universal agreement that this was in fact an existing public health crisis, not a "potential" or future threat.

Thus, the pandemic influenza domain was marked by two distinct agenda setting stages. Whereas the first resulted in considerable policy change, the second largely focused on policy maintenance, program oversight, and crisis management. The following section investigates the implementation stage, once again comparing and contrasting the distinctive approaches to dealing with avian versus swine influenza.

## Policy Implementation

Just as agenda setting was marked by two distinct but interrelated stages, so too can we distinguish between the implementation processes surrounding avian versus swine influenza. The first stage of activity, which began in roughly 2006, included all of the activities needed to prepare the country for a potential avian influenza pandemic, including the implementation of the *National Strategy*, PREPA, and PAHPA. The second stage, by contrast, focused not on preparing, but on managing the swine flu crisis. Crisis management often drew heavily on the sprawling preparedness framework established in anticipation of the avian influenza. The following section examines both stages in detail.

### Implementation and Avian Influenza Preparedness

Implementation efforts were grounded in the *National Strategy for Pandemic Influenza Implementation Plan* (*Implementation Plan* hereafter), a 234-page document outlining scores of activities needed to achieve preparedness.[118] The absence of a clearly identifiable problem logically undermined bureaucrats' ability to examine the effectiveness of various policy interventions. In turn, the Bush Administration seemed to take great pains to "construct" measures or benchmarks of implementation success. One could argue these benchmarks represented surrogates or proxies for an actual problem and allowed agencies to gauge productivity and effectiveness in the absence of a tangible social ill.

In March 2006, HHS offered the first of what would be six *Pandemic Planning Update* reports. The updates provided a detailed account of the progress of the *Implementation Plan*, offering updates on everything from the status of surveillance efforts to vaccine development. By presenting implementation as a proverbial "laundry list" of distinct but interrelated actions, bureaucrats were, again, able to demonstrate measurable success in the face of a most uncertain problem.[119] For every box checked, it seemed, another element of preparedness was achieved.

The government website "FLU.gov" was yet another example of policymakers' desire to develop constructed measures of implementation progress. Flu.gov presented a "Summary of Progress" in December of 2006 that listed each of the various actions outlined in the *Implementation Plan*. Next to each, a designation of "complete" or "in progress" was provided, indicating whether or not a particular goal was achieved. Similar planning updates were offered in conjunction with the implementation of other laws and plans. HHS, for example, provided a series of "progress reports" on the status of PAHPA implementation.[120] In sum, although government progress reports are by no means unique to anticipatory policymaking, there seemed to be a special emphasis on disseminating these types of updates.

The *Implementation Plan* established four critical implementation areas, notably surveillance, the procurement of antiviral drugs, vaccine development, and state and local preparedness. Despite the *Implementation Plan*'s presentation of international surveillance as a core pandemic planning activity, it is not entirely possible to separate H5N1 surveillance from disease surveillance in general. The global surveillance network for emerging diseases consists of a "bewildering mix" of agencies, programs, institutions, and initiatives.[121] Indeed, it was this very system that first identified avian influenza in the 1990s, years before Congress or the President initiated any policymaking. The *National Strategy* did, however, inject money into this system, incentivizing these institutions and programs to pay more attention to H5N1.[122] Still, some critics maintained this funding made little tangible difference, as many nations lack the capacity to track and monitor diseases within their borders.[123]

Disease surveillance is complex and can engender thorny international controversies. Countries must report disease outbreaks and, in the event that an outbreak occurs, agree to give international experts access to contaminated sites in order to investigate the threat. Neither of these critical steps are a given. In June 2008, Indonesia announced it would no longer report avian influenza deaths as they occurred. Indonesian officials worried broadcasting their country's high death rate would undermine their international standing. Moreover, they refused to submit samples of the virus to the WHO and questioned the logic of sharing samples that could be used to create a vaccine that they themselves could not afford.[124] Concerns with reporting were magnified by China's cover-up of Severe Acute Respiratory Syndrome (SARS) in 2002. The deadly respiratory virus, which originated in China, infected 27 Americans and 251 Canadians—44 of whom died. Many observers believed China's failure to report contributed to the disease's spread.[125]

Income inequalities further exacerbated these difficulties. Many Asian and African countries simply could not afford to develop a strong public health infrastructure and struggled to discern—let alone contain—novel disease outbreaks.[126] A number of U.S. government agencies set out to strengthen the surveillance capabilities of poorer countries, particularly those with high rates of avian influenza infection. Among other measures, USAID implemented more than $55 million worth of surveillance programs abroad, including disease monitoring, laboratory diagnoses, and disease containment programs in Laos, Cambodia, China, Vietnam, and Indonesia; distributed more than 30,000 personal protective-equipment sets (decontaminating sprayers, protective boots, protective coveralls, masks); and deployed infectious disease and animal experts across Asia, Africa, and Eastern Europe. The CDC spent over $100 million on international surveillance initiatives through FY2006.[127] The CDC's Global Disease Detection (GDD) Initiative, which aims to rapidly mobilize resources to identify and contain outbreaks, launched a number of surveillance programs, which set out to help countries suffering from persistent outbreaks test a vaccine; develop clinical trials and a research infrastructure in southeast Asia; and survey animal outbreaks in Asia.[128]

Domestic surveillance efforts were more straightforward. The USDA, working in conjunction with the DOI and other agencies, launched a multi-pronged program aimed at detecting H5N1 in wild migratory birds. Participating agencies reported testing more than 35,000 wild migratory birds by 2006. Participating agencies also created a GPS-monitoring system capable of tracking flocks across the globe.[129]

Of all the goals outlined in the *National Strategy*, antiviral stockpiling was perhaps the most quantifiable. The federal government hoped to stockpile 80 million doses of antiviral drugs, enough to treat roughly 25 percent of the American population. Stockpiling was a "shared responsibility," meaning 60 percent of the stockpile would be acquired by the national government while the remaining 40 percent would be acquired by the states at a reduced rate (three-quarters of the original cost). Two companies, Roche and GlaxoSmithKline, produced the vast majority of these drugs—generic drugs were not readily available. Countries and even private companies interested in building their own anti-viral cache were advised to place an order and assume their place in line.[130]

However, stockpiling efforts bottlenecked in the spring of 2006, a testament to the considerable concern surrounding avian influenza. In March, the Bush Administration announced it was buying 14 million doses of antiviral drugs, bringing the total number of purchased antiviral drugs to about 19.5 million—less than half of the original goal. Democrats scoffed at the announcement, questioning why the stockpiling process was taking so long. "We appropriated this money in December, and now it's March. Why are they waiting so long, and where are the rest of the orders?" asked Allison Dobson, a spokeswoman for Senator Tom Harkin (D-IA).[131]

The cause of the delay was multifaceted. For one, the drug manufacturers themselves were overwhelmed. Roche reported more than 65 countries placed orders to fill their stockpiles. At the same time, the Bush Administration consciously delayed spending $101 million of the original appropriation because it was trying to determine "how much of the remaining medicines should be delivered as liquids, which are easier for children and the elderly to take."[132] Magnifying these concerns, reports indicated wildly different purchasing habits on the part of states. In 2008, the *Washington Post* reported: "Fifteen states purchased less than 50 percent of the amount recommended by the federal government."[133] Commenting on the situation, Senator Charles Schumer (D-NY) stated: "Despite the urgency and the need in the event of an avian flu pandemic, the administration is slow-walking the much-needed dollars because of their overall budget problems. That is wrong."[134] The antiviral controversy eventually subsided, owing in part to H5N1's failure to mutate into a pandemic strain. It was not until January of 2009 that the Administration declared it had reached its goal of stockpiling 81 million courses. The Administration also reported states purchased more than 22 million treatment courses.[135]

The Bush Administration worked to stockpile enough "pre-pandemic" vaccine to inoculate at least 20 million Americans.[136] In September, HHS signed a contract for more than $160 million with pharmaceutical companies Sanofi Pasteur and Chiron to create a pre-pandemic vaccine.[137] By January 2009, the U.S. government stockpiled 2.2 million pre-pandemic vaccine courses, enough to cover first responders and healthcare workers at the onset of a pandemic.[138]

However, a pre-pandemic vaccine, which was based on circulating H5N1 viruses, would offer minimal protection against a mutated pandemic strain. With this in mind, the Bush Administration set a target of being able to produce 600 million doses of pandemic vaccine—enough to inoculate every American twice—within six months of a pandemic. HHS set out to vastly expand domestic manufacturing capacity. The program provided private drug manufacturers contracts to improve their facilities, encouraged expansion of both egg-based and cell-based vaccine production methods, and funded the creation of "next generation" vaccine development techniques.[139]

Determining the actual value of these programs was obviously difficult prior to the 2009 pandemic. In the *Pandemic Planning Update VI*, HHS reported more than $120 million was filtered into domestic vaccine manufacturing facilities, which updated their existing egg-based production systems to allow for faster vaccine development. The improvements led to a record output of seasonal influenza vaccine, a promising sign for those hoping for rapid production during a pandemic. Another $1 billion was allocated for contracts with private companies to develop advanced cell-based technology, which was projected to produce approximately 240 million doses of vaccine within six months of the pandemic. HHS reported that "five manufacturers are on target for reaching the milestones required in their contracts, and all are moving toward FDA approval of cell-based influenza vaccines."[140]

Finally, the *Implementation Plan* requested state governments take steps to ready for a pandemic. State planning included everything from outlining travel restrictions to determining drug distribution protocols. By the fall of 2006, the Bush Administration reported every state had, at the very least, developed a draft plan. The federal government provided states funding to support their preparedness effort, including $225 million to conduct "tabletop" exercises, which were essentially crisis management simulations conducted by HHS representatives.[141]

In the absence of a manifest problem, the Bush Administration claimed implementation success. As early as July 2007, Dr. Rajeev Venkayya, Special Assistant to the President for Biodefense, revealed the Administration was well on its way to achieving "preparedness":

Of the actions that were due at 12 months, we assess that 86 percent of those actions have been completed. That's to be compared to a score of around 92 percent that we released at the six-month mark. There are about 14 percent of the actions that are not yet completed. We document those in the action-by-action detailed report. We anticipate those being completed in the 18-month time frame at the next six-month report.[142]

Of course, many goals took years to complete and public health agencies continued implementing the plan well into 2009. Yet, while public health agencies and officials remained vigilant, something of a malaise settled over the rest of the policymaking community. At a 2007 Council on Foreign Relations meeting on avian flu preparedness, Laurie Garrett indicated that she was surprised by the waning interest: "Let me just start by saying that we had a meeting 17 months ago here. The turnout was about five times what we see in the room here today."[143] By 2007, the number of H5N1 cases and deaths had in fact declined, although the virus maintained an exceptionally high lethality rate—59 deaths resulted from 88 human cases. The virus continued to taper in 2008 (36 cases and 28 deaths). However, this indifference would be short lived after a new virus emerged in 2009, reigniting interest in pandemic influenza.

## Implementation and Swine Influenza Response

The 2009 outbreak of the swine influenza was a critical test of the Bush-era preparedness regime. Upon identifying the outbreak, President Obama quickly turned to the preparedness programs established by his predecessor.[144] In fact, he praised the Bush model in saying "I think the Bush administration did a good job of creating the infrastructure so that we can respond."[145] Within days, President Obama requested influenza reports be integrated into his daily intelligence briefing. He also updated the public regularly.[146] Pursuant to the National Response Framework, Secretary of Homeland Security Janet Napolitano assumed her role as the Principal Federal Official (PFO) for coordinating the national response on April 26, 2009. A day earlier, HHS declared the outbreak a "public health emergency," enabling the FDA to allow the use of unapproved medical treatments in the face of an emergency. The declaration also meant HHS would assume oversight of all federal public health and medical responses. President Obama later elevated the threat level to a "national emergency" designation in October 2009. The President also requested $1.5 billion to fund preparedness programs in the summer of 2009.[147] In other words, the Administration transitioned quickly from anticipatory policymaking to emergency management.

The CDC originally recommended 159 million swine flu vaccine courses be made available at the onset of a pandemic, which was enough to cover America's "high risk" population.[148] But as late as November 2009, only 32.3 million doses were available. Concern and frustration abounded. By November, there were upwards of 30 million H1N1 cases in the U.S. alone.[149] Dr. Jonathan Fielding, Los Angeles County's health commissioner, griped: "The vaccine situation is quite frustrating. The numbers we were asked to anticipate and plan for have turned out to be gross overestimates of what has been supplied to us."[150]

The vaccine shortage constituted nothing short of an implementation failure and Republican politicians and pundits laid the blame squarely on the Obama Administration, deeming the debacle President's "Katrina" moment.[151] Republican Senator Susan Collins of Maine stated:

The fact that there are vaccine shortages is a huge problem. I believe the administration took the pandemic seriously, but I also believe administration officials were so determined to show that everything was under control that they sent the wrong signals about the adequacy of supplies of the vaccine.[152]

Administration officials countered that the private companies contracted to make the vaccine should be held accountable, not them. Nicole Lurie, Deputy to HHS Secretary Kathleen Sebelius, criticized the drug manufacturers, noting that when they "hit some stumbling blocks, they sometimes thought the fix was around the corner and didn't always feel the need to tell us, and then sometimes the fix wasn't around the corner."[153] According to representatives of the vaccine companies, however, the Obama Administration was well aware of the situation and was updated "on a daily basis."[154] HHS later conceded that the vaccine development programs mandated by PAHPA and other statutes were underway, but not fully functional by the fall of 2009.[155]

The task of investigating this misstep fell on Congress, which convened a number of hearings to determine the cause of the vaccine shortage. On October 21, 2009, the Senate Committee on Homeland Security held a hearing entitled *H1N1 Flu: Monitoring the Nation's Response* where it peppered HHS Secretary Sebelius with questions relating to the cause of the shortage and how the Administration intended to remedy this situation. One particularly pointed exchange saw Senator McCain (R-AZ) repeatedly ask the Secretary when the Administration would be caught up in terms of vaccine supplies, to which Secretary Sebelius eventually replied: "Senator, I have no idea ... I will try to get that."[156] Senator Joseph Lieberman (I-CT), Chairman of the Committee, later echoed McCain's inquiry: "My concern now is that the spread of the disease has gone beyond the government's ability to take actions to prevent and respond to them."[157] The vaccine crisis was eventually resolved and by December 2009 ample swine flu vaccine was available.[158]

While elements of the all-hazards approach were integrated into response efforts, including the national emergency declaration as well as the application of the National Response Framework, the swine flu pandemic did not trigger the militarized response feared by many. For example, despite calls by Republican politicians to close the borders, the President remained wedded to public health best practices. Heeding the advice of CDC representatives and other public health officials, the President rejected the idea of isolation and argued border closure would be "akin to closing the barn door after the horses are out."[159] Thus, while previous policymaking efforts unquestionably altered certain aspects of the pandemic domain, there is little evidence to suggest they diminished the public health community's power.

Nor did the federal government usurp state power. Even in the face of a national emergency declaration, the Obama Administration respected the traditional authority of states over preparedness and emergency response. States maintained control over vaccine distribution, despite the fact that some of their decisions amounted to public relations blunders. In one notable case, the New York City Health Department was lambasted for sending early vaccines to Wall Street firms, like Goldman Sachs. The Obama Administration, however, refrained from publically second-guessing their decision. What is more, decisions about school closures and quarantine measures were guided by the states without federal interference, although the CDC provided guidance documents.[160]

When all was said and done, the Administration's swine flu response was largely seen as a success, save the vaccine blunder. Unlike the Bush Administration, the effectiveness of the Obama Administration's response was not evaluated based on "constructed" metrics of preparedness, but much more concrete terms, notably the number of American lives lost. Roughly 10,000 Americans died from the outbreak, far less than the 30,000 to 90,000 forecasted by the President's Council of Advisors on Science and Technology.[161] "I would give them a B for performance," indicated Dr. Eric Toner, a senior associate at the Center for Biosecurity at the University of Pittsburgh Medical Center.[162] Yanzhong Huang, Senior Fellow for Global Health at the Council on Foreign Relations and associate professor at Seton Hall University, argued the response was quite effective and "informed by science and epidemiology."[163] Of course, many observers concluded that the relatively small death toll was more a reflection of the relative "mildness" of the H1N1 strain, than the Obama Administration's efforts.[164] In other words, the Obama Administration might have been lucky the much-feared avian influenza virus never mutated into an airborne strain.

## Conclusion

By the end of 2009, the swine flu pandemic had abated. The virus was highly infectious, but nowhere near as lethal as expected. In the wake of the swine influenza pandemic, pandemic influenza has all but disappeared from the institutional agenda and the 112th and 113th Congresses have recorded very little policy activity.

The avian influenza case provides a compelling illustration of anticipatory policymaking. To sidestep the inherent uncertainty associated with the pandemic issue, problem definers relied heavily on analogy and other forms of implied comparison. Uncertainty also influenced the implementation process and, in the absence of an actual problem, bureaucrats were forced to manufacture measures of programmatic success. Indicators drove the agenda setting process, which lasted for almost two Congress, although Hurricane Katrina helped compel decisive policy action. Policy change preceded the onset of pandemic influenza, and many of the laws and plans created in anticipation of avian influenza helped the federal government and the states navigate the swine flu pandemic.

While congressional concerns with avian influenza have diminished, the U.S. public health apparatus continues to survey the globe for other emerging pandemic threats. In fact, a new variant of avian influenza, H7N9 bird flu, appeared in China in March 2013. By February 2015, the virus had infected close to 571 individuals across Southeast Asia. Public health experts estimated 212 of these cases resulted in death.[165] The concerns associated with H7N9 are the same as those with H5N1, which are that the highly pathogenic and lethal virus will mutate into an airborne strain. Assuming these fears are eventually confirmed, how might the prior policymaking conflict impact future avian influenza policymaking? Can H7N9—or any other novel influenza—induce another round of anticipatory policy change or have policymakers become desensitized to this threat? What lessons might be gleaned from the swine influenza case? Unfortunately, we likely will not be able to answer these questions until pandemic influenza rears its ugly head again.

## Notes

1  World Health Organization. 2015. "Cumulative Number of Confirmed Cases for Avian Influenza (AH5N1) Reported to WHO, 2003–2015." Available at www.who.int/influenza/human_animal_interface/EN_GIP_20150126Cumulat iveNumberH5N1cases.pdf?ua=1 (accessed March 28, 2015).
2  An earlier version of this case was published in *Crisis, Risk, and Hazards in Public Policy*. See: Rob A. DeLeo. 2010. "Anticipatory-Conjectural Policy Problems: A Case Study of the Avian Influenza." *Crisis, Risk, and Hazards in Public Policy* 1(1): 147–184.
3  Center for Disease Control and Prevention. 2010. "Key Facts About Avian Influenza (Bird Flu) and Highly Pathogenic Avian Influenza A (H5N1) Virus." Available at www.cdc.gov/flu/avian/gen-info/facts.htm (accessed March 28, 2015).
4  World Health Organization (WHO). 2011. "Avian Influenza." April. Available at www.who.int/mediacentre/factsheets/avian_influenza/en/ (accessed March 28, 2015).
5  Michael T. Osterholm. 2005. "Preparing for the Next Pandemic." *The New England Journal of Medicine* 352: 1839–1842.
6  World Health Organization 2011; Peter Doshi. 2011. "The Elusive Definition of Pandemic Influenza." *Bulletin of the World Health Organization* 89: 532–538.
7  Mike Davis. 2006. *The Monster at Our Door: The Global Threat of Avian Flu*. New York: Henry Holt.
8  Robert G. Webster and Elena A. Govorkova. 2006. "H5N1 Influenza—Continuing Evolution and Spread." *New England Journal of Medicine* 355: 2174–2177.
9  Raymond A. Strikas, Gregory S. Wallace, and Martin G. Myers. 2002. "Influenza Pandemic Preparedness Action Plan for the United States: 2002 Update." *Clinical Infectious Diseases* 5 (September 1): 590–596; John M. Barry. 2004. *The Great Influenza: The Story of the Deadliest Pandemic in History*. New York: Viking Press; Davis 2006.
10  Philip Elliott. 2009. "Obama: Swine Flu A National Emergency." *Huffington Post*, October 25. Available at www.huffingtonpost.com/2009/10/24/obama-declares-swine-flu_n_332617.html (accessed March 28, 2015).
11  Gregory A. Poland. 2006. "Vaccine against Avian Influenza—A Race against Time." *The New England Journal of Medicine* 354(13): 1411–1413, 1411.

12 Ibid.
13 Centers for Disease Control and Prevention. 2008. "Avian Influenza A Virus Infections of Humans." Available at www.cdc.gov/flu/avian/gen-info/avian-flu-humans.htm/ (accessed March 28, 2015); World Health Organization. 2011b. "H5N1 Avian Influenza: Timeline of Major Events." December 13. Available at www.who.int/influenza/human_animal_interface/avian_influenza/H5N1_avian_influenza_update.pdf (accessed March 28, 2015).
14 Davis 2006.
15 Daniel DeNoon. 2005b. "WHO: Threat of Bird Flu Pandemic Rises." *Fox News*, March 23. Available at www.foxnews.com/story/0,2933,157218,00. html (accessed March 28, 2015); Webster and Govorkova 2006.
16 Daniel DeNoon. 2005. "Bird Flu Threat Rise." *WebMD Health News*, May 20. Available at www.webmd.com/cold-and-flu/news/20050520/bird-flu-threat-rises (accessed March 28, 2015).
17 Webster and Govorkova 2006, 2174.
18 Anthony S. Fauci. 2006. "Pandemic Influenza Threat and Preparedness." *Emerging Infectious Diseases* 12(1): 73–77, 73.
19 Ibid.
20 Edward A. Gargan. 1997. "Avian Flu Strain Spreads to Humans." *New York Times*, December 9, 9.; Arkansas Department of Health. 2011. "Epidemiology." Available at www.healthy.arkansas.gov/programsServices/epidemiology/Pages/default.aspx (accessed March 28, 2015).
21 Bryan D. Jones and Frank R. Baumgartner. 2005. *The Politics of Attention: How Government Prioritizes Problems*. Chicago: The University of Chicago Press.
22 Davis 2006; Gargan 1997.
23 Erik Larson. 1998. "The Flu Hunters." *Time*, February 23: 56.
24 World Health Organization. 2012. "Cumulative Number of Confirmed Human Cases of Avian Influenza A(H5N1) Reported to WHO." May 2. Available at www.who.int/influenza/human_animal_interface/H5N1_cumulative_table_archives/en/index.html (accessed March 28, 2015).
25 Minnesota Department of Public Health. 2012. "Avian Flu and Pandemic Flu: The Difference—and the Connection." Available at www.health.state.mn.us/divs/idepc/diseases/flu/avian/avianpandemic.html (accessed March 28, 2015).
26 City of Long Beach California Department of Health and Human Services. 2012. "Avian/Pandemic Influenza Information Page." Available at www.longbeach.gov/health/influenza.asp (accessed March 28, 2015).
27 Richard J. Webby and Robert G. Webster. 2003. "Are We Ready for Pandemic Influenza?" *Science* 302(5650): 1519–1522, 1522.
28 Declan Butler. 2012. "Death-rate Row Blurs Mutant Flu Debate." *Science*, February 13. Available at www.nature.com/news/death-rate-row-blurs-mutant-flu-debate-1.10022 (accessed March 28, 2015).
29 Lawrence Altman. 1997. "When a Novel Flu is Involved, Health Officials Get Jumpy." *New York Times*, December 30, F7.
30 Geoffrey Crowley. 1998. "Assessing the Threat: Why 'Bird Flu' Isn't Likely to Turn into a Global Killer," *Newsweek* 12: 41.
31 David A. Rochefort and Roger W. Cobb. 1994. "Problem Definition: An Emerging Perspective." In *The Politics of Problem Definition: Shaping the Policy Agenda*, eds. David A. Rochefort and Roger W. Cobb. Kansas: University of Kansas Press: 1–31.
32 Davis 2006, 24.
33 Martin Enserink. 2004. "WHO Adds More '1918' to Pandemic Predictions." *Science* 5704: 2025.
34 Congressional Budget Office (CBO). 2006. *A Potential Influenza Pandemic: Possible Macroeconomic Effects and Policy Issues*. Washington, D.C.: The

Congress of the United States. Available at www.cbo.gov/sites/default/files/12–08-birdflu.pdf (accessed March 28, 2015).

35  Marc Lallanilla. 2005. "Spanish Flu of 1918: Could It Happen Again?" *ABC News*, October 5. Available at: http://abcnews.go.com/Health/AvianFlu/story?id=1183172 (accessed March 28, 2015).

36  Osterholm 2005.

37  Bernadine Healy. 2006. "The Young People's Plague." *U.S. News & World Report* 140(16): 63.

38  Erik Larson. 1998. "The Flu Hunters." *Time*, February 23: 54–64; Rochefort and Cobb 1994.

39  Fred Levit. 1998. "Asian 'Bird Flu' Isn't Similar to 1918." *New York Times*, January 5, A18.

40  Richard E. Neustadt and Harvey V. Fineberg. 1978. *The Swine Flu Affair: Decision-Making on a Slippery Disease*. Washington, D.C.: U.S. Department of Health, Education, and Welfare.

41  Altman 1997.

42  Marc Siegel. 2006. "A Pandemic of Fear." *Washington Post*, March 26, B07.

43  Elisabeth Rosenthal. 2006. "On the Front: A Pandemic is Worrisome But 'Unlikely.'" *New York Times*, March 2006, F1.

44  Using ProQuest Congressional, I searched for all mentions of the terms "pandemic influenza," "pandemic flu," "avian influenza," "avian flu," "bird flu," "bird influenza," "h5n1," "swine influenza," "swine flu," and "h1n1" in the *Congressional Record*.

45  All committee hearing data was compiled used ProQuest Congressional. I repeated the search string used for my *Congressional Record* data. Hearing data was derived by searching in all fields *except* full text. Hearings were then manually reviewed and all hearings focusing on topics other than pandemic influenza were removed.

46  Data used in Figure 4.3 came from the source: World Health Organization 2012.

47  "CNN Takes In-depth Look at Bird Flu." 2005. *CNN*, November 2. Available at http://articles.cnn.com/2005–11–02/health/birdflu.tv_1_h5n1-bird-flu-human-to-human?_s=PM:HEALTH. (accessed March 28, 2015).

48  U.S. Congress. House. 2005. "The Threat of and Planning for Pandemic Flu." Subcommittee on Health. May 26. 109th Cong., 1st sess. Washington, D.C.: Government Printing Office: 65.

49  Christina E. Mills, James M. Robins, Carl T. Bergstrom, and Marc Lipsitch. 2006. "Pandemic Influenza: Risk of Multiple Introductions and the Need to Prepare for Them." *PLoS Medicine* 3(6): 769–773, 769. Available at http://octavia.zoology.washington.edu/publications/MillsEtAl06.pdf (accessed March 8, 2015).

50  Sarah A. Lister. 2007. *Pandemic Influenza: Appropriations for Public Health Preparedness and Response*. Congressional Research Service January 23. Washington, D.C.: Library of Congress. Available at https://fas.org/sgp/crs/misc/RS22576.pdf (accessed March 8, 2015).

51  Lister 2007; Stewart Simonson. 2010. "Reflections on Preparedness: Pandemic Planning in the Bush Administration." *Saint Louis University Journal of Health Law & Policy* 4(1): 5–33. Available at www.slu.edu/Documents/law/SLUJHP/Simonson_Article.pdf (accessed March 28, 2015).

52  Simonson 2010.

53  Sarah A. Lister. 2005. *Pandemic Influenza: Domestic Preparedness Efforts*. Congressional Research Service November 10. Washington, D.C.: Library of Congress.

54  Ibid.

55  Ibid.

56  Lister 2005; Andrew Lakoff. 2007. "From Population to Vital System:

National Security and the Changing Object of Public Health." *ARC Working Paper*, No. 7. Available at http://anthropos-lab.net/wp/publications/2007/08/workingpaperno7.pdf (accessed March 28, 2015).
57 United States Department of Health and Human Services. 2005. *HHS Pandemic Influenza Plan*. November. Washington. Available at www.flu.gov/planning-preparedness/federal/hhspandemic influenzaplan.pdf. (accessed March 28, 2015); Lister 2005.
58 Betsey Gilman Duane. 2004. "ABSA Influenza Pandemic Plan Position Paper." *ABSA News*: 231. Available at www.absa.org/abj/abj/040904influenza.pdf. (accessed March 28, 2015): 231.
59 White House Homeland Security Council. 2005. *National Strategy for Pandemic Influenza*. November 5. Washington. Available at www.flu.gov/planning-preparedness/federal/pandemic-influenza.pdf (accessed March 28, 2015); Lister 2005.
60 White House Homeland Security Council 2005.
61 Christine Craig and Laurence Hecht. 2005. "Will Vaccine Funds Be In Time for Pandemic?" *Executive Intelligence Review*, November 4. Available at www.larouchepub.com/eiw/public/2005/2005_40–49/2005_40–49/2005–44/pdf/32–33_43_eco.pdf (accessed March 28, 2015).
62 "Bush Unveils $7.1 Billion Plan to Prepare for Flu Pandemic." 2005. *CNN*, November 2. Available at http://articles.cnn.com/2005–11–01/health/us.flu.plan_1_pandemic-strain-vaccine-makers-flu-pandemic?_s=PM:HEALTH. (accessed March 28, 2015); James T. Matthews. 2006. "Egg-Based Production of Influenza Vaccine: 30 Years of Commercial Experience." *The Bridge* 36(3): 17–24.
63 Craig and Hecht 2005.
64 "Bush Pandemic Flu Strategy: Detection, Treatment, Response." 2005. *Environmental News Service*, November 1. Available at www.ens-newswire.com/ens/nov2005/2005–11–01–01.html (accessed March 28, 2015).
65 Lister 2005.
66 Ricardo Alonso-Zaldivar. 2005. "Bush's Flu Plan Stresses Vaccine." *Los Angeles Times*, November 2, 2.
67 Ibid.
68 Lister 2005, 16.
69 Lister 2005.
70 Kim Coghill. 2005. "Greenwood Praises Passage of Pandemic Influenza Plan." *Biotechnology Industry Organization*, December 22. Available at www.bio.org/media/press-release/greenwood-praises-passage-pandemic-influenza-plan (accessed March 9, 2015).
71 Nellie Bristol. 2005. "US President Releases Long-awaited Preparedness Plan." *The Lancet* 366(9498): 1683.
72 Allen, Mike. 2005. "Bush v. Bird Flu." *Time*, November 1. Available at www.time.com/time/nation/article/0,8599,1125104,00.html (accessed March 9, 2015).
73 Nellie Bristol. 2005. "US President Releases Long-awaited Preparedness Plan." *The Lancet*. 366(9498): 1683. Available at: www.thelancet.com/journals/lancet/article/PIIS0140–6736(05)67678–7/abstract (accessed March 10, 2015).
74 Jane Parry. 2007. "Ten Years of Fighting Bird Flu." *Bulletin of the World Health Organization* 85(1): 3–6, 4.
75 It is exceedingly difficult to account for all pandemic appropriations, as many programs are designed to address various public health needs not just influenza preparedness. These numbers capture annual appropriations and supplemental appropriation that were *specifically* allocated for pandemic influenza between FY2004 and FY2011.
76 Laurie Garrett. 2005. "The Next Pandemic?" *Foreign Affairs* 84(4): 3–23.
77 Ibid., 5.

78  "Avian Flu, Bioterror and Washington." 2005. *Washington Times*, September 29, A22.
79  Christine Gorman. 2005. "Avian Flu: How Scared Should We Be?" *Time*, October 17: 30–34, 30.
80  Richard Morin. 2005. "Bush Approval Rating at All-Time Low." 2005. *Washington Post*, September 12. Available at www.washingtonpost.com/wpdyn/content/article/2005/09/12/AR2005091201158_pf.htm (accessed March 28, 2015).
81  Nancy Shute. 2005. "A Man With An Antiflu Plan." *U.S. News & World Report* 139(18): 32, 32.
82  "Democrats Work to Protect Americans From Avian Flu." 2005. United States Senate Democrats. October 5. Available at http://democrats.senate.gov/2005/10/05/democrats-work-to-protect-americans-from-avian-flu/#.VP85_0trWlI (accessed March 28, 2015).
83  Lakoff 2007, 21.
84  Michael Mair, Beth Maldin, and Brad Smith. 2006. "Passage of S. 3678: The Pandemic and All-Hazards Preparedness Act." Center for Biosecurity, University of Pittsburgh Medical Center. Available at www.upmchealthsecurity.org/our-work/pubs_archive/pubs-pdfs/archive/2006–12–20-allhazardsprepact.pdf (accessed March 12, 2015).
85  James G. Hodge, Lawrence Gostin, and Jon S. Vernick. 2007. "The Pandemic and All-Hazards Preparedness Act: Improving Public Health Emergency Response." *The Journal of the American Medical Association* 297(15): 1708–1711.
86  Thomas A. Birkland and Sarah Waterman. 2008. "Is Federalism the Reason for Policy Failure in Hurricane Katrina?" *Publius* 38(4): 692–714.
87  B. Kurt Cooper. 2006. "High and Dry? The Public Readiness and Emergency Preparedness Act and Liability Protection for Pharmaceutical Manufacturers." *Journal of Health Law* 40(1): 2–3.
88  Jeffrey H. Birnbaum. 2005. "Vaccine Funding Tied to Liability." *Washington Post*, November 17. Available at www.washingtonpost.com/wp-dyn/content/article/2005/11/16/AR2005111602238.html (accessed March 28, 2015).
89  The Model State Health Emergency Powers Act. 2001.The Center for Law and the Public's Health at Georgetown and Johns Hopkins Universities. Available at www.publichealthlaw.net/ModelLaws/MSEHPA.php (accessed March 28, 2015).
90  American Civil Liberties Union. 2002. "Model State Health Emergency Powers Act." January 1. Available at www.aclu.org/technology-and-liberty/model-state-emergency-health-powers-act (accessed March 28, 2015).
91  Sue Bevins. 2002. "The Model State Health Emergency Powers Act: An Assault on Civil Liberties in the Name of Homeland Security." The Heritage Foundation. June 10. Available at www.heritage.org/research/lecture/the-model-state-emergency-health-powers-act (accessed March 28, 2015).
92  The Model State Health Emergency Powers Act 2001.
93  Peter J. May, Ashley E. Jochim, and Joshua Sapotichne. 2011. "Constructing Homeland Security: An Anemic Policy Regime." *Policy Studies Journal* 39(2): 285–307.
94  Jeremy Youde. 2008. "Who's Afraid of Chicken? Securitization and Avian Flu." *Democracy and Security* 4(2): 148–169; David Brown. 2005. "Military's Role in a Flu Pandemic." *Washington Post*, October 5, A05.
95  Peter J. May and Chris Koski. 2013. "Addressing Public Risks: Extreme Events and Critical Infrastructures." *Review of Policy Research* 30(2): 139–159, 139; Thomas A. Birkland. 2006. *Lessons of Disaster: Policy Change After Catastrophic Events*. Washington, D.C.: Georgetown University Press.

96  United States Department of Health and Human Services. 2006. *Department of Health and Human Services Pandemic Planning Update.* March 13. Washington, D.C.: 2. Available at www.cap.org/apps/docs/committees/microbiology/panflu_final3_13_(2).pdf (accessed March 28, 2015).

97  Shirley S. Ho, Dominique Brossard, and Dietram A. Scheufele. 2007. "Public Reactions to Global Health Threats and Infectious Diseases." *Public Opinion Quarterly* 4 (Winter): 671–692.

98  Ibid., 678.

99  Jo Tuckman and Robert Booth. 2009. "Four-year-old Could Hold Key in Search for Source of Swine Flu Outbreak." *Guardian*, April 27. Available at www.guardian.co.uk/world/2009/apr/27/swine-flu-search-outbreak-source (accessed March 28, 2015).

100  Centers for Disease Control and Prevention. 2009b. "WHO Pandemic Declaration." Available at www.cdc.gov/h1n1flu/who/ (accessed March 28, 2015).

101  United States Department of Health and Human Services. 2009. "Obama Administration Calls on Nation to Begin Planning and Preparing for Fall Flu Season & the New H1N1 Virus." July 9. Available at www.hhs.gov/news/press/2009pres/07/20090709a.html (accessed March 28, 2015).

102  See, for example: Brendan Maher. 2010. "Crisis Communicator." *Nature* 463(14): 150–152; Xana O'Neill. 2009. "The Little Boy Behind the Swine Flu Crisis." *NBC Miami*, April 28. Available at www.nbcmiami.com/news/archive/NATL-The-Little-Boy-Behind-the-Swine-Flu-Crisis-.html (accessed March 28, 2015).

103  Robert Pear and Gardiner Harris. 2009. "Obama Seeks to Ease Fears on Swine Flu." *New York Times*, April 28, 1.

104  "Study: Swine Flu Resembles 1918 Virus." 2009. *Fox News*, July 13. Available at www.foxnews.com/story/0,2933,532020,00.html. (accessed March 28, 2015).

105  Rush Limbaugh. 2009. "Obama's Swine Flu Vaccine Fiasco." *The Rush Limbaugh Show*, November 3. Available at www.rushlimbaugh.com/daily/2009/11/03/obama_s_swine_flu_vaccine_fiasco (accessed March 28, 2015).

106  Martin Schram. 2009. "Will Swine Flu be Obama's Katrina?" *Washington Times*, August 29. Available at www.washingtontimes.com/news/2009/aug/29/will-swine-flu-be-obamas-katrina/ (accessed March 28, 2015).

107  U.S. Congress. Senate. 2009. "Swine Flu: Coordinating the Federal Response." Committee on Homeland Security and Government Affairs. April 29. 111th Cong., 1st sess. 3. Washington, D.C.: Government Printing Office.

108  U.S. Congress. House. 2009. "The Administration's Flu Vaccine Program: Health, Safety, and Distribution." Committee on Oversight and Government Reform. September 29. 111th Cong., 1st sess. Washington, D.C.: Government Printing Office.

109  United States Department of Health and Human Services 2009.

110  Jackie Calmes and Donald G. McNeil Jr. 2009. "H1N1 Widespread in 46 States as Vaccines Lag." *New York Times*, October 24, A1.

111  Calmes and McNeil 2009.

112  Patricia Zengerle. 2009. "Obama Declares Swine Flu a National Emergency." *Reuters*, October 24. Available at http://mobile.reuters.com/article/topNews/idUSTRE59N19E20091024 (accessed March 28, 2015).

113  Calmes and McNeil 2009.

114  American Civil Liberties Union. 2009. "Maintaining Civil Liberties Protections in Response to the H1N1 Flu." An ACLU White Paper. November: 7. Available at www.aclu.org/files/assets/H1N1_Report_FINAL.pdf (accessed March 28, 2015).

115  Ibid.

116  Perhaps the one important item worth noting is the 2013 reauthorization of PAHPA. The reauthorized Act increased preparedness funding while bolstering

the Food and Drug Administration's capacity to respond to public health emergencies.
117 Arjen Boin and Paul 't Hart. 2003. "Public Leadership in Times of Crisis: Mission Impossible?" *Public Administration Review* 63(5): 544–553.
118 Homeland Security Council. 2006. *National Strategy for Pandemic Influenza Implementation Plan*. The White House. May. Available at www.flu.gov/planning-preparedness/federal/pandemic-influenza-implementation.pdf (accessed March 28, 2015).
119 United States Department of Health and Human Services. 2009b. *Pandemic Planning Update VI*. A report from Secretary Michael O. Leavitt, January 8. Available at www.flu.gov/pandemic/history/panflureport6.pdf (accessed March 28, 2015).
120 United States Department of Health and Human Services. 2008. *Pandemic and All-Hazards Preparedness Act: Progress Report on the Implementation of Provisions Addressing High Risk Individuals*. August. Available at www.phe.gov/Preparedness/legal/pahpa/Documents/pahpa-at-risk-report0901.pdf (accessed March 28, 2015).
121 Ian Scoones and Paul Forster. 2010. "Unpacking the International Response to Avian Influenza: Actors, Networks and Narratives." In *Avian Influenza: Science, Policy, and Politics*, ed. Ian Scoones. Washington, D.C.: Earthscan: 19–64, 45.
122 Scoones and Forster 2010.
123 Declan Butler. 2012b. "Flu Surveillance Lacking." *Nature*, March 28. Available at www.nature.com/news/flusurveillance-lacking-1.10301 (accessed March 28, 2015).
124 Bryan Walsh. 2007. "Indonesia's Bird Flu Showdown." *Time*, May 10. Available at http://content.time.com/time/health/article/0,8599,1619229,00.html (accessed March 28, 2015).
125 "China Accused of SARS Cover-up." 2003. *BBC News*, April 9. Available at http://news.bbc.co.uk/2/hi/health/2932319.stm (accessed March 28, 2015).
126 Robert F. Breiman, Abdulsalami Nasidi, Mark A. Katz, and John Vertefeuille. 2007. "Preparedness for Highly Pathogenic Avian Influenza Pandemic in Africa." *Emerging Infectious Diseases* 13(10): 1453–1458.
127 Tiaji Salaam-Blyther. 2011. *Centers for Disease Control and Prevention Global Health Programs: FY2001-FY2012 Request*. June 27. Congressional Research Service. Washington, D.C.: Library of Congress. Available at www.fas.org/sgp/crs/misc/R40239.pdf (accessed March 28, 2015).
128 Ibid.
129 United States Department of Health and Human Services. 2006b. *Department of Health and Human Services Pandemic Planning Update III*. A report from Secretary Michael O. Leavitt, November 13. Available at www.flu.gov/pandemic/history/panflureport3.pdf (accessed March 28, 2015).
130 United States Department of Health and Human Services, 2006; Gardiner Harris. 2006. "U.S. Stockpiles Antiviral Drugs, but Democrats Critical of Pace." *New York Times*, March 2, A21.
131 Harris 2006, A21.
132 Ibid., A21.
133 "Antiviral Stockpiles." 2009. *Washington Post*. Available at www.washingtonpost.com/wpdyn/content/graphic/2009/05/01/GR2009050100352.html (accessed March 28, 2015).
134 Harris 2006, A21.
135 United States Department of Health and Human Services. 2009b. *Pandemic Planning Update VI*. A report from Secretary Michael O. Leavitt, January 8. Available at www.flu.gov/pandemic/history/panflureport6.pdf (accessed March 28, 2015).

136 Caleb Hellerman. 2006. "Bird Flu Vaccine Eggs All In One Basket." *CNN*, March 20. Available at http://articles.cnn.com/2005–12–08/health/pdg.bird. flu.vaccine_1_vaccine-targets-sanofi-pasteur-pandemic-strain?_s=PM:HEALTH (accessed March 28, 2015).
137 Ibid.
138 Infectious Disease Society of America (IDSA). 2011. "Vaccine Development." June 16. Available at http://biodefense.idsociety.org/idsa/influenza/panflu/ biofacts/panflu_vax.html (accessed March 28, 2015).
139 United States Department of Health and Human Services 2009b.
140 Ibid., 9.
141 United States Department of Health and Human Services 2006b.
142 Office of the Press Secretary. 2007. "Press Briefing on National Strategy for Pandemic Influenza Implementation Plan One Year Summary." The White House, July 17. Available at http://georgewbush-whitehouse.archives.gov/ news/releases/2007/07/20070717–13.html (accessed March 28, 2015).
143 Council on Foreign Relations (CFR). 2007. "Is The Bird Flu Threat Still Real and Are We Prepared? [Rush Transcript; Federal News Service]." April 12. Washington, D.C.: Federal News Service, Inc. Available at www.cfr.org/ public-health-threats/bird-flu-threat-still-real-we-prepared-rush-transcript-federal-news-service/p13115 (accessed March 28, 2015).
144 Scott Wilson Scott and Spencer S. Hsu. "A Bush Team Strategy Becomes Obama's Swine Flu Playbook." *Washington Post*, May 1. Available at www. washingtonpost.com/wp dyn/content/article/2009/04/30/AR2009043003910. html (accessed March 28, 2015).
145 Brian Naylor. 2009. "Obama Flu Response Relied On Bush Plan." *NPR*, May 7. Available at www.npr.org/templates/story/story.php?storyId=103908247 (accessed March 28, 2015).
146 Robert Pear and Gardiner Harris. 2009. "Obama Seeks to Ease Fears on Swine Flu." *New York Times*, April 28, A1.
147 Sarah A. Lister and C. Stephen Redhead. 2009. *The 2009 Influenza Pandemic: An Overview*. August 6. Congressional Research Service. Washington, D.C.: Library of Congress. Available at http://fpc.state.gov/documents/organiza-tion/128854.pdf (accessed March 28, 2015).
148 Richard Knox. 2009. "Swine Flu Vaccine Shortages: Why?" *NPR*, October 26. Available at www.npr.org/templates/story/story.php?storyId=114156775 (accessed March 28, 2015).
149 Centers for Disease Control and Prevention. 2009. "CDC Estimates of 2009 H1N1 Influenza Cases, Hospitalizations, and Deaths in the United States, April–November 14, 2009." December 10. Available at www.cdc.gov/h1n1flu/ estimates/April_November_14.htm (accessed March 28, 2015).
150 Knox 2009.
151 Limbaugh 2009.
152 Sheryl Gay Stolberg. 2009. "Vaccine Shortage Is Political Test for White House." *New York Times*, October 28, A04.
153 Michael D. Shear and Rob Stein. 2009. "Administration Officials Blame Shortage of H1N1 Vaccine on Manufacturers, Science." *Washington Post*, October 27, A11.
154 Ibid. A11.
155 United States Department of Health and Human Services. 2012. "An HHS Retrospective on the 2009 H1N1 Influenza Pandemic to Advance All Hazards Preparedness." June 15. Available at www.phe.gov/Preparedness/mcm/h1n1-retrospective/Documents/h1n1-retrospective.pdf (accessed March 28, 2015).
156 U.S. Congress. Senate. 2009. "H1N1 Flu: Monitoring the Nation's Response." Committee on Homeland Security and Governmental Affairs. October 21. 111th Cong., 1st sess. Washington, D.C.: Government Printing Office: 100.

157 Shear and Stein 2009, A11.
158 "Obamas Get Their Swine Flu Shots." 2009. *New York Times*, December 21. Available at www.nytimes.com/2009/12/22/us/politics/22brfs-OBAMASGET-THE_BRF.html (accessed March 28, 2015).
159 Yanzhong Huang. 2010. "Comparing the H1N1 Crises and Responses in the US and China." Centre for Non-Traditional Security Studies Working Paper Series. Working Paper No. 1, November: 8 Available at www.rsis.edu.sg/NTS/resources/research_papers/NTS%20Working%20Paper1.pdf. (accessed March 28, 2015).
160 Donald G. McNeil. 2010. "U.S. Reactions to Swine Flu Apt and Lucky." *New York Times*, January 2, A1.
161 Ibid.
162 Stolberg 2009, A04.
163 Huang 2010, 11.
164 McNeil 2010.
165 World Health Organization. 2014. "WHO Risk Assessment Of Human Infections With Avian Influenza A(H7N9) Virus". February 23. Available at www.who.int/influenza/human_animal_interface/influenza_h7n9/RiskAssessment_H7N9_23Feb20115.pdf?ua=1 (accessed March 28, 2015).

# 5   Too Hot to Handle
## Mitigation, Adaptation, and the Quest for Climate Change Policy

Climate change refers to significant change in measures of the Earth's climate, which last for an extended period of time. Among these changes is a marked rise in the Earth's temperature, a phenomenon called "global warming." Global warming is primarily caused by an accumulation of certain "greenhouse gases" in the atmosphere that allow light to enter but not exit, in turn trapping heat—a so-called "greenhouse effect." Most scientists believe the acceleration of global warming in recent decades is caused or, at least, exacerbated by human activities (e.g., factories, the automobile, large electricity grids). The implications of global warming are many, including warming oceans, melting of polar glaciers, rising sea levels, destruction of ecological systems, diminished food security, and a proliferation of zoological and tropical diseases.[1]

Concerns with global warming center on the possibility that humankind is approaching a "tipping point," which is a threshold that, once crossed, will initiate dire and irreversible changes in the Earth's climate. Such changes may include permanent alterations in weather patterns, irregular ocean currents, and eradication of entire forests and species. Because these various systems are interconnected, radical change in one will likely alter others, further accelerating warming trends. Although the precise temperature threshold likely to bring about this tipping point has not been definitely determined, most agree that, while humankind has yet to reach this point of no return, we are dangerously close—perhaps fewer than 50 years away.[2]

This case chronicles efforts in the U.S. to anticipate, mitigate, and, most recently, adapt to climate change. The climate change case differs from the avian influenza and nanotechnology cases in at least two critical respects. First, decades of high emission rates mean many of the dire effects of global warming are well underway. However, environmentalists and other proponents of policies to address climate change argue these risks will multiply if anticipatory action is not taken to both *mitigate* emissions and *adapt* to existing and future environmental changes, such as rising tides, heat waves, and the increased spread of disease. Second, global warming is a far more politicized problem than either of the two previous cases. Attempts to regulate climate change can be traced to the Nixon Administration, and the subject has been addressed by planks in both major parties' platforms. Climate change's extended life-span as a policy issue has given rise to a fairly well-established interest group structure that has effectively gridlocked policy change and innovation at the national level.

This chapter focuses on the climate change debate from 2000 to present, although an overview of the climate change's policy history is provided. What factors drove global warming's rise to prominence as an agenda item? In the absence of federal law, what steps have national bureaucracies taken to combat climate change? To what extent have state and local governments shaped the climate change domain? These questions, as well as others, are answered below.

## Dimensions of the Problem

The United Nations' Intergovernmental Panel on Climate Change (IPCC), which is responsible for providing comprehensive scientific and technical assessments of the risks associated with climate change as well as the human causes of climate change, estimate temperatures have warmed roughly 0.74 degrees Celsius over the last century. The United States National Oceanic and Atmospheric Administration's (NOAA) National Climate Data Center (NCDC) reports that more than half (11-years) of the top-20 warmest years on record occurred in the twenty-first century. Six of the top-25 warmest years occurred in the 1990s, while the remaining two occurred in the late 1980s.[3] NASA and NOAA scientists recently proclaimed 2014 the single hottest year on record.[4]

Most climate scientists attribute increases in average temperatures to the successive accumulation of infrared heat-trapping gases, a phenomenon commonly referred to as the "greenhouse effect."[5] Carbon dioxide is the greatest single contributor to global warming, responsible for upwards of 60 percent of greenhouse emissions, with methane contributing another 15–20 percent, and the remaining 20 percent composed of nitrogen oxide, chlorofluorocarbons, and ozone. The U.S. today is the second largest emitter of greenhouse gases, after China, which became the biggest emitter in 2006. Together, these two countries account for more than 40 percent of all greenhouse gas emissions worldwide. Burning of fossil fuels is the primary source of carbon dioxide emissions. The U.S. Environmental Protection Agency (EPA) estimates that electricity generation accounts for nearly 41 percent of all carbon dioxide emissions in the U.S.[6] Other important sources are the manufacturing, construction, and mining industries; residential and commercial practices (e.g., home heating, lighting, cooling, and appliance use); and transportation (e.g., automobile use).[7] Agricultural deforestation, or the clearing of large segments of forest to make space for crops or cattle, also contributes to climate change. Not only does barren deforested land expose the Earth to further heating, but trees remove toxic impurities and gases from the air.[8]

One of the most dramatic implications of climate change is rising sea levels. According to some estimates, seas have risen by roughly 15–20 centimeters since the start of the twentieth century and are calculated to rise another 17 to 28 inches by 2100.[9] One report estimates roughly 3.7 million Americans to be at risk from rising sea levels. Higher atmospheric temperatures have warmed ocean waters, accelerating the melting and erosion of much of the globe's polar ice sheets and shelves.[10] Sea level increases also can be highly disruptive to ecological systems and settlement patterns will be interrupted as people are forced to retreat from coasts and islands.[11]

As much as one-third of all wildlife specifies could be extinct by 2050 if current warming trends continue. Moreover, an increase in tropical diseases is anticipated because warmer weather and moister climates facilitate the spread of vector organisms (e.g., insects, snails, and rodents), which are the primary transmitters of malaria, dengue fever, yellow fever, viral encephalitis, and other deadly diseases.[12] Some researchers have even linked global warming to extreme weather events, such as hurricanes, tsunamis, wildfires, and floods.[13] And, if unabated, climate change could cost as much as 3.2 percent of global GDP by 2030, with the world's least developed countries suffering loses as high as 11 percent of their GDP. Many estimates hold climate change already costs trillions of dollars a year.[14]

Not all of these dangers are inevitable, however. Scientists fear the possibility of an approaching tipping point—a temperature threshold that, once breached, will bring about irreversible changes. The concept of a tipping point actually denotes a number of watershed events (a number of tipping points), each of which is expected to have enormously dire consequences. The Earth's various systems are interconnected. Once one is disrupted others will soon follow. Scientists often cite a disruption of the Atlantic "conveyor belt" as a critical tipping point. The Atlantic conveyor belt is the process through which warm water is sent northward and colder water is sent southward in the Atlantic Ocean. Drastic influxes of coldwater due to melting ice caps threatens to permanently disrupt this cycle by preventing warm water from moving north. Scientists worry that this occurrence will spawn stronger hurricanes and hinder the Earth's capacity to cool itself. Another tipping point would occur if Greenland become entirely ice free, which would also cause a 20 to 23-foot rise in sea levels. Additionally, the destruction of more than half the Amazon rain forest, which could occur by 2200, would eradicate entire species and deprive the Earth of one of its most important air purifying systems.[15]

Scientific recognition of global warming dates to 1896, when Swedish scientist Svante Arrhenius argued that influxes in carbon dioxide in the atmosphere could alter the Earth's surface temperature. Systemic research throughout the 1950s confirmed Arrhenius' hypothesis, revealing a connection between carbon dioxide and rising temperatures. Concerns were magnified in 1985 when a team of British scientists discovered a hole in the ozone layer. The discovery captured widespread media attention and was presented as tangible evidence of the connection between human activity and atmospheric destruction.[16]

The 1978 National Climate Act (P.L. 95–367), an interagency research endeavor authorized to gather climate information and consider the policy implications of climate change, represented the first piece of climate change legislation in the U.S. However, climate change did not solidify its place on the U.S. policy agenda until 1988, a year when scorching heat waves caused severe droughts and stifled agriculture production across the country. The 1988 drought lent "proximity" to the issue, showing Americans that rising temperatures could have an immediate impact on their lives.[17]

Representative Albert Gore (D-TN), who convened the first congressional hearing on climate change in 1976, emerged as one of the foremost advocates of climate change policy. During his time in Congress and later as Vice President, he tirelessly advocated for climate change legislation. In fact, his efforts, which continued after his career in public service, earned him the 2007 Nobel Peace Prize.[18]

Outside Congress, a number of important issue entrepreneurs set out to raise awareness of climate change as well. Few received more attention than James Hansen, Director of NASA's Goddard Institute for Space Studies, who published a number of highly influential papers modeling the long-term implications of climate change. In what has come to be seen as a watershed moment in the history of climate change policymaking, Hansen testified at a 1988 U.S. Senate hearing that he was "99 percent" confident long-term warming trends were underway and they would likely bring frequent storms, flooding, and life-threatening heat waves.[19] In addition, an army of environmental interest groups continue to call for domestic and international interventions, including Greenpeace, Earth First!, and the National Wildlife Federation.

Conversely, a bevy of industry groups construe climate change legislation, and especially emissions caps, as a threat to profits. In 1989 the Global Climate Coalition (GCC) was formed to represent a diversity of corporate interests, including Chevron, Chrysler, Shell Oil, Texaco, Ford, General Motors, the American Petroleum Institute, and the American Forest & Paper Association, amongst others. The GCC aggressively disputed the contention that climate change posed a threat and set out to influence policy at the domestic and international levels.[20] In the years to follow, a variety of industry groups sought to strategically undermine the science of global warming by highlighting the inherent uncertainties surrounding the forecasting of future trends.[21]

In 1987, Congress passed The Global Climate Protection Act (P.L. 100–204), which tried to strengthen the Climate Program by authorizing the EPA and Department of State (DOS) to develop climate change policy. Despite signing the Act, President Ronald Reagan objected to vesting policymaking authority in the EPA and DOS, arguing it would interfere with the White House Office of Science and Technology Policy's jurisdiction over science issues.[22]

The United Nations established the IPCC in 1988. In the years to follow, IPCC reports were seen as arguably the most authoritative documents in climate change research, aggregating "indicators"—scientific findings—of climate change in a single text.[23] The movement for policy change has since progressed along two parallel, but at times intersecting, tracks—the domestic and international arenas. In 1990, the IPCC released its first report, which confirmed that climate change was underway and would continue in the absence of government intervention. The platform for contemporary international climate change treaty negotiations was established at the 1992 Earth Summit in Rio de Janiero, otherwise known as the United Nations Conference on Environment and Development (UNCED). The result was the creation of the United Nations Framework Convention on Climate Change (UNFCCC), a global treaty aimed at stabilizing greenhouse gas in the atmosphere and reversing global warming trends. Although the UNFCCC lacked legally binding limits on greenhouse gas emissions and enforcement mechanisms, it called for a number of updates or "protocols" that would, presumably, result in mandatory emissions standards. The first protocol established a broad goal to stabilize greenhouse gases at 1990 levels by the year 2000. The U.S. was one of the first nations to formally ratify the UNFCC, which was supported by the entire European Union and 167 other nations.[24]

Over the course of his four years in office, President George H. W. Bush enacted laws and established programs that set out to reduce American dependence on foreign oil,[25] provide states with various incentives to establish more energy efficient transportation systems,[26] and provide technical assistance to states creating their own climate change plans.[27,28] The Clean Air Act Amendments of 1990 (P.L. 101–549) was the most significant policy of the Bush era. To be sure, it was not a direct response to climate change, but instead sought to reduce sulfur dioxide emissions. Concerns with sulfur dioxide can be traced to the 1970s when scientists established a link between the atmospheric pollutant and acid rain. Acid rain was shown to contaminate water supplies and terrestrial ecosystems while eroding historical buildings, monuments, and other materials.[29]

The Clean Air Act Amendments of 1990 utilized a market-based, "cap-and-trade" approach to limit the amount of sulfur dioxide that could be emitted by manufacturing industries (e.g., automobile, paper, steel, rubber, and chemical). Broadly speaking, cap-and-trade designs place a ceiling or "cap" on the annual amount of sulfur dioxide (or any other gas, for that matter) industry can emit in a given year. Industries then receive emissions allowances that can be sold or bought. Industries with more advanced factories are better positioned to sell their allowances, as they will easily meet the yearly cap. On the other hand, transitioning industries struggling to meet the cap can purchase additional credits, which allow them to over-emit. The overall "cap" on annual emissions is lowered each year thereby requiring industries to gradually improve their factories or continue purchasing allowances.[30]

The cap-and-trade approach represented an important departure from "command and control" instruments, which imposed uniform standards across all firms. By implementing a market-based approach, the Clean Air Act Amendments of 1990 allowed industries to gradually reduce emissions without assuming significant economic losses. This design was politically appealing to environmentalists and industry alike, although some observers frowned at the so-called "moral hazard" associated with allowing firms to pollute.[31] The Act, which passed the House and Senate with *overwhelmingly* majorities, showed that, with the right policy design, bipartisan support for climate change policy was possible.[32]

Upon assuming office in 1993, Democratic President Bill Clinton and Vice President Al Gore sought to maintain year 2000 emissions at 1990 levels. But the majority of their domestic climate initiatives were rejected by the Republican-controlled Congress.[33] Instead, the thrust of the Clinton Administration's climate change agenda occurred at the international level. Calls for an international emission standard came to a head in 1997 at the negotiation meetings in Kyoto, Japan. The Kyoto Protocol set binding emissions targets for all signatories. Years earlier, in 1995, the Clinton Administration accepted the "Berlin Mandates," a precursor to Kyoto that called for legally binding emissions targets but exempted developing nations from the requirements. Many observers balked at the exemption, arguing it was unfair to require some nations to invest in the environment but not others.[34]

Months before Kyoto, the U.S. Senate unanimously passed a resolution preemptively rejecting any future climate change agreements without formal commitments from developing countries. Despite this opposition, the Clinton Administration formally signed the Protocol in November 1998. Senators berated the President's wanton disregard of their previous resolution.[35] Senator Larry Craig (R-ID) succinctly captured elected officials concerns, stating that the treaty was "designed to give some nations a free ride, it is designed to raise energy prices in the United States and it is designed to perpetuate a new U.N. bureaucracy to manage global resource allocation."[36] Even Senator John Kerry (D-MA), a proponent of emission controls, concluded the treaty was "not ratifiable in the Senate."[37]

Clinton left office in 2001 without having submitted the treaty for Senate confirmation. By 2000, not only had emissions failed to stabilize to 1990 levels, but increased by nearly 15 percent over the decade.[38] The policy divisions evidenced in the 1990s crystallized into a clear ideological split between environmentalists and industry, Democrats and Republicans. It is against this backdrop that the following section examines the global warming problem definition debate.

## Discourse of Conflict

To be sure, scientists "know," or are fairly certain, the Earth is warming and that these changes are caused by human activity, although, as described below, both claims have been refuted by skeptics.[39] Still, they cannot conclusively determine how much warming will occur in the future, and estimates range from 2.1 degrees and 11 degrees Fahrenheit by 2100. Nor can scientists predict how much carbon dioxide or other greenhouse gas emissions will be released in the future; how different ecosystems and species will respond to warmer temperatures; or the human, health, and economic implications of climate change.[40] And, like all anticipatory problems, climate change is very abstract and many individuals struggle to observe its immediate relevance to their day-to-day lives despite being asked to make immediate economic sacrifices.[41]

Climate change policymaking has always relied heavily on problem indicators. While a number of metrics have been presented (e.g., measurements of sea level rise, counts of extreme weather events), measures of average annual temperatures are arguably the most prominent indicators of climate change. Similar to the other cases, indicators are packaged as part of larger narratives that work to mitigate perceived uncertainties and frame global warming as an immediate threat. As early as 1988, NASA Scientist James Hansen declared "the earth is warmer in 1988 than at any time in the history of instrumental measurement."[42] Decades later, President Barack Obama cited similar indicators in his 2015 State of the Union Address:

> 2014 was the planet's warmest year on record. Now, one year doesn't make a trend, but this does—14 of the 15 warmest years on record have all fallen in the first 15 years of this century.
>
> I've heard some folks try to dodge the evidence by saying they're not scientists; that we don't have enough information to act. Well, I'm not a scientist, either. But you know what—I know a lot of really good scientists at NASA, and NOAA, and at our major universities. The best scientists in the world are all telling us that our activities are changing the climate, and if we do not act forcefully, we'll continue to see rising oceans, longer, hotter heat waves, dangerous droughts and floods, and massive disruptions that can trigger greater migration, conflict, and hunger around the globe. The Pentagon says that climate change poses immediate risks to our national security. We should act like it.[43]

Indicators of increased warming helped convey an emerging threat, if not a full-blown crisis. At a Field Hearing of the Energy and Natural Resources Subcommittee, Senator John McCain (R-AZ), one of the few proponents of emissions legislation in his party, declared: "A common misperception is that this is a crisis that is down the road. Climate change is real. It's happening now."[44] In a 2007 address before the United Nations General Assembly, Secretary-General Ban Ki-moon also evoked the rhetoric of crisis, stating:

> Today, war continues to threaten countless men, women and children across the globe. It is the source of untold suffering and loss. And the majority of the UN's work still focuses on preventing and ending conflict. But, the danger posed by war to all of humanity—and to our planet—is at least matched by the climate crisis and global warming.[45]

Public discourse also relies heavily on the use of analogy as a means of communicating this complex phenomenon. The best example is, of course, the "greenhouse effect" metaphor, which constitutes *the* formative depiction and explanation of how emitted gases are heating the Earth. And while the greenhouse effect did not constitute a politically motivated narrative per se, its importance in communicating an easily understandable explanation of climate change was instrumental in making it a salient political issue.

One of the most pervasive strategic political analogies was the contention that climate change was tantamount to World War II. This analogy held that climate change not only constituted a problem of epic proportions, but that "[a]ddressing the root causes of global warming will require a level of national and international cooperation not seen since the Allied nations' response during World War II."[46] This analogy was so important that *Time* magazine adorned the cover of its April 28, 2008 "Special Environment Issue" with a doctored image of the iconic World War II photograph depicting six American Marines raising the U.S. flag in Iwo Jima. However, in this instance the Marines are raising a tree, and the magazine cover bears the title "How to Win the War on Global Warming." *Time* correspondent Bryan Walsh argued that a World War II-type effort on the part of the U.S. is not only our best—and perhaps only—hope for reversing climate change trends, but will likely pay considerable economic dividends in the long run. Writes Walsh:

> Forget precedents like the Manhattan Project, which developed the atom bomb, or the Apollo program that put men on the moon—single-focus programs both, however hard they were to pull off. Think instead of the overnight conversion of the World War II-era industrial sector into a vast machine capable of churning out 60,000 tanks and 300,000 planes, an effort that not only didn't bankrupt the nation but instead made it rich and powerful beyond its imagining and—oh, yes—won the war in the process.[47]

Stone notes that analogies and metaphors to war help legitimize long term and large-scale policy commitments.[48] A nation at war will leverage all of its resources—financial and human—to ensure victory. While global warming would not require human sacrifice, it would demand considerable government spending, international cooperation, and significant economic changes. Former Vice President Gore went so far as to quote an address by Winston Churchill on the eve of Britain's entrance into World War II. Quoting Churchill, Gore writes: " 'The era of procrastination, of half measures, of soothing and baffling expedience of delays, is coming to a close. In its place we are entering a period of consequences.' "[49] In his own words, Gore adds: "We are facing a global climate crisis. It is deepening. We are entering a period of consequences."[50]

Proponents of climate change legislation also note a moral obligation to future generations who will suffer dire consequences if corrective steps are not taken soon. Environment America, a federation of state-based advocacy organizations, claimed: "If we want to spare our children and grandchildren the worst consequences of global warming, we must dramatically reduce the carbon pollution that we pump into the atmosphere. And, as most scientists agree, we better do it soon."[51] The State of the World Forum, an organization that helps world leaders find solutions to critical global challenges, echoed these sentiments: "Make no mistake: this will impact you and *certainly* all of our children."[52] Such a narrative lends proximity to the issue by depicting climate change as a threat to "our children."[53]

Those expressing skepticism about climate change have also set out to frame the debate. Countering claims that indicators of rising global temperatures represent "proof" of a crisis, opponents leverage an analogy of their own, arguing that the current warming period is similar to previous warming periods in Earth's history. They assert that, although current temperature increases are indisputable, the extent to which mankind is responsible for this change is open to debate. Critics often cite studies demonstrating a Medieval Warm Period which occurred between roughly AD 950 and 1250, during which the Earth's temperatures rose sharply, followed by a stark downturn in temperatures from 1400 to 1700. Commenting on these findings, James Inhofe (R-OK), current chair of the Senate Committee on Environment and Public Works and arguably the most outspoken climate change skeptic, wrote, "if the earth was warmer during the Middle Ages than the age of coal-fired power plants and SUVs, what role do manmade emissions play in influencing climate?"[54]

But the most prevalent narrative proposed by opponents of stringent emissions standards frames climate change policies as a threat to the American economy. For example, at a 2003 House hearing on the Kyoto Protocol's potential impact on the coal industry, Representative Robert Ney (R-OH) declared: "The Kyoto Protocol, beyond any question, will eliminate thousands of jobs across the country. Jobs will be lost throughout the industry in all coal producing states and many other regions in our country."[55] Similarly, Representative Richard Pombo (R-CA) stated: "At a time when people in this room have personally experienced the pain and trauma of massive plant closings in recent years, the Kyoto Treaty would add further insult to injury. In short, Kyoto means pink slips in French."[56]

The jobs argument is multifaceted, integrating a variety of concerns relating to domestic employment and international competiveness. Updating factories to meet these standards promises to be a costly undertaking and, in order to cover the costs of these conversions, many fossil-fuel-dependent industries anticipate layoffs. This narrative also lends proximity to the debate while, again, identifying a sympathetic problem population, namely American workers. In an op-ed appearing on the conservative website *RedState.com*, then-Republican Presidential hopeful Rick Santorum propounded these themes:

> [I]n Washington, blocking the American dream has become political sport. The Washington Establishment would rather fight global warming than fight for American jobs.
> We are the collateral damage of the war against global warming.[57]

The jobs argument closely dovetails with the argument that international emissions standards are simply unfair. Representative Joe Knollenberg (R-MI) summarized these concerns in 1998:

> This fatally-flawed agreement [the Kyoto Protocol] is blatantly unfair because it exempts developing nations from making any commitment to reduce their emissions of greenhouse gases. As a result, nations like China, India, Mexico, and Brazil, [...] will be given a free pass while the United States is forced to struggle with the Kyoto treaty's stringent mandates.[58]

A more extreme narrative has argued climate change policy is objectionable not necessarily because it threatens American jobs, but rather because global warming does not, in fact, exist. No popular figure better exemplified this trend than techno-thriller writer Michael Crichton. In 2004, Crichton published the best-selling book *State of Fear*, which describes a group of eco-terrorists attempting to create widespread fear as a result of a series of environmental catastrophes in order to forward their climate change agenda. The book, which sold more than 1.5 million copies worldwide, represented a harsh critique of what Crichton sees as misleading climate science.[59] The biggest proponent of this "conspiracy theory" in Congress is Senator Inhofe. Inhofe, who published a book entitled *The Greatest Hoax: How the Global Warming Conspiracy Threatens Your Future*, is quoted as saying: "With all the hysteria, all the fear, all the phony science, could it be that manmade global warming is the greatest hoax ever perpetrated on the American people? I believe it is."[60]

This complex collection of definitional claims were staples of the climate change debate. In many resprects, the polarized narratives described above foreshadowed the subsequent agenda setting and policy formulation stages, as the national government has been effectively gridlocked on the issue of climate change for more than a decade. The following section chronicles this struggle for policy change.

## Agenda Setting and Policy Formulation

Since 2000, policymaker attention to climate change has spiked on two distinct but overlapping occasions. The first occurred in 2007 (110th Congress), a year that saw the Democratic Party gain control of both chambers of the legislative branch for the first time since 1995. By this time, the "cap-and-trade" approach represented the dominant mechanism for combating greenhouse gas emissions. The 110th Congress was also much more receptive to the importance of climate change adaptation, at least relative to previous sessions. Whereas emissions policies aim to mitigate climate change by reducing greenhouse gases, adaptation redesigns and prepares communities for impending and unavoidable changes in the environment (e.g., heat waves, rising tides, new diseases). Adaptation provisions were integrated into most of the emission proposals introduced in 2007. A number of other policies also focused exclusively on adaptation. However, none of the major climate change policies—mitigation or adaptation—were enacted.

The election of Democrat Barack Obama as President ushered in a second wave of activity in 2009. Of all the policies developed during this period, the American Clean Energy and Security Act of 2009 (H.R. 2454), which narrowly passed the House, and, in the Senate, the Clean Energy Jobs and American Power Act (S. 1733), were the strongest contenders for enactment. Adaptation also remained an important component of these laws as well as a number of other, more targeted policies. Yet, despite tremendous optimism, Congress was once again unable to pass national climate change legislation.

This pattern of congressional attention is clearly illustrated in Figures 5.1 and 5.2. Figure 5.1, which shows the number of times "climate change" and "global warming" were entered in the *Congressional Record*, indicates 2007 was a breakthrough year for the climate change domain.[61] The number of entries skyrocketed between 2006 (424 entries) and 2007 (1,058 entries). The number of entries recorded in 2007 eclipsed the previous high of 818 entries in 1990, arguably the most productive year for climate change legislation to date. Climate change remained an important topic in 2008 (906 entries) before peaking in 2009 (1,265 entries). A precipitous decline in attention occurred in subsequent years.[62]

Figure 5.2 depicts a very similar trend. The number of congressional hearings on the topics "global warming" and "climate change" *exploded* from 2006 (25 hearings) to 2007 (119 hearings).[63] Prior to 2007, only one year, 1989 (34 hearings), had more than 25 hearings. A noticeably high number of hearings were held in 2008 (80 hearings) and the number of hearings on climate change once again eclipsed "100" in 2009 (102 hearings). These numbers dropped considerably from 2010 to 2014, averaging close to 22 hearings per year during this period.

To be sure, the climate change domain has yet to see the sort of dramatic national policy change evidenced in the previous cases. Still, the absence of a singular federal statute does not mean the climate change domain has been devoid of change. Instead, a diverse mix of state actions, regulatory changes, and international agreements have collectively shaped and altered subsystem dynamics. This "patchworking" of policies has occurred simultaneously in the areas of *both* mitigation and adaption,

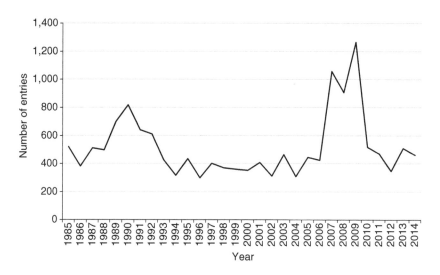

*Figure 5.1* Number of Climate Change Entries in the *Congressional Record*, 1985–2014 (source: search on ProQuest Congressional for the terms "climate change" and "global warming").

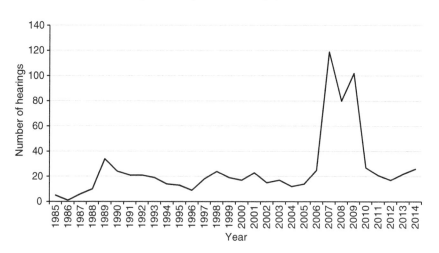

*Figure 5.2* Number of Congressional Hearings on Climate Change, 1985–2014 (source: search on ProQuest Congressional for the terms "climate change" and "global warming").

although emissions capitation remains the proverbial "Holy Grail" of climate change policy. While it is certainly beyond the scope of this book to review *all* of these changes, this section considers not only attempts to broker a national policy regime but also the multilevel dynamics of climate change policy. The first part of this section will focus on climate change mitigation, while the second part will focus on adaptation.

### Climate Change Mitigation

Months after assuming office, President George W. Bush called the Kyoto Protocol "unrealistic," adding that he would oppose any treaty or law that threatened the U.S. economy.[64] The Heritage Foundation, a conservative think tank, praised the new President, writing:

> President Bush is right to walk away from the Kyoto Protocol. It is a flawed agreement for addressing the issue of global temperature changes and their impact on the environment. Considerable uncertainty remains about the science of climate change and mankind's contribution to it.[65]

Highlighting the uncertainties of climate change was an important part of opposition groups' strategy during this period. A leaked 1998 memorandum from the American Petroleum Institute (API) read: "Victory will be achieved when [a]verage citizens 'understand' (recognize) uncertainties in climate science; [and] recognition of uncertainties becomes part of the 'conventional wisdom'."[66] Various points of evidence imply this thinking resonated with the Bush Administration. A controversial memorandum from Republican strategist Frank Luntz to President George W. Bush leaked in 2002 indicated:

> There is still a window of opportunity to challenge the science. Voters believe that there is no consensus about global warming within the scientific community. Should the public come to believe that the scientific issues are settled, their views about global warming will change accordingly. Therefore, you need to continue to make the lack of scientific certainty a primary issue in the debate.[67]

A 2007 report by the House Committee on Oversight and Government Reform confirmed "There was a systematic White House effort to minimize the significance of climate change by editing climate change reports."[68] The report added "a systematic White House effort to censor climate scientists by controlling their access to the press and editing testimony to Congress."[69]

However, outside of the most conservative political circles, "doing nothing" was not a politically viable option. In turn, President Bush announced his Clear Skies Initiative on February 14, 2002. Clear Skies established caps on sulfur dioxide, nitrogen oxides, and mercury. The Bush

Administration estimated greenhouse gas intensity would decline by nearly 18 percent by 2012 under the Clear Skies Initiative. Yet, overall greenhouse gas emissions were expected to increase during this period and the Bush Administration made no attempt to return emissions to 1990s levels.[70] Nor did the plan regulate carbon dioxide, the primary source of warming. Environmentalists and many Democrats dismissed the proposal as a half-hearted attempt to combat climate change.[71]

Senator George Voinovich (R-OH), Senator Inhofe (R-OK), Representative Billy Tauzin (R-LA), and Representative Joe Barton (R-TX) introduced a version of the Clear Skies Initiative in Congress. The bill provoked a number of competing proposals, including the Clean Air Planning Act of 2003 (S. 843), the Clean Power Act of 2003 (S. 366), and the Climate Stewardship Act of 2003 (S. 139). By the 109th Congress (2005–2006), a number of competing proposals were circulating the legislative branch.[72]

Notions of time and temporality factored prominently into the design of these proposals. Policies set out to curb emissions 10, 25, even 50 years into the future.[73] While cap-and-trade provisions were typically the centerpiece of these plans, most policies also worked to reduce energy use, promote energy efficiency, and advance renewable energy technologies. Proposals were compared based on to the extent to which *future* emissions totals (caps) would remain in line with the levels recorded in *previous years*. A 2008 document by the PEW Center on Global Climate Change used a number of criteria to compare the seven leading proposals, including their "2010–2019 Cap," "2020–2029 Cap," and the "2030–2050 Cap." For example, the Climate Stewardship and Innovation Act (S. 280) was projected to bring emissions to a "2004 level in 2012" and a "1990 level in 2020," whereas the Low Carbon Economy Act (S. 1766) would "Start at [a] 2012 level in 2012" but bring emissions to a "2006 level in 2020."[74]

Such long-range planning, projections, and estimations epitomized conjectural policymaking.[75] In order to determine the efficacy of these policies, policymakers and stakeholders essentially had to forecast future warming trends. This was hardly a trivial task. Indeed, the IPCC's 2000 *Special Report on Emissions Scenarios* offered more than 40 different future emissions scenarios. Each scenario was influenced by a variety of different variables, including, but by no means limited to, the rate of economic growth, population increases, future technological developments, and projected land use.[76] Change in any one of these factors could impact whether a policy reached its stated goals.

Thus, similar to nanotechnology R&D and influenza preparedness, planning was a staple climate change policymaking. Of course, any preoccupation with the future is bound to be rife with political pitfalls. In fact, Richard J. Lazarus, professor of law at Georgetown University, argues it represents the *greatest* impediment to successful climate change policy. Writes Lazarus:

Climate change legislation is peculiarly vulnerable to being unraveled over time for a variety of reasons, but especially because of the extent to which it imposes costs on the short term for the realization of benefits many decades and sometimes centuries later.[77]

Climate change planning is difficult because it militates against the politician's desire—*need*—to enact policies that produce immediate outputs. Compounding this problem, climate change policy is often construed as entailing short-term job loss, a tough sell for any elected official.

Of the various proposals offered in 2003, only the Climate Stewardship Act of 2003 (S. 139) received a floor vote. Co-sponsored by Senators McCain (R-AZ) and Joseph Lieberman (I-CT), S. 139 set *mandatory* caps on carbon dioxide emissions, requiring that they remain at 2000 levels by 2012. Companies were allowed to pursue a myriad of carbon trading options. The Climate Stewardship Act was defeated in the Senate by a vote of 43 to 55. However, proponents of climate change legislation claimed victory, arguing the narrow vote was a promising sign for future climate change legislation. In fact, an anonymous staffer from the Environment and Public Works Committee suggested the Act was actually closer to passing than the vote tally implied: "We've heard that the opposition was surprised and intimidated by the results, especially given that in the days before the vote it looked awfully close to passing."[78]

Climate change emerged as a top agenda item in the 110th Congress after Democrats regained control of Congress for the first time since 1995. Ironically, the uptick in national attention was partially stimulated by the states, which were busy enacting mitigation programs of their own. In 2007, 12 states, a handful of cities, and a number of environmental organizations sued the EPA, charging carbon dioxide and other greenhouse gases constituted air pollutants under the Clean Air Act. Therefore, they argued, the EPA was required to regulate emissions from motor vehicles. In a 5–4 decision, the U.S. Supreme Court ruled in favor of the states.[79] The ruling set off a feeding frenzy within Congress, "[catalyzing] calls for more comprehensive federal climate change legislation ... that covers sectors other than transportation as well as non-$CO_2$ [carbon dioxide] greenhouse gases."[80] By July 2008, more than 200 bills, resolutions, and amendments were introduced in the 110th Congress.[81]

Warning signs—problem indicators—abounded during this period. The IPCC released its long-awaited Fourth Assessment Report in 2007, which provided a wealth of evidence demonstrating with certainty that the planet was, in fact, warming at an alarming rate and that this trend was directly attributable to human activity.[82] It concluded by offering a blunt warning to policymakers: "Delayed emission reductions significantly constrain the opportunities to achieve lower stabilisation levels and increase the risk of more severe climate change impacts."[83]

The House Committee on Energy and Commerce and the Senate Committee on Environment and Public Works, both of which have jurisdiction over environmental issues, were important venues during this period, convening a number of investigatory hearings and stewarding legislation along the institutional pathways of Congress. Debate within these venues was polarized. At a 2005 hearing, James Jeffords (I-VT), who once called global warming his number one priority, called for stringent emissions standards:

> [W]e cannot legislate responsibly and ignore manmade global warming completely. The U.S. power sector emits one-tenth of the world's total carbon dioxide emission. To ignore this fact defies reason, logic and the peer review work of the National Academy of Sciences, the American Geophysical Union and the International Panel on Climate Change.[84]

Conversely, at a 2009 hearing on Clean Energy Jobs and American Power Act (S. 1733), one of the various cap-and-trade bills circulating at this time, Brett A. Vassey, President & CEO of Virginia Manufacturers Association testified that his group wanted:

> Congress to develop responsible policies that protect domestic jobs and the environment. We are concerned that these bills will "cap" industrial competitiveness and "trade" domestic manufacturing jobs abroad for an entirely undefined environmental benefit. We can do better and we must do better.[85]

While climate change was largely framed as an environmental issue, the perceived magnitude of the problem coupled with its uncertainty accommodated narratives linking it to other policy areas, a dynamic observed in previous cases as well. Global warming was also portrayed as a foreign policy threat,[86] a byproduct of American's dependence on foreign oil,[87] a public health problem,[88] and an opportunity to develop a new "green" economy.[89] As such, hearings were convened in a variety of committees including the House Committee on Foreign Affairs; the Senate Committee on Foreign Relations; House Committee on Transportation and Infrastructure; Senate Committee on Banking, Housing, and Urban Affairs; and the House Committee on Small Business amongst others.

By the 110th Congress, scores of proposals set out to establish emissions standards, but, by the end of the 110th Congress, only one bill was poised for explicit consideration.[90] The America's Climate Security Act of 2007 (S. 2191), which was co-sponsored by Senator Lieberman and Senator John Warner (R-VA), promised to reduce greenhouse gas emissions 19 percent below 2005 levels by 2020, and 63 percent below 2005 levels by 2050. Once again, a complex trading scheme was developed and the bill capped overall emissions in the electricity, industrial, transportation, and natural gas sectors. Despite being approved by the Committee on Environment and Public Works, it was defeated in the Senate by a vote of 48–3.[91] Industry groups aggressively contested the measure, threatening to run television advertisements targeting senators representing districts with large numbers of emitting facilities.[92]

Democratic gains in both chambers positioned the 111th Congress to break the deadlock. Equally important, the election of President Barack Obama gave Democrats control of the White House for the first time in eight years. During his campaign, President Obama proposed an ambitious domestic cap-and-trade initiative that would reduce emissions by 80 percent by 2050.[93] Climate change was one of the President's top priorities, second only, perhaps, to health care.[94]

Months into President Obama's first term, Representatives Henry Waxman (D-CA) and Edward Markey (D-MA) introduced the American Clean Energy and Security Act of 2009 (H.R. 2454), which used a cap-and-trade approach to reduce greenhouse emissions to 17 percent below 2005 levels by 2020 and 80 percent below by 2050. These emissions standards were applied to "covered entities," which included most of the highest emitting industries (e.g., electricity, oil, manufacturing, etc.). These entities were required to obtain "tradable emissions allowances" for every ton of carbon dioxide emitted in the prior year. Beyond the cap-and-trade program, the Act promoted energy efficiency, renewable electricity, green building and job training, and more efficient appliances and transportation. It also offered grants to support state mitigation programs.[95]

The proposal enjoyed the support of a handful of automotive companies, gas companies, and labor unions in addition to the usual army of environmental groups.[96] President Obama worked to recast emissions policy—and H.R. 2454—as an economic opportunity. At a 2009 speech, the President proclaimed:

> The nation that leads in the creation of a clean energy economy will be the nation that leads the twenty-first century's global economy. That's what this legislation seeks to achieve. It's a bill that will open the door to a better future for this nation and that's why I urge members of Congress to come together and pass it.[97]

The American Clean Energy and Security Act of 2009 *narrowly* passed the Democratic House by a vote of 219 to 212 (with three abstentions).[98]

Meanwhile, in the Senate, the Clean Energy Jobs and American Power Act (S. 1733) proposed a gradual easing of emissions to 17 percent below 2005 levels by 2020 and 80 percent below 2005 by 2050.[99] Although the bill aligned with American Clean Energy and Security Act of 2009, co-sponsoring Senators John Kerry (D-MA), Joseph Lieberman (I-CT), and Lindsey Graham (R-SC) were open to compromise.[100] Commenting on the need to remain flexible, Senator Kerry stated: "For climate, it's the bottom of the ninth inning and the bases are loaded if we can just push these runs across the plate."[101]

Two grand concessions would be required if the bill had any hope of passing the Senate. First, the policy expanded subsidies for nuclear power plants, a provision that was especially important to Senator Graham (R-SC), a proponent of nuclear energy. Second, to appeal to Republicans representing coastal states, the bill provided subsidies for offshore oil drilling.[102] Even President Obama was willing to support offshore drilling if it meant passing mitigation legislation, declaring: "It turns out, by the way, that oil rigs today generally don't cause spills. They are technologically very advanced. Even during Katrina, the spills didn't come from the oil rigs, they came from the refineries onshore."[103]

In a horrific twist of fate, an explosion on a British Petroleum rig on April 20, 2010, sent oil streaming into the Gulf of Mexico, killing 11 workers and blanketing the water in a toxic plume. The *Deepwater Horizon* oil spill released nearly five million barrels of oil into the water, devastating entire ecosystems and crippling the economy of the Gulf Coast.[104] In the wake of the disaster, few Senators were willing to support legislation subsidizing offshore drilling. In a last-ditch effort to appease those concerned with the drilling provision, Senator Kerry inserted a provision allowing states to veto offshore drilling up to 75 miles off their coast.[105] Kerry's attempt was futile, however, and on July 22, 2010, Senate Majority Leader Harry Reid (D-NV) formally announced the Senate would not persue comprehensive mitigation legislation before the August break.

The 112th Congress, which saw Republicans regain the House and gain seats in the Senate, shelved the climate change debate and did not attempt to pass emissions legislation.[106] Although most sober observers acknowledge Senate approval of Kyoto is unlikely, President Obama has taken modest steps to reengage the international community. In November 2014, the President struck a climate deal with the Chinese government. According to the deal, China promised to cap emissions by 2030, while the U.S. promised to cut emissions by 26 to 28 percent of 2005 levels by 2025. Climate scientists called the agreement an important step in the right direction, but warned it still fell short of solving the climate change problem.[107]

Yet, in the absence of a national standard, state governments have adopted emissions programs of their own. Schreurs notes one of the "unintended consequences of the Bush Administration's obstructionist attitude on climate action was to mobilize sub-national governments into taking action independently of the federal government."[108] By 2006, virtually every state in the country had taken some steps to address climate change. Long-term and anticipatory planning was important to these policy designs and states adopted a dizzying array of policies, including cap-and-trade arrangements, carbon capture and sequestration programs, automobile emissions standards, clean and sustainable energy programs, and electricity regulations.[109]

California has been a leader in this area for decades, establishing the nation's first comprehensive greenhouse gas inventory in 1988 as well as the first greenhouse gas regulation for automobile emissions in 2003.[110] In 2006, California adopted legislation requiring that greenhouse gas emissions be reduced to 1990 levels by 2020.[111] The Global Warming Solutions Act of 2006 (AB 32), which covers all of the gases included in the Kyoto Protocol, uses a comprehensive "cap and trade" program and requires participation from all large businesses that emit more than 25,000 metric tons of carbon dioxide a year.[112,113]

The Regional Greenhouse Gas Initiative (RGGI), the nation's first mandatory cap-and-trade program for greenhouse gases, encompasses the Northeastern states of Connecticut, Delaware, Maine, Maryland, Massachusetts, New Hampshire, New York, Rhode Island, and Vermont.[114] RGGI initially sought to cap carbon dioxide emissions at projected 2009 levels until 2015. After 2015, the program would implement gradual reductions, eventually achieving a 10 percent reduction from 2009 levels by 2019. Participants revised the program in 2012 to include a more stringent emissions cap, in part because actual 2012 emissions were far lower than originally anticipated. The new emissions cap reduces the amount of allowable emissions from 165 metric tons of carbon dioxide to 91 metric tons.[115] Governors in Arizona, California, New Mexico, Oregon, and Washington, adopted a similar initiative in 2007, the Western Climate Initiative (WCI). The Canadian provinces of Ontario, Manitoba, and Quebec joined the WCI, which aims to bring emissions 15 percent below 2005 levels by 2020.[116]

Local governments have been very active as well. Boulder, Colorado became the first city in the U.S. to experiment with a "carbon tax" system in 2006, charging a direct fee to residents and businesses based on how much electricity they use.[117] San Francisco, California adopted a similar carbon tax program in 2008.[118] Local governments have also entered into cooperative initiatives. The Cities for Climate Protection (CCP) program, which encompasses more than 650 municipal governments across more than 30 countries, allows participating cities to establish their own emissions targets.[119]

Environmental groups never abandoned the national government, as evidenced in the 2009 push for national legislation. Instead, states were "added" as an additional policy venue.[120] Powerful environmental groups, including Environmental Defense Fund (EDF) and the Natural Resources Defense Council, co-sponsored California's 2006 legislation but remained actively engaged in the fight for a national standard.[121] What is more, many of these "early adopter" states had long records of environmental activism and were populated by policymakers who were very supportive of emissions legislation.[122] Soon after California's passage of AB 32, Assembly Speaker Fabian Núñez underscored his state's rich history of environmental leadership: "We feel that California has always been a leader in protecting the environment. We now have moved it to the next level. We'd all like to see California one day be carbon free."[123]

Venue adding allowed environmentalists and other proponents of reform to reframe the issue of climate change. One of the more prominent definitions presented climate change policy not as a burden, but as an opportunity to promote economic growth through "green jobs." Hess et al. note: "Part of the attractiveness of 'green jobs' is that it is a wide tent where many constituencies can gather, ranging from antipoverty constituencies that want job training for persons with employment barriers to high-tech venture capitalists and entrepreneurs."[124] In other words, the green jobs narrative helped facilitate issue expansion. Massachusetts, for example, adopted a handful of laws between 2007 and 2013 that set out to curb emissions, invest in clean energy industries and jobs, and promote energy efficiency.[125] Instead of presenting these reforms as separate actions, policymakers—and especially Massachusetts Governor Deval Patrick—pitched these actions as part of a larger effort to upgrade the state's economy and promote "green industries and jobs."[126] Other states, including Michigan, Wisconsin, Florida, Colorado, and Washington, employed similar strategies and profited from the green jobs frame.[127] Uncertainty about the future implications of climate change not only allowed problem definers to imagine the dangerous consequences of failing to act, it also allowed them to envision the economic opportunities created by what was otherwise a most unfortunate situation.

Still, evolutionary theory implies state and federal policymaking are not mutually exclusive, but collectively shape the broader climate change domain.[128] Indeed, Carlson argues federal laws and regulations created pathways for states to pursue emissions policy—a phenomena she refers to as "iterative federalism."[129] For example, the Clean Air Amendments of 1990 created commissions that explicitly encouraged cooperation between states in the Northeast region. This skeletal framework later served as the basis for the RGGI's multi-state structure. Similarly, California's emission policy would not have been possible had it not been for the EPA's 2009 decision to grant the state a waiver allowing it to regulate tailpipe emissions.[130]

And just as national policy helped facilitate state innovation, so too did state policy encourage national policy change. Environmentalists readily cited state programs, like RGGI, to justify their claims that a national reform was feasible. Others believed decentralized state regulatory standards would motivate carbon producers to support a uniform national system, if only to reduce confusion.[131] State programs also provided the Obama Administration with a backup plan in that it allowed the President to marshal his administrative powers to support state policies in the face of congressional gridlock. As described in this chapter's implementation section, in June 2014, the President proposed a rule requiring that states cut carbon pollution from power plants by 30 percent below 2005 levels by 2030. In order to comply with the rule, EPA officials encouraged states to participate in existing regional markets, like the RGGI. The President called the rule a "sensible, state-based plan."[132]

Of course, industry groups have worked to counter the addition of states as venues. "We're in an era where state legislatures and statewide races are just more important than congressional races to the long-term policy," said Michael Davis, CEO of Enterprise Washington, a coalition of businesses that campaigns against state emissions laws.[133] In 2014, the American Legislative Exchange Council (ALEC), an industry-financed group that advocates for conservative state policies, poured millions of dollars into state campaigns and even drafted model environmental legislation designed to block state and EPA power plant emission programs.[134]

Although it is difficult to determine the overall effectiveness of these programs, a number of major emitting states, including California, New York, and Michigan, have managed to keep emissions growth rates in the single digits.[135] Despite these efforts, the current intergovernmental arrangement in the U.S. is unlikely to halt current warming trends. These concerns are magnified by the fact that many Kyoto signatories have also struggled to meet their emissions targets—and many other major emitters remain unwilling to commit to binding targets. In the face of these shortcomings, governments have been forced to begin adapting to climate change.

## Agenda Catalysts

More than any other case in this book, mitigation policy has been shaped by Kingdon's "politics stream."[136] One of the best predictors of institutional attention, it seemed, were changes in party control of both Congress and, later, the White House, from Republicans to Democrats. This partisan divide is reinforced by a protracted battle between industry and environmental lobbyists. Thus far, industry has succeeded in blocking national policy change.

This chapter implies climate change's so-called "wickedness"—the extent it has eluded policy change—is in a large part attributable to its temporal dimensions or the fact that "it imposes costs on the short term for the realization of benefits many decades and sometimes centuries later."[137] Complicating matters further, costs are incurred by some of the most powerful groups in the country, who not only absorb these financial damages but can pass them onto employees and consumers—voters. Thus, while all anticipatory problems necessitate a degree of short-term investment to procure a future output, the perceived magnitude of these immediate costs—the scope of redistribution—is much greater in the case of climate change.

Yet, the climate change case still shared many similarities with the previous cases. Pralle argues, and this chapter shows, indicators play an important agenda setting function with respect to climate change.[138] IPCC findings drew policymaker and public attention to this issue, especially the 2007 Fourth Assessment report. Running measures of annual temperatures lent further credence to the notion that a "tipping point" loomed and anticipatory action was in order. Not surprisingly, opponents contested the meaning of these indicators as well as the methodologies used to derive them.

Extreme weather events had a similar "nudging" effect in the climate change domain. Policy activity was triggered not by a singular event but by a number of events—hurricanes, droughts, floods—that *collectively* symbolized the urgency of climate change. Ironically, the most important "singular" event might have been the Gulf Oil Spill, which effectively *prevented* policy change. While it is impossible to determine whether the Clean Energy Jobs and American Power Act would have secured the 60 votes needed in the Senate, this disaster unquestionably undermined the bill's progress.

Figure 5.3 shows the number of stories in the *New York Times* on the topic "climate change."[139] For the most part, it demonstrates a growing media interest in climate change beginning in 2000 and peaking in 2007. The number of stories jumped from 2006 (928 stories) to 2007 (1,609 stories). Coverage remained significant in 2008 (1,285 stories) and 2009 (1,315 stories), before dipping below 1,000 stories in 2010 (856 stories). However, media coverage of climate change—even in "slow" years— dwarfed peak coverage of pandemic influenza and nanotechnology, a testament to the fact that the attention baseline was much higher in the climate change case.

Despite the inability of national institutions to enact emissions policy, the last 15 years have not been devoid of policy change. State governments have enacted their own mitigation programs, creating a virtual mosaic of laws within the climate domain. Engaging select states as "additional" policy venues allowed environmentalists to access more favorable political and institutional arenas. California as well as many of the Northeastern states partnering in the RGGI are comprised of policymakers who are much more receptive to the need for climate change legislation. On the other hand, the liberal coasts are hardly the only regions participating in

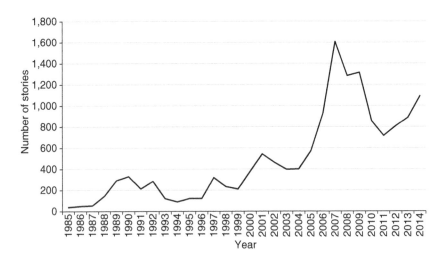

*Figure 5.3* Number of Stories on Climate Change in the *New York Times*, 1985–2014 (source: Search on Lexis-Nexis for the terms "climate change" and "global warming" in the *New York Times*).

this "bottom up" regime and some unlikely states (e.g., Texas) have taken substantive steps to combat climate change.[140]

While it is impossible to devise a "one size fits all" characterization describing policy change at the state level, evidence suggests many states used the uncertainty surrounding climate change to their advantage. The most obvious example is the visioning of climate change as an economic opportunity. What is more, states borrowed many of the same policy designs experimented with at the federal level, including cap-and-trade schemes. Thus, a number of subnational governments have proven quite adept at fashioning anticipatory policy. This activity has not gone unnoticed by the federal government. This chapter will later explore recent attempts by executive agencies to promote a "bottom up" approach to emissions policy in the absence of national legislation. But first, let us review the politics of climate change adaptation.

## Climate Change Adaptation

Adaptation seeks to reduce vulnerabilities in the face of actual or expected climate and climate-induced environmental changes.[141] Adaptation requires communities adjust future living conditions in the face of anticipated changes. While successful adaptation might include elements of resilience and preparedness, the principle of adaptation calls for fundamentally redesigning modern society—where we live, our access to natural resources, and even how we travel—in order to reduce and avoid vulnerabilities.

A great deal of adaptation policymaking has revolved around water, as rising tides already threaten coastal communities. Governments have considered or developed regulations dictating the types of structures that can be built along the coasts, established new insurance and zoning provisions for high risk areas, fortified coastlines, and restored wetlands. Soaring temperatures demand investments in new types of cooling systems, force public health practitioners to grapple with new types diseases, and threaten crop yields. Equally alarming, climate change already endangers many animal species. Saving them may require proactive government intervention.[142] Adaptation thus cuts across a variety of policy domains.

Adaptation is very costly. Klaus Jacob, a geophysicist at Colombia University, notes that climate proofing the New York City transit system alone "will probably cost billions [of dollars], maybe tens of billions."[143] Most of the subway is already below sea level and flooding during Hurricane Sandy cost the city more than $8 million.[144] While it is impossible to determine these costs on a national scale, some estimates hold adaptation will cost tens or even hundreds of billions of dollars per year by 2050 in the U.S. alone.[145]

As evidenced in Figures 5.4 and 5.5, adaptation attracted very little policymaker attention before 2007. Figure 5.4 shows the number of times the term "climate change adaption" was entered in the *Congressional Record*, and indicates growing attention from 2007 (55 entries) to 2009 (69 entries).[146] These numbers dwarf the 2006 total of six entries as well as the previous

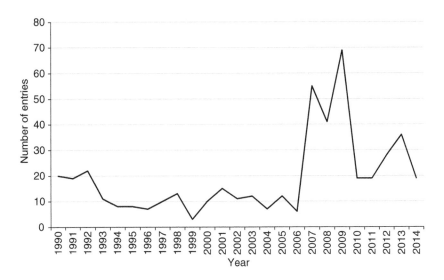

*Figure 5.4* Number of Climate Change Adaptation Entries in the *Congressional Record*, 1990–2014 (source: search on ProQuest Congressional for the terms "climate change adaptation").

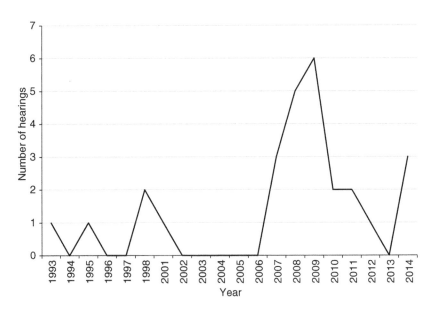

*Figure 5.5* Number of Congressional Hearings on Climate Change Adaptation, 1993–2014 (source: search on ProQuest Congressional for the terms "climate change adaptation").

high of 22 entries in 1992. They also coincide with the broader push for mitigation legislation.

Figure 5.5 is demonstrative of a similar pattern, as the number of hearings that discussed or focused on the topic of adaptation increased dramatically in 2007 (three hearings), 2008 (five hearings), and 2009 (six hearings).[147] Indeed, 2007 marked the first hearing on adaptation in nearly six years.

Prior to 2007, adaptation was widely seen as a "taboo." Whereas climate deniers objected the notion of adapting to a threat that did not exist, environmentalists and liberals worried introducing adaptation as a policy option would undermine their argument that the best—indeed only—way to address climate change was through emissions reductions. Even former Vice President Al Gore warned, "The advice to adapt is an obstacle to the correct political response, which is prevention."[148] At the international level, by contrast, adaptation has represented an important component of the UNFCC since 2000, in part because island nations suffered greatly at the hands of rising seas.[149]

As noted above, this uptick in attention coincided with the push for emissions legislation. Many of the mitigation policies introduced during the 110th Congress included provisions to improve resilience and promote adaptation. Take, for example, the all important American Clean Energy and Security Act of 2009 (Waxman–Markey), which called for both a "National Adaptation Strategy" to address the impacts of climate change on ecological systems as well as a more comprehensive "National Adaptation Plan" to identify vulnerabilities, coordinate state and national activities, prioritize high risk regions, and begin to develop estimates of the costs of adaptation.

A number of factors converged to promote this sudden uptick in attention. For one, the scientific community's ability to understand and identify the risks associated with climate change had improved dramatically by 2000, a testament to decades of sustained research. In its 2007 Fourth Assessment Report, the IPCC concluded the era for procrastination was over and "additional adaptation measures will be required at regional and local levels to reduce the adverse impacts of projected climate change and variability, regardless of the scale of mitigation undertaken over the next two to three decades."[150] The urgency to adapt was further magnified by what Pielke et al. describe as a "timescale mismatch" between mitigation policy and the time it would take to actually *reverse* the processes already set in motion. The authors write:

Whatever actions ultimately lead to the decarbonization of the global energy system, it will be many decades before they have a discernible effect on the climate. Historical emissions dictate that climate change is unavoidable. And even the most optimistic emissions projections show global greenhouse-gas concentrations rising for the foreseeable future.[151]

In other words, government's failure to *anticipate* and act quickly effectively foreclosed the possibility of avoiding many climate change's most negative consequences.

A number of extreme weather events, including Hurricanes Rita and Katrina, floods in the Northeast and Midwest, and nationwide droughts, also stimulated policymaker interest in adaptation. In 2008, the U.S. Climate Change Science Program warned the frequency of these events would increase:

> In the future, with continued global warming, heat waves and heavy downpours are very likely to further increase in frequency and intensity. Substantial areas of North America are likely to have more frequent droughts of greater severity. Hurricane wind speeds, rainfall intensity, and storm surge levels are likely to increase. The strongest cold season storms are likely to become more frequent, with stronger winds and more extreme wave heights.[152]

The onus was therefore on government to help "at risk" regions avoid—or at least minimize—the human and economic toll associated with these types of extreme events.

In light of this reality, many policymakers began advocating for multifaceted climate change policies integrating elements of both mitigation and adaptation. As early as 2003, Senator Mark Udall (D-CO) proclaimed:

> we are going to have to deal with climate change with some mix of mitigation and adaptation. We must acknowledge the interdependence of our social, economic and environmental systems and learn to anticipate and adjust to changes that will inevitably occur.[153]

Years later, in 2013, Senator Chris Coons (D-DE) echoed these sentiments stating: "The second part of climate change policy is adaptation. It is based on accepting the reality our climate is changing and that it will have real effects on our planet and our communities."[154]

So prominent was this shift that some hearings and laws from this period focused *exclusively* on adaptation measures. In March 2009, the House Committee on Energy and Commerce's Subcommittee on Energy and Environment convened a hearing, titled "Preparing for Climate Change: Adaptation Policies and Programs," to assess the various policy options for adapting to climate change. Subcommittee Chair Markey (D-MA) opened the hearing by stating: "Mitigation, the act of reducing greenhouse gas emissions, will not be enough. Our country and other nations must also implement adaptation policies to respond to changes in our climate, in our ecosystems, and in our infrastructure."[155] Hearings were held in other committees as well, including the House Committee on Energy Independence and Global Warming,[156] the Senate Committee on Commerce, Science, and Transportation,[157] the Senate Committee on Foreign Relations,[158] and the

House Committee on Natural Resources.[159] Much like emissions policy, adaptation cut across various venues.

Building on this momentum, a number of policies from this period worked to improve the nation's capacity to adapt to climate change. Planning was central to these proposals. The Climate Change Adaptation Act (S. 2355) required the President to develop a five-year strategy for adapting to climate change, mandated planning for coastal and oceanic changes, and offered grants to states to support adaptation and mitigation efforts. The Investing in Climate Action and Protection Act (H.R. 6186) also offered a comprehensive framework, creating a "National Climate Change Adaptation Council" to coordinate planning and implementation across federal agencies, requiring the NOAA assess regional vulnerabilities, and allocating money to support the UNFCC adaptation program.

Other policies from 110th and 111th Congresses focused on more specific areas of adaptation, such as maintaining clean drinking water,[160] ensuring coastal adaptation and preparedness,[161] protecting natural resources and wildlife systems,[162] increasing the capacity of the public health community to address new diseases,[163] mitigating the effects of climate change in the Arctic,[164] supporting international adaptation,[165] and assisting Native Americans in their adaptation efforts.[166] None of these policies were enacted and only two laws passed during this time included *any* adaptation provisions—neither of which amounted to a significant policy change.[167]

Adaptation was a peripheral issue for much of the 112th Congress.[168] However, federal interest was reignited by Hurricane Sandy, which resulted in 117 American deaths and billions of dollars in damage across the U.S.'s East Coast. "Sandy was a wake-up call, not just for the eastern seaboard, but for communities all over the country that we need to start preparing for climate change now," indicated Brian Holland, Climate Program Director at ICLEI-Local Governments for Sustainability, an environmental association of cities and counties.[169] Months after the event, Representative Lois Capps (D-CA) indicated: "we cannot afford to sit back and wait for the next Hurricane Sandy to devastate American lives and property. Especially in these tight economic times, I think we can all agree that reducing the cost of extreme weather events is a good idea. And one of the most effective ways to reduce these costs is to plan ahead."[170]

Once again, a variety of adaptation policies were introduced in the 113th Congress, but none of them were enacted.[171] Adaptation did, however, factor prominently in the Disaster Relief Appropriation Act of 2013 (P.L. 113–2), which appropriated $5.38 billion to the Federal Transit Administration to help the agency prepare for climate-related disasters, $2.9 billion to the Army Corps of Engineers to address flood risks in coastal communities, and required FEMA to provide Congress a series of recommendations for reducing future vulnerabilities.[172]

By 2007 there was a great deal of optimism that by casting adaptation as part of climate change policy, proponents of policy change could mobilize

Republican support. Specifically, adaptation was perceived as being *less threatening* to industry.[173] Thus far, Republicans have rejected almost any policy aimed at addressing climate change, including a request by the Department of Defense (DOD) to allocate money to help the armed services adapt to rising seas, extreme weather, and widespread droughts.[174] David McKinley (R-WV), who co-sponsored an amendment barring the DOD request, stated: "The climate is obviously changing; it has always been changing. With all the unrest around the [world], why should Congress divert funds from the mission of our military and national security to support a political ideology?"[175]

States and localities have been important drivers of adaptation policy as well. As of this writing, 15 states have already completed adaptation plans, outlining specific goals, targets, and requirements. Another five states are currently planning while seven states intend to integrate adaptation planning within their larger climate change mitigation and action plans.[176] A number of state legislatures have adopted more specific statutes, including Florida, Washington, Hawaii, New York, Louisiana, North Carolina, Rhode Island, and Virginia.[177]

Local governments have also developed fairly robust adaptation measures. In the wake of Hurricane Sandy, New York City created a 250 page plan that proposed close to $19.5 billion worth of adaptation measures, including creating more resilient hospitals, homes, and nursing facilities; elevating the electrical infrastructure; improving and strengthening subway systems; and restoring sand dunes.[178] Climate change adaptation also factored prominently in Chicago's 2007 Climate Action Plan (CCAP), which sought to update the city's heat response plan and encourage the development of innovative ideas for cooling the city.[179] The Center for Climate and Energy Solutions reports *at least* ten other cities have completed substantive adaptation plans while many others are currently in the process of planning.[180]

The U.S. has yet to enact a large scale, national adaptation policy. Still, many observers remain optimistic that progress will be made in the coming years. In the meantime, however, the Obama Administration has taken matters into their own hands, devising a series of regulations and executive orders to promote adaptation. These measures are explored in detail below, in the implementation section of this chapter.

## Agenda Catalysts

Once banished as an objectionable alternative to emissions policy, adaptation provisions were integrated into many of the mitigation policies introduced between 2007 and 2009. In fact, a number of bills introduced during this period focused exclusively on creating a national adaptation program and strengthening certain areas of vulnerability, like coastlines, natural resources, and marine ecosystems. What triggered this uptick in attention? The so-called "timescale mismatch" between the time it would take to reverse warming trends and the actual onset of climate change-related problems (e.g., rising seas and extreme weather) necessitated policymakers at least consider adaptation programs.[181] By 2007, many believed climate change was already well underway and changes were inevitable and unavoidable. In this respect, the emergence of adaptation is partially an outgrowth of the temporal dimensions of climate change policy—indeed of all anticipatory problems. Eventually, it seems, these once "future" threats can evolve into imminent dangers.

Indicators of increased warming were another important driver of institutional attention. For example, the 2007 IPCC Assessment Report, explicitly called for nations to initiate adaptation planning. Additionally, weather related disasters, most notably Hurricane Sandy, were also cited as evidence of the need to adapt to climate change at the national, state, and local levels of government.

Adaptation and mitigation policy have been coupled. To be sure, a comprehensive climate change regime is most desirable, as both approaches set out to remedy the same problem. In theory, however, climate change adaptation should not be *as* politicized as mitigation policy. Unlike mitigation policy, which pits short-term economic loss against long-term gain, adaptation imposes far fewer costs on industry and even promises immediate policy outputs, like fortified coasts, sturdier buildings, and improved transportation infrastructures. Thus far, however, Republicans and industry groups have objected to any form of climate change legislation—mitigation or adaption. Past experience suggests this marriage will prove disadvantageous to the promotion of climate change adaptation and policy change might necessitate separating these ends.

## Policy Implementation

Unable to generate policy change through the legislative process, President Obama has instructed federal agencies, most notably the EPA, to establish binding greenhouse gas limits for mobile and, most recently, stationary emitters. Instead of imposing a proverbial 'one-size-fits-all' national framework, the EPA has often allowed states to craft their own distinctive mitigation designs, particularly with respect to stationary sources, so long as they meet a number of targets established by the federal government. If implemented, this flexible framework promises to fundamentally reshape the American climate change domain, establishing a robust "bottom up" regime in the absence of congressional policy.

The Obama Administration has also taken steps to advance adaptation. A variety of interagency task forces and working groups have initiated planning across the federal government while promoting great cooperation with the states. While these adaptation efforts are very much in their early stages, the Administration recently requested a significant increase in funding for the President's adaptation plan, suggesting a sincere commitment to these goals. The following section considers the implementation of both mitigation and adaption programs.

## Mitigating Climate Change

Enacted in 1963, the Clean Air Act authorizes the EPA to develop regulations for airborne contaminants hazardous to human health. The law currently allows the EPA to regulate greenhouse gas emissions from stationary pollution sources, such as factories or manufacturing facilities, as well as mobile pollution sources, such as automobiles and trucks.[182] The EPA did not always recognize the Clean Air Act's applicability to greenhouse gases. In 1999, the International Center for Technology Assessment (ICTA), an advocacy group focusing on science and technology issues, filed a petition for rulemaking that requested the EPA regulate greenhouse gases emitted from new motor vehicles, arguing that greenhouses gases fit the definition of an "air pollutant" described in the Clean Air Act.[183]

After collecting public commentary for nearly two years, the EPA formally denied the ICTA's petition on September 8, 2003. The EPA concluded Congress did not intend for the Clean Air Act to regulate emissions, that such regulation would be inconsistent with the Bush Administration's climate change policy, and that there was simply too much uncertainty surrounding the climate issue to determine whether greenhouse gases could be covered under the Act.[184] It was not until *Massachusetts et al.* v. *EPA* in 2007 that the agency reemerged as a legitimate policy venue. To be sure, the ruling did not require the EPA to regulate greenhouse gases. Instead, the Supreme Court held the Clean Air Act's definition of air pollutants was sufficiently broad to encompass greenhouse gases and that the EPA's decision to abstain from regulation was arbitrary, political, and not based on criteria outlined by the Act. Denying regulation on the basis of "uncertainty" would not suffice. Instead, EPA had to provide a logical explanation and judgment as to whether or not greenhouse gas emissions contribute to climate change.[185]

However, procuring a positive decision was unlikely so long as George W. Bush remained in Office. In a July 2008 Advance Notice of Proposed Rulemaking on Regulating Greenhouse Gas Emissions Under the Clean Air Act, which was used to alert the public and stakeholders that the agency was considering a new rule, the EPA Administrator wrote that "the ANPR demonstrates [that] the Clean Air Act, an outdated law originally enacted to control regional pollutants that cause direct health effects, is ill-suited for the task of regulating global greenhouse gases."[186]

Soon after President Obama assumed office in 2009, the EPA began reversing the Bush-era environmental policies. On December 15, 2009, the EPA published an Endangerment Finding, which emphatically concluded the six greenhouse gases (including carbon dioxide) met the Clean Air Act's definition of an air pollutant and could be reasonably anticipated to endanger public health and welfare. In crafting its decisions, the EPA explicitly addressed the enormous uncertainties associated with global warming, arguing that warming trends were real and that scientists have clearly established the link between greenhouse gas emissions and climate change.

In fact, the EPA drafted a 500-page response to public commentary and grounded its final decision in a sweeping analysis of existing scientific findings on climate change.[187] This decision was later challenged, but upheld, in court.[188]

Similar to the nanotechnology case, the EPA's attention to detail was partially an outgrowth of the enormous uncertainty engendered by this problem and partially an outgrowth of its politically charged environment. Commenting on the complexity of the situation, one administration official, who chose to remain anonymous, indicated:

> If the administration gets it wrong, we're looking at years of litigation, legislation and public and business outcry. If we get it right, we're facing the same thing. Can we get it right? Or is this just too big a challenge, too complex a legal, scientific, political and regulatory puzzle?[189]

Questions about the applicability of existing statutes versus the need for new regulation often arise during anticipatory policymaking, as evidenced in previous cases. It is exceedingly difficult to determine the best or most appropriate policy response to a problem that does not exist in the present or, at least, is not fully developed. Of course, the decision to regulate emissions via the Clean Air Act was also politically motivated and not just the product of "sound" scientific decision making. President Obama, it seemed, was determined to govern climate change with or without congressional approval.[190]

Industry groups, as well as some members of Congress, took offense to the proposed regulatory standards. Fred Upton (R-MI), Chairman of the House Energy and Commerce Committee, said: "This move represents an unconstitutional power grab that will kill millions of jobs—unless Congress steps in."[191] A contingent of Republican legislators, most of whom represented coal, oil, and gas states, mounted a multiple-pronged legislative attack against EPA's proposed regulation, which included a series of resolutions to block the EPA from regulating greenhouse gases, a bill to formally amend the Clean Air Act to prohibit greenhouse gas regulation, and budget stipulations indicating the EPA could not use any of its funding to regulate greenhouse gas emissions.[192] None of these proposals were enacted however.[193]

EPA regulations have grown increasingly ambitious with every passing year. One of the earliest rules, the *Light-Duty Vehicle Greenhouse Gas Emission Standards and Corporate Average Fuel Economy Standards Rule* (LDV rule), was established in 2010 through a partnership between the EPA and the National Highway Safety Administration. The LDV rule required automakers to meet increasingly stringent motor vehicle emission standards between 2012 and 2016. Although it primarily relies on command-and-control instruments, the LDV rule included a number of offsets to accommodate industry needs and allow for flexibility. For one, it is graduated and thus provides manufacturers time to prepare for increasingly stringent requirements. It also includes a "bank and trade" system. While all automakers are required to meet certain standards, which peak at a fuel economy of roughly 34.1 miles per gallon in model year 2016, they can bank emissions credits should a particular fleet's emissions fall below that standard. Banked credits can be saved for future use or sold to laggard manufacturers.[194] The agency also granted waivers to states wanting to establish more stringent fuel economy standards than those imposed by the national government. As indicated above, California was one of the first beneficiaries of this change in 2009.[195]

The EPA also targeted stationary sources, such as factories, manufacturing facilities, and electric power plants. The Clean Air Act originally required that any new or recently modified stationary source emitting (or potentially emitting) 100 to 250 tons of pollutants per year (e.g., lead, sulfur dioxide, nitrogen dioxide) undergo a formal review and permitting process. Facilities must install the "Best Available Control Technology" (BACT) or "an emissions limitation which is based on the maximum degree of control that can be achieve."[196]

Retrofitting this statute to cover greenhouse gases proved arduous. With respect to greenhouse gases, a threshold of 100 to 250 tons annually is exceptionally low because these gases are emitted in far greater bulk than other pollutants. In fact, this baseline would require most small commercial businesses to obtain a permit.[197] The EPA conceded this standard was not feasible and chose to establish a new rule, a "tailoring rule," that specifically applied to greenhouse gases.[198] The EPA chose to implement the new rule in a series of phases, beginning with facilities emitting (or projected to emit) more than 75,000 tons of greenhouse gases per year and eventually covering projects emitting 100,000 tons of greenhouse gases per year as well as existing facilities that increase emissions by at least 75,000 tons per year. The agency also promised to produce a multi-year study examining greenhouse gas permitting for small facilities. The agency proudly touted stationary facilities responsible for 70 percent of the country's emissions would be regulated under this rule.[199]

Upon completing the tailoring rule, EPA Administrator Lisa Jackson indicated: "There is no denying our responsibility to protect the planet for our children and grandchildren. It's long past time we unleashed our American ingenuity and started building the efficient, prosperous clean energy economy of the future."[200] Environmentalists applauded the new rule. Frank O'Donnell, President of Clean Air Watch, stated:

> It's clear that EPA is trying to fine-tune it and make sure that the permit requirements are truly limited to the biggest sources. I think the EPA is trying to act very responsibly, and they're trying to say to the Congress and the public, "We're not the green monsters you think we are,"[201]

Industry groups argued the measure constituted an inappropriate application of the Clean Air Act. Howard Feldman, Director of Regulatory and

Scientific Affairs at the American Petroleum Institute, stated: "The Clean Air Act is not designed to regulate greenhouse gas emissions, and this tailoring rule doesn't fix the problems with the Clean Air Act doing it."[202]

Determining a BACT for greenhouse gases was also incredibly complex. In 2009, the EPA convened a special panel to help determine the type of pollution-reducing technology that stationary sources should be required to install. However, the panel, which included an eclectic mix of industry representatives, environmentalists, and state regulatory agencies, was unable to reconcile "divergent points of view" and failed to produce a final recommendation.[203] In turn, the EPA opted to provide industry substantial flexibility, allowing BACT to be determined on a case-by-case basis instead of requiring one specific technology. Industry representatives, who feared the EPA would impose a uniform standard, welcomed this decision.[204]

But the agency's most significant actions, it seems, have yet to come. In 2013, President Obama released his Climate Action Plan, which promised to *significantly* expand agency efforts in both the areas of adaptation and mitigation. In conjunction with the Action Plan, the EPA released the "Clean Power Plan," which caps carbon pollution from power plants. Environmentalists, including the Union of Concerned Scientists, have widely applauded the proposal, calling it a "game changer."[205] The draft plan sets out to reduce carbon emissions by 30 percent below 2005 levels by 2030. Capitalizing on the state emission programs, the proposed rule works to "reinforce the actions already taken by states" by providing them flexibility to chart their own course in meeting this goal. It emphasizes a series of "building blocks" (e.g., renewable energy, nuclear power, efficiency improvements at fossil fuel plants, etc.), which provide states a menu of options for achieving their emissions targets. Mixing and matching different building blocks, each state can chart its own, unique path to federal compliance. State plans are evaluated based on the extent to which they can feasibly implement their desired building blocks.[206]

While the Clean Power Plan remains a work in progress (the final rule is slated for a June 2015 publication), industry has already expressed concerns. The electricity industry warned of "widespread rolling blackouts," as the new rule will require significant reductions in their fossil fuel generating capacity.[207] Senator Inhofe expressed similar concerns about the electricity industry and wrote the "Administration's aim of this rulemaking is less about reducing greenhouse gas emissions and more about taking over another sector of the economy. This is wholly inappropriate, and I urge the withdrawal of this rule."[208]

If implemented, the rule will represent arguably the most significant mitigation action to date, potentially paving the way for a "bottom up" and state driven national emissions regime. Yet even the most stringent emission reductions cannot reverse the warming trends already set in motion by years of emissions and inaction. To this end, the Obama Administration has initiated adaptation planning.

*An Emerging Adaptation Regime?*

Executive Order 13514, signed by President Obama in 2009, established the Climate Change Adaptation Task Force, an interagency panel responsible for developing a national climate change adaptation strategy. The Order also instructed individual agencies to develop their own adaptation plans, a process that culminated in 2013. Similar to pandemic preparedness, the Task Force codified a number of "strategic goals," which could guide long-term adaptation. These goals included a variety of elements, such as interagency planning, greater integration of science and administrative decision making, identification of cross cutting jurisdictional issues, enhancement of international adaptation efforts, and better coordination between the federal government and the states.

With this broad infrastructure in place, President Obama later integrated adaptation into the 2013 Climate Action Plan.[209] The President codified many of the items listed in the Action Plan with "Executive Order 13653: Preparing the United States for the Impacts of Climate Change," which established a new interagency panel on Climate Preparedness and Resilience to replace the Task Force; created a State, Local, and Tribal Leaders Task Force on Climate Preparedness and Resilience; mandated a full assessment and inventory of federal government land and water use policy; expanded data sharing mechanisms; and required a second round of agency planning for climate change.[210]

While plans, strategies, task forces, and councils are often construed as a form of political obfuscation, President Obama's FY2016 budget proposal allocated millions of dollars to support the Action Plan's adaptation projects. Among some of the most notable items were a $50 million grant program for coastal communities, a $50 million grant program to restore ecosystems at risk of flooding, a $175 million increase in FEMA's flood protection program, and a $184 million increase in the national government's flood insurance program.[211] Commenting on this plan, President Obama indicated: "The failure to invest in climate solutions and climate preparedness does not just fly in the face of the overwhelming judgment of science—it is fiscally unwise."[212]

Adaptation policymaking is still very much in its early stage and individual agencies have only begun to implement programs, including the Action Plan. A 2012 report by the Center for Climate and Energy Solutions revealed a fairly broad collection of initiatives spanning virtually every federal department. To be sure, most of this activity has centered on planning initiatives, interagency working groups, and data collection efforts. However, a number of agencies have implemented more substantive programs. For example, the United States Department of Agriculture (USDA) launched a pest detection program to identify new pests and potentially invasive species that may emerge in the face of a changing climate. The Department of Health and Human Services (HHS) developed a number of programs that disseminate information about the dangers of climate change to health professionals. The CDC published toolkits to help local officials prepare for heat waves. FEMA updated its community rating system, which provides financial incentives for building practices that mitigate flood risks, to include language specifically addressing climate change.[213]

As evidenced in the avian influenza and nanotechnology cases, determining the effectiveness of anticipatory policy interventions is exceedingly difficult in the absence of an existing problem. This dynamic is less prominent with respect to mitigation policy, which establishes fairly rigid and measurable goals—reductions in annual emissions. Adaptation, however, strives to prepare for events and changes that may or may not appear until years if not decades in the future. To this end, bureaucracies have taken great pains to compile progress reports of their planning efforts, highlighting important developments and achievements. This practice dates back to the original 2009 Climate Change Adaptation Task Force, which issued two successive reports on the federal government's adaptation efforts. In the first report, released in 2010, the Task Force scripted a series of broad strategic goals, which would collectively improve the government's adaptation planning.[214] The second report, released one year later in 2011, detailed agency progress and spotlighted specific programs designed and implemented according to the Task Force's original recommendations.[215] More recently, the government website "Performance.gov," which provides publically accessible updates and evaluations of agency performance, has emerged as an important repository of this information. The website provides a collection of federal adaptation plans and updates, allowing visitors to "track" the progress of different measures.[216]

The implementation stage has been an important arena for establishing substantive policy outputs. Whereas the EPA's efforts to mitigate greenhouse gases are now several years underway and potentially poised for a significant breakthrough, the federal government has only recently initiated adaptation policy. With the 2016 presidential election looming, Republican nominees have already threatened to scale back President Obama's initiatives in both areas. In January 2015, Republican presidential hopeful Senator Ted Cruz (R-TX) called the EPA "locusts" and lamented the fact that "you can't use pesticide against them."[217] Former Arkansas Governor Mike Huckabee also took aim at the President's plan arguing that recent beheadings by the Islamic State are far more important "than a sunburn."[218] Time will tell whether the ambitious Climate Action Plan actually comes to fruition.

## Conclusion

Climate change policymaking has been shaped by many of the same explanatory factors identified in prior cases. An acute level of uncertainty engendered a political discourse rife with analogy, claims of crisis, proximity, and attempts to identify a sympathetic problem population. Uncertainty prompted narratives "linking" climate change to other problems—national security threats, extreme weather, emerging diseases—as well as narratives framing climate change as an economic opportunity. Agenda setting and policymaking was fairly protracted and two "surges" in activity occurred over a four-year period. Agenda setting was also indicator driven, but was helped along by a series of extreme weather events, which symbolized the dangers of failing to act on climate change. The need to project decades into the future gave rise to an array of models, scenarios, and plans. And regulatory agencies had to recalibrate existing policy to accommodate climate change's novel features, particularly with respect to greenhouse gas emissions. These patterns were, more or less, observed in all three cases.

Despite its similarities, the climate change case also represents somewhat of an outlier. Specifically, while the above-described factors provide a well-rounded explanation of the policymaking dynamics associated with this case, "politics," it seems, functioned as a proverbial "trump card." Climate change is unquestionably the most politicized anticipatory problem examined in this book, as neither the pandemic nor nanotechnology domains include such powerful and entrenched interest groups. Indeed, global warming in many ways represents a "wedge-issue" and is perhaps tantamount to problems like health care reform, the economy, and even national security, at least with respect to levels of issue attention.[219] It is, without question, the defining environmental issue of our times.

What is more, this noticeably charged political environment fueled an equally divisive problem definition debate, which pitted mitigation policy against jobs and the economy. This narrative proved incredibly difficult to overcome, as proponents of emissions standards struggled to convince policymakers and the public that climate change legislation did not represent a threat to their self-interest—to their wallet. Neither of the previous problems was framed as such a direct threat to the economy.

It is precisely this feature—this remarkable politicization—that best explains the second distinctive feature of the global warming case: policy change has come from subnational initiatives and, most recently, regulatory activity. The U.S. has yet to commit to a uniform national emissions reduction regime, and is often perceived as an impediment to international policy change. Gridlock at the national level necessitated that environmentalists explore alternative venues, a trend that is likely to continue in coming years. The EPA's latest attempts to unify the states toward a common emissions goal is promising, but will likely come under intense scrutiny in coming years.

## Notes

1  Andrew E. Dessler and Edward A. Parson. 2006. *The Science and Politics of Global Climate Change: A Guide to the Debate*. Cambridge: Cambridge University Press; Loren R. Cass. 2006. *The Failures of American and European Climate Policy: International Norms, Domestic Politics, And Unachievable Commitments*. Albany: State University of New York Press;

James E. Hansen. 2006. "Can We Still Dangerous Human-Made Climate Change?" *Social Research* 73(3): 949–971.

2 Hansen 2006.

3 "The Hottest Years on Record." 2010. *The Economist*, December 3. Available at www.economist.com/blogs/dailychart/2010/12/climate_change (accessed March 28, 2015).

4 The National Aeronautics and Space Administration. 2015. "NASA, NOAA Find 2014 Warmest Year in Modern Record." January 16. Available at www.nasa.gov/press/2015/january/nasa-determines-2014-warmest-year-in-modern-record/#.VRKlFEtrWlJ (March 28, 2015).

5 Intergovernmental Panel on Climate Change (IPCC). 2007. *Climate Change 2007: Synthesis Report*. Geneva: IPCC. Available at www.ipcc.ch/pdf/assessment-report/ar4/syr/ar4_syr.pdf (March 28, 2015); Dessler and Parson 2006.

6 Environmental Protection Agency (EPA). 2012. "Greenhouse Gas Emissions." Available at www.epa.gov/climatechange/ghgemissions/gases/co2.html (accessed March 28, 2015).

7 Cass 2006; Environmental Protection Agency (EPA). 2013. Greenhouse Gas Emissions Data. September. Available at www.epa.gov/climatechange/ghgemissions/global.html (accessed March 28, 2015).

8 Gordon B. Bonan. 2008. "Forests and Climate Change: Forcing, Feedbacks, and the Climate Benefits of Forests." *Science* 320 (5882): 1444–1449; Deborah Lawrence and Karen Vandecar. 2015. "Effects of tropical deforestation on climate and agriculture." *Nature Climate Change* 5: 27–36.

9 Climate Institute. 2010. "Oceans & Sea Level Rise." Available at www.climate.org/topics/sea-level/ (accessed March 28, 2015); Intergovernmental Panel on Climate Change (IPCC) 2007.

10 Charley Cameron 2012. "New Report Finds 3.7 Million U.S. Residents at Risk from Rising Sea Levels." *Inhabitat*, March 14. Available at http://inhabitat.com/new-report-finds-3-7-million-u-s-residents-at-risk-from-rising-sea-levels/ (accessed March 28, 2015).

11 Tiffany Stecker. 2011. "Identifying 'Hot Spots' of Future Food Shortages Due to Climate Change." *Scientific American*, June 3. Available at www.scientificamerican.com/article.cfm?id=identifying-future-food-shortage-hot-spots-due-to-climate-change (accessed March 28, 2015); Hansen 2006.

12 Union of Concerned Scientists. 2003. "Early Warning Signs of Global Warming: Spreading Disease." Available at www.ucsusa.org/global_warming/science_and_impacts/impacts/early-warning-signs-of-global-9.html (accessed March 28, 2015); Stephen Leahy. 2004. "Grim Signs Mark Global Warming." *Wired*, November 10. Available at www.wired.com/science/discoveries/news/2004/11/65654 (accessed March 28, 2015).

13 Holli Riebeek. 2005. "The Rising Cost of Natural Hazards." The National Aeronautics and Space Administration, March 28. Available at http://earthobservatory.nasa.gov/Features/RisingCost/rising_cost.php (accessed March 28, 2015).

14 DARA. 2012. *2nd Edition Climate Vulnerability Monitor: A Guide to the Cold Calculus of a Hot Planet*. Available at http://daraint.org/wp-content/uploads/2012/09/CVM2ndEd-FrontMatter.pdf (accessed March 28, 2015).

15 Christine Dell'Amore. 2010. "Five Global Warming 'Tipping Points'." *National Geographic* October 28. Available at http://news.nationalgeographic.com/news/2009/03/photogalleries/tipping-points-climate-change/index.html (accessed March 28, 2015); Hansen 2006.

16 Anthony Giddens. 2011. *The Politics of Climate Change, Second Edition*. Oxford: Polity; Richard D. Besel. 2013. "Accommodating Climate Change Science: James Hansen and the Rhetorical/Political Emergence of Global Warming." *Science in Context* 26(1): 135–152.

17 Nathaniel Massey. 2012. "1988 vs. 2012: How Heat Waves and Droughts Fuel Climate Perception." *Environmental Change in the News*, University of Maryland. Available at www.climateneeds.umd.edu/climatewire-08-28-12/article-02. php. (accessed March 28, 2015); Basel 2013; David A. Rochefort and Roger W. Cobb. 1994. "Problem Definition: An Emerging Perspective." In *The Politics of Problem Definition: Shaping the Policy Agenda*, eds. David A. Rochefort and Roger W. Cobb. Kansas: University of Kansas Press: 1–31; Roger A. Pielke. 2000. "Policy History of the US Global Change Research Program: Part I. Administrative Development." *Global Environmental Change* 10: 9–25; Roger A Pielke. 2000b. "Policy History of the US Global Change Research Program: Part II. Legislative Process." *Global Environmental Change* 10: 133–144.
18 Jessica Aldred. 2007. "Timeline: Al Gore." *Guardian*, October 12. Available at www.guardian.co.uk/environment/2007/oct/12/climatechange1 (accessed March 28, 2015).
19 Clair L. Parkinson. 2010. *Coming Climate Crisis? Consider the Past Beware the Big Fix*. Plymouth: Rowan & Littlefield.
20 Robert Cox. 2013. *Environmental Communication and the Public Sphere*. Thousand Oaks: Sage.
21 Naomi Oreskes and Erik M. Conway. 2010. *Merchants of Doubt: How a Handful of Scientists Obscured the Truth on Issues from Tobacco to Global Warming*. New York: Bloomsbury Press.
22 Pielke 2000, 2000b.
23 Cass 2006.
24 Barry Rabe. 2004. *Statehouse and Greenhouse: The Emerging Politics of American Climate Change Policy*. Washington, D.C.: Brookings Institution Press.
25 The 1992 Energy Policy Act (P.L. 102–486).
26 Intermodal Surface Transportation Efficiency Act of 1991 (P.L. 102–240).
27 This program was called "The State and Local Climate Change Program."
28 Rabe 2004; Cynthia J. Burbank and Parson Brinckerhoff. 2009. *Strategies for Reducing the Impact of Surface Transportation on Global Climate Change: A Synthesis of Policy Research and State and Local Mitigation Strategies*. American Association of State Highway and Transportation Officials (AASHTO). Available at http://climatechange.transportation.org/pdf/nchrp_2024_59_final_report_031309.pdf (accessed March 28, 2015).
29 United States Environmental Protection Agency (EPA). 2014. "Reducing Acid Rain." October 28. Available at www.epa.gov/airquality/peg_caa/acidrain.html (accessed March 28, 2015).
30 Glen Anderson and David Sullivan. 2009. "Reducing Greenhouse Gas Emissions: Carbon Cap and Trade and the Carbon Tax." *National Conference of State Legislatures*, July. Available at www.ncsl.org/documents/environ/captrade.pdf (accessed March 28, 2015); Gerald F. Talbert. 2012. "Conservation Marketplace." National Association of Conservation Districts. Available at www.nacdnet.org/resources/reports/marketplace.phtml (accessed March 28, 2015).
31 Richard Connif. 2009. "The Political History of Cap and Trade." *Smithsonian Magazine*, August. Available at www.smithsonianmag.com/air/the-political-history-of-cap-and-trade-34711212/?no-ist (accessed March 28, 2015).
32 See: www.govtrack.us/congress/bills/101/hr3030.
33 Giddons 2009; Rabe 2004.
34 Rabe 2004.
35 Angela Antonelli and Bett D. Schaefer. 1998. "Why Kyoto Signing Signals Disregard For Congress." The Heritage Foundation, November 23. Available at www.heritage.org/research/reports/1998/11/why-the-kyoto-signing-signals-disregard-for-congress (accessed March 28, 2015).

36 Helen Dewar and Kevin Sullivan. "Senate Republicans Call Kyoto Pact Dead." *Washington Post*, December 11, A37.
37 Ibid.
38 Rabe 2004.
39 Union of Concerned Scientists. 2012. "Manipulation of Global Warming Science." Available at www.ucsusa.org/scientific_integrity/abuses_of_science/manipulation-of-global.html (accessed March 28, 2015).
40 David Beillo. 2007. "Climate Change's Uncertainty Principle." *Scientific American*, November 29. Available at www.zcommunications.org/climate-changes-uncertainty-principle-by-david-biello (accessed March 28, 2015); John Quiggin. 2008. "Uncertainty and Climate Change Policy." *Economic Analysis & Policy* 38(2): 203–210.
41 Daniel Sarewitz and Roger Pielke Jr. 2000. "Breaking the Global Warming Deadlock." *The Atlantic*, July. Available at www.theatlantic.com/past/issues/2000/07/sarewitz.htm (accessed March 28, 2015).
42 U.S. Congress. Senate. 1988. "Greenhouse Effect and Global Climate Change: Hearing before the Committee on Energy and Natural Resources." Committee on Energy and Natural Resources. June 23. 100th Cong., 2nd sess. Washington, D.C.: U.S. Government Printing Office: 39.
43 Barack Obama. 2015. "2015 State of the Union Address." Available at www.npr.org/2015/01/20/378680818/transcript-president-obamas-state-of-the-union-address (accessed March 28, 2015).
44 Kristen Wyatt. 2009. "Senators Tour US Park, Hear about Global Warming." *Guardian*, August 24. Available at www.guardian.co.uk/world/feedarticle/8672910 (accessed March 28, 2015).
45 Ibid.
46 Brendan Mackey and Song Li. 2007. "Stand Up for the Earth Community: Win the Struggle Against Global Warming." *Pacific Ecologist* (Summer): 10–13, 10.
47 Bryan Walsh. 2008. "How to Win the War on Global Warming." *Time*, April 17. Available at: www.time.com/time/specials/2007/article/0,28804,1730759_1731383_1731363,00.html (accessed March 28, 2015).
48 Deborah Stone. 2002. *Policy Paradox: The Art of Political Decision Making.* New York: W.W. Norton.
49 Al Gore. 2007. *The Assault on Reason.* New York: Penguin Books: 210.
50 Ibid.
51 Environment America. 2011. *Annual Report: Recapping our work in 2011 for our members.* Boston: 7. Available at http://environmentamerica.org/sites/environment/files/ANN_AME_FY11_WEB.pdf (accessed March 28, 2015).
52 State of the World Forum. 2012. Available at www.worldforum.org/ (accessed March 28, 2015).
53 Rochefort and Cobb 1994.
54 James Inhofe. 2004. *The Facts and Science of Climate Change.* Washington, D.C.: U.S. Senate: 18. Available at www.epw.senate.gov/repwhitepapers/ClimateChange.pdf (accessed March 28, 2015).
55 U.S. Congress. House. 2003. "Kyoto Global Warming Treaty's Impact on Ohio's Coal-Dependent Communities." Committee on Resources. May 13. 108th Cong., 1st sess. Washington, D.C.: U.S. Government Printing Office: 4.
56 Ibid. 3.
57 Rick Santorum. 2012. "Blown and Tossed by the Winds of Political Correctness." *Redstate.com*, March 10. Available at www.redstate.com/rjsantorum/2012/03/10/blown-and-tossed-by-the-winds-of-political-correctness/ (accessed March 28, 2015).
58 Benito Müller. 2001. "Fatally Flawed Inequity: Kyoto's Unfair Burden on the United States & the Chinese Challenge to American Emission Dominance."

Presented at World Bank Climate Change Day, Washington D.C., June 14, 2001 and Climate Strategies Review, Brussels, June 19: 1. Available at www.oxford climatepolicy.org/publications/documents/ffi.pdf (accessed March 28, 2015).

59  Michael Crichton. 2009. *State of Fear*. New York: Harper; Chris Mooney. 2005. "Some Like It Hot." *Truthout*, May/June. Available at http://archive. truthout.org/article/chris-mooney-some-like-it-hot (accessed March 28, 2015).

60  Quoted in the *Congressional Record* 149 (July 28, 2003): S 19943.

61  All *Congressional Record* data was derived from ProQuest Congressional's online database. I searched the *Congressional Record* for all mentions of the terms "climate change" and "global warming."

62  Please note, *Congressional Record* and hearing data in this chapter extends back to the 1980s, which is much earlier than the previous cases. This reflects the fact that climate change has, in fact, percolated on the public and institutional agendas much longer than H5N1 avian influenza and nanotechnology. What is more, climate change is a much more prominent national issue, so the attention baseline is much higher.

63  All committee hearing data was derived from ProQuest Congressional. My initial search included the terms "climate change" or "global warming." In keeping with previous chapters, hearing data was derived by searching all fields *except* full text. This approach helps minimize the retrieval of hearings that only make a passing reference to climate change or global warming. Hearings were then manually reviewed in order to ensure climate change or global warming were, in fact, focal topics.

64  Larry Parker, John Blodgett, and Brent D. Yacobucci. 2011. "U.S. Global Climate Change Policy: Evolving Views of Cost, Competiveness, and Comprehensiveness." February 24. Congressional Research Service. Washington, D.C.: Library of Congress: 7. Available at www.fas.org/sgp/crs/misc/RL30024.pdf (accessed March 28, 2015).

65  Charli E. Coon. 2001. "Why President Bush is Right to Abandon the Kyoto Protocol." *The Heritage Foundation Backgrounder*. No. 1437, May 11. Available at www.grida.no/geo/GEO/Geo-2-011.htm (accessed March 28, 2015).

66  Kirsty Hamilton. 1998. *The Oil Industry and Climate Change*. A Greenpeace Briefing. August: 37.

67  Quiggin 2008, 207.

68  U.S. Congress. House. 2007. *Political Interference With Climate Change Science Under the Bush Administration*. Committee on Government Reform. Report: ii. Available at http://earthjustice.org/sites/default/files/library/reports/house-of-representative-2007-majority-report-on-climate-change-science.pdf (accessed March 28, 2015).

69  Ibid., 32; Similar reports were released by other organizations. See: Union of Concerned Scientists 2004. *Scientific Integrity in Policymaking: An Investigation into the Bush Administration's Misuse of Science*. March. Cambridge, MA. Available at www.ucsusa.org/our-work/center-science-and-democracy/promoting-scientific-integrity/scientific-integrity-in.html#.VRdGx0trWlI (accessed March 28, 2015).

70  Parker et al. 2011.

71  Gordon Prather. 2002. "Clear Skies, But No Carbon Dioxide Cap." *WND Commentary*, February 16. Available at www.wnd.com/2002/02/12813/ (accessed March 28, 2015).

72  Center for Climate and Energy Solutions. 2012. "Legislation in the 109th Congress Related to Global Climate Change." Arlington. Available at www. c2es.org/federal/congress/109 (accessed March 28, 2015); Center for Climate and Energy Solutions. 2012b. "109th Congress Index of Proposals." Arlington. Available at www.c2es.org/federal/congress/109/bills_index (accessed March 28, 2015).

73 John Larsen. 2006. "Global Warming Legislation in the 109th Congress." World Resources Institute, November 3. Available at www.wri.org/stories/2006/11/global-warming-legislation-109th-congress (accessed March 28, 2015).

74 PEW Center on Global Climate Change. 2008. "Economy-wide Cap-and-Trade Proposals in the 110th Congress." December. Available at www.c2es.org/docUploads/Chart-and-Graph-120108.pdf (accessed March 28, 2015).

75 Thomas A. Birkland. 1997. *After Disaster: Agenda Setting, Public Policy, and Focusing Events*. Washington, D.C.: Georgetown University Press.

76 Intergovernmental Panel on Climate Change (IPCC). 2000. *IPCC Special Report on Emissions Scenarios*. Summary for Policymakers. Available at www.ipcc.ch/pdf/special-reports/spm/sres-en.pdf (accessed March 28, 2015).

77 Richard J. Lazarus. 2010. "Super Wicked Problems and Climate Change: Restraining the Present to Liberate the Future." *Environmental Law and Policy Annual Review* 40: 10749–10756, 10749.

78 Amanda Little. 2003. "The climate bill lost out, but the environment may yet prove the winner." *Grist*, November 5. Available at http://grist.org/article/thrill/ (accessed March 28, 2015).

79 *Massachusetts et al.* v. *Environmental Protection Agency*. 2007. April 2. United States Supreme Court. Case brief available at: www.oyez.org/cases/200 0–2009/2006/2006_05_1120/ (accessed March 28, 2015).

80 Center for Climate and Energy Solutions. 2014. "Clean Air Act Cases." Available at: www.c2es.org/federal/courts/clean-air-act-cases (accessed March 28, 2015).

81 Center for Climate and Energy Solutions. 2008. "Legislation in the 110th Congress Related to Global Climate Change." Available at www.c2es.org/federal/congress/110 (accessed March 28, 2015).

82 It is also worth noting that former-Vice President Al Gore's Academy Award winning film *An Inconvenient Truth* was released in 2006. While the film unquestionably brought increased public attention to the issue of climate change, studies question whether it actually *helped* the movement to enact substantive emissions caps. Some observers contend the film further polarized the debate. See: Deborah Lynn Guber. 2013. "A Cooling for Climate Change? Party Polarization and the Politics of Global Warming." *American Behavioral Scientist* 57(1): 93–115.

83 Intergovernmental Panel on Climate Change (IPCC). 2007b. "Working Group 1: The Scientific Basis." Summary for policymakers. Geneva: 66. Available at www.ipcc.ch/ipccreports/tar/wg1/index.php?idp=5 (accessed March 28, 2015).

84 U.S. Congress. Senate. 2005. "The Need for Multi-Emissions Legislation." Committee on Environment and Public Works. January 29. 109th Cong., 1st sess. Washington, D.C.: U.S. Government Printing Office: 16–17.

85 Brett A. Vassey. 2009. "Testimony for Legislative Hearing on S. 1733, Clean Energy Jobs and American Power Act, U.S. Senate Committee on Environment and Public Works." October 28: 6. Available at http://epw.senate.gov/public/index.cfm?FuseAction=Files.View&FileStore_id=30d85d3d-d032–4293-a9da-d239e06d040a (accessed March 28, 2015).

86 Rebecca Lefton and Daniel J. Weiss. 2010. "Oil Dependence is a Dangerous Habit." Center for American Progress, January 13. Available at www.americanprogress.org/issues/2010/01/oil_imports_security.html (accessed March 28, 2015).

87 U.S. Congress. House. 2005. "Energy Demand in the 21st Century: Are Congress and the Executive Branch Meeting the Challenge?" Subcommittee on Energy and Resources. March 16. 109th Cong., 1st sess. Washington, D.C.: U.S. Government Printing Office.

88 World Health Organization. 2014. "Climate Change and Infectious Diseases." Available at www.who.int/globalchange/climate/summary/en/index5.html (accessed March 28, 2015).

89 Biello 2014.
90 Notable examples include the Lieberman–Warner Climate Security Act (S. 2191), Low Carbon Economy Act (S. 1766), Climate Stewardship and Innovation Act (S. 280), Global Warming Pollution Reduction Act (S. 309), Global Warming Reduction Act (S. 485), Climate Stewardship Act (H.R. 620), and Safe Climate Act of 2007 (H.R. 1590).
91 James E. McCarthy and Larry Parker. 2010. *EPA Regulation of Greenhouse Gases: Congressional Responses and Options.* Congressional Research Service. June 8. Washington, D.C.: Library of Congress. Available at www.fas.org/sgp/crs/misc/R41212.pdf (accessed March 28, 2015).
92 "Global Warming Bill Faces Stiff GOP Opposition." 2008. *CNN*, July 2. Available at: http://articles.cnn.com/2008–06–02/politics/senate.greenhousegas_1_global-warming-bill-energy-costs-democratic-backed-bill?_s=PM:POLITICS (accessed March 28, 2015).
93 Parker et al. 2011.
94 Ryan Lizza. 2010. "As the World Burns." *The New Yorker*, October 11. Available at www.newyorker.com/reporting/2010/10/11/101011fa_fact_lizza (accessed March 28, 2015).
95 U.S. Green Building Council. 2009. "Highlights: American Clean Energy and Security Act of 2009 (H.R. 2454)." Available at www.usgbc.org/ShowFile.aspx?DocumentID=6070 (accessed March 28, 2015).
96 Nicholas Loris and Ben Lieberman. 2009. "Cap and Trade (American Clean Energy and Security Act): Talking Points." The Heritage Foundation, Distributed by the Sumter County Republican Executive Committee. Available at www.sumterrepublicans.com/images/talkingpoints/Talking%20Points%20%20American%20Clean%20Energy%20and%20Security%20Act.pdf (accessed March 28, 2015); Tony Dutzik and Emily Figdor. 2009. *The Clean Energy Future Starts Here: Understanding the American Clean Energy and Security Act.* Environment America, Fall. Available at www.environmentamerica.org/sites/environment/files/reports/clean-energy-future-starts-here.pdf (accessed March 28, 2015).
97 Press Conference by the President. 2009. The White House. June 23. Available at www.whitehouse.gov/the-press-office/press-conference-president-6–23–09 (accessed March 28, 2015).
98 See: www.opencongress.org/bill/hr2454–111/actions_votes.
99 Other notable proposals included the American Clean Energy Leadership Act of 2009 (S. 1462), Practical Energy and Climate Plan (S. 3464), Carbon Limits and Energy for America's Renewal Act (S. 2877), and Clean Energy Jobs and American Power Act of 2009 (S. 1733).
100 Chaddock 2010; Franz Matzner, and Jim Presswood. 2010. "An Energy Bill Without a Carbon Cap Could Do More Harm than Good." National Resources Defense Council. Available at http://docs.nrdc.org/legislation/files/leg_10030901a.pdf (accessed March 28, 2015); Center for Climate and Energy Solutions 2010. "Lugar Practical Energy and Climate Plan (S.3464)." Available at www.c2es.org/federal/congress/111/lugar-practical-energy-climate-plan (accessed March 28, 2015).
101 Matthew Daly and Frederic J. Frommer. 2010. "States could veto neighbor's offshore drilling under energy-climate bill." *Cleveland.com*, May 11. Available at www.cleveland.com/nation/index.ssf/2010/05/states_could_veto_neighbors_of.html (accessed March 28, 2015).
102 Daly and Frommer 2010; Lizza 2010.
103 Brad Johnson. 2010. "Citing Katrina Myth, Obama Claimed 'Oil Rigs Today Don't Generally Cause Spills.'" *Think Progress*, April 28: 456. Available at http://thinkprogress.org/climate/2010/04/28/174651/obama-katrina-spill/ (accessed March 28, 2015).

104 Cheryl Dybas. 2012. "Gulf of Mexico Oil Spill's Effects on Deep-Water Corals." National Science Foundation, March 26. Available at www.pbs.org/newshour/rundown/2010/08/new-estimate-puts-oil-leak-at-49-million-barrels.html (accessed March 28, 2015); Maureen Hoch. 2010. "New Estimates Puts Gulf Oil Leak at 205 Million Gallons." *PBS Newshour*, August 2. Available at www.pbs.org/newshour/rundown/2010/08/new-estimate-puts-oil-leak-at-49-million-barrels.html (accessed March 28, 2015).

105 Daly and Frommer 2010.

106 Center for Climate and Energy Solutions. 2012c. "Climate Debate in Congress." Available at www.c2es.org/federal/congress (accessed March 28, 2015).

107 Emily Atkin 2014. "What Climate Scientists Have to Say About Obama's Deal With China." *Climate Progress*, November 12. Available at: http://thinkprogress.org/climate/2014/11/12/3591341/china-climate-deal-scientists-reaction/ (accessed March 28, 2015); David Jackson. 2014. "Obama Clears the Air on Climate Deal with China." *USA Today*, November 12. Available at www.usatoday.com/story/news/nation/2014/11/12/obama-china-xi-jinping-climate-change-agreement-coal/18901537/ (accessed March 28, 2015).

108 Miranda A. Schreurs. 2010. "Climate Change Politics in the United States: Melting the Ice." *Analyse & Kritik* 32(1): 177–189, 183.

109 Rabe 2004; Rabe 2008. "States on Steroids: The Intergovernmental Odyssey of American Climate Policy." *Review of Policy Research* 25(2): 105–128.

110 Ann E. Carlson. 2009. "Iterative Federalism and Climate Change." *Northwestern University Law Review* 103(3): 1097–1162; Weiss and Mark Sarro. 2009. *The Economic Impact of AB 32 on California Small Businesses.* Prepared for the Union of Concerned Scientists. December. Available at www.ucsusa.org/sites/default/files/legacy/assets/documents/global_warming/AB-32-and-CA-small-business-report.pdf (accessed March 28, 2015); State of California. 2006. "State of California's Actions to Address Global Climate Change." Available at www.climatechange.ca.gov/climate_action_team/reports/2006report/2005–12–08_STATE_ACTIONS_REPORT.PDF (accessed March 28, 2015).

111 Environmental Defense Fund. 2012. "California is Leading the Climate Change Fight." Available at www.edf.org/climate/AB32 (accessed March 28, 2015); California Assembly Bill No. 32. 2006. Chapter 488. Available at: www.leginfo.ca.gov/pub/05–06/bill/asm/ab_0001–0050/ab_32_bill_20060927_chaptered.pdf (accessed March 28, 2015).

112 Under the program, which was designed by the California Air Resources Board (CARB), state agencies give away most of the emissions credits for free and the remainder of permits are auctioned off quarterly.

113 Dana Hull. 2013. "13 Things to Know About California's Cap-And-Trade Program." *San Jose Mercury News*, February 22. Available at www.mercurynews.com/ci_22092533/13-things-know-about-california-cap-trade-program (accessed March 28, 2015); Katherin Hsia-Khung and Erica Morehouse. 2015. *Carbon Market in California: A Comprehensive Analysis of the Golden State's Cap-and Trade Program.* Environmental Defense Fund. Available at www.edf.org/sites/default/files/content/carbon-market-california-year_two.pdf (accessed March 28, 2015).

114 New Jersey Governor Chris Christie pulled his state out of the program in 2011.

115 Jonathan L. Ramseur. 2014. *The Regional Greenhouse Gas Initiative: Lessons Learned and Issues for Policy Makers.* Congressional Research Service. November 14. Washington, D.C.: Library of Congress. Available at www.fas.org/sgp/crs/misc/R41836.pdf (accessed March 28, 2015); Anderson and Sullivan 2009.

116 Center for Climate and Energy Solutions. 2014. "Western Climate Initiative." Available at www.c2es.org/us-states-regions/regional-climate-initiatives/western-climate-initiative (accessed March 28, 2015).

117 "City Residents Vote to Tax Selves for Carbon Use." 2006. *MSNBC News*, November 10. Available at www.nbcnews.com/id/15651688/ns/us_news-environment/t/city-residents-vote-tax-selves-carbon-use/#.VRNcC0trWlJ (accessed March 28, 2015); Anderson and Sullivan 2009.

118 "S.F. Bay Area Passes Carbon Tax." 2008. *Environmental Leader*, May 22. Available at www.environmentalleader.com/2008/05/22/sf-bay-area-passes-carbon-tax/ (accessed March 28, 2015).

119 Philipp Pattberg and Johannes Stripple. 2008. "Beyond the Public and Private Divide: Remapping Transnational Climate Governance in the 21st century." *International Environmental Agreements* 8: 367–388.

120 Sarah Pralle. "The Mouse That Roared: Agenda Setting in Canadian Pesticide Politics." *Policy Studies Journal* 34(2): 171–194.

121 Environmental Defense Fund 2012.

122 Paul Posner. 2005. "The Politics of Preemption: Prospects for the States." *PS: Political Science and Politics* 38(3): 371–374.

123 Felicity Barringer. 2006. "Officials Reach California Deal to Cut Emissions." *New York Times*, August 30. Available at www.nytimes.com/2006/08/31/washington/31warming.html?pagewanted=print (accessed March 28, 2015).

124 David J. Hess, David A. Banks, Bob Darrow, Joseph Datko, Jaime D. Ewalt, Rebecca Gresh, Matthew Hoffmann, Anthony Sarkis, and Logan D.A. Williams. 2010. *Building Clean-Energy Industries and Green Jobs: Policy Innovations at the State and Local Government Level.* Science and Technology Studies Department, Rensselaer Polytechnic Institute: 11.

125 Notable bills include The Clean Energy and Biofuels Act, The Green Jobs Act, and The Global Warming Solutions Act.

126 Jon Chesto. 2014. "Deval Patrick's Renewable Revolution." *Boston Business Journal*, September 26. Available at www.bizjournals.com/boston/print-edition/2014/09/26/deval-patrick-s-renewable-revolution.html?page=all (accessed March 28, 2015).

127 Hess et al. 2010.

128 Peter John. 1999. "Ideas and Interests; Agendas and Implementation: An Evolutionary Explanation of Policy Change in British Local Government Finance." *British Journal of Politics and International Relations* 1(1): 39–62.

129 Carlson. 2009.

130 Ibid.

131 Hess et al. 2010.

132 Coral Davenport and Peter Baker. 2014. "Taking Page From Health Care Act, Obama Climate Plan Relies on States." *New York Times*, June 3, AO16.

133 Darren Goode and Andrew Restuca. 2014. "Greens Take 2014 Fight to States." *Politico*, September 30. Available at www.politico.com/story/2014/09/environmental-groups-spending-states-2014-elections-111435.html (accessed March 28, 2015).

134 Tom Hamburger. 2014. "Fossil-Fuel Lobbyists, Bolstered by GOP Wins, Work to Curb Environmental Rules." *Washington Post*, December 7. Available at www.washingtonpost.com/politics/fossil-fuel-lobbyists-bolstered-by-gop-wins-work-to-curb-environmental-rules/2014/12/07/3ef05bc0–79b9–11e4–9a27–6fdbc612bff8_story.html (accessed March 28, 2015).

135 Rabe 2008.

136 Kingdon 2004.

137 Richard J. Lazarus. 2010. "Super Wicked Problems and Climate Change: Restraining the Present to Liberate the Future." *Environmental Law and Policy Review* 40: 10749–10767, 10749.

138 Sarah Pralle. 2009. "Agenda Setting and Climate Change." *Environmental Politics* 18: 781–789.
139 This data was derived using Lexis-Nexis' online database. I searched within the *New York Times* for the terms "climate change" or "global warming."
140 Rabe 2004.
141 Marc Landy. 2010. "Climate Adaptation and Federal Megadisaster Policy: Lessons from Katrina." Resources for the Future, February. Available at www. rff.org/RFF/Documents/RFF-IB-10–02.pdf (accessed March 28, 2015).
142 United States Environmental Protection Agency. 2014. "Adaptation Overview" Available at www.epa.gov/climatechange/impacts-adaptation/adapt-overview.html (accessed March 28, 2015).
143 Eric Klinberg. 2013. "Adaptation: How Can Cities Climate Proof?" *The New Yorker*, January 7: 33.
144 Alexis Petru. 2014. "Two Years After Sandy: NYC Plans for Transportation Resiliency." *Triple Pundit*. September 29. Available at www.triplepundit. com/2014/09/two-years-sandy-nyc-plans-transportation-resiliency/ (accessed March 28, 2015).
145 Fran Sussman, Nisha Krishnan, Kathryn Maher, Rawlings Miller, Charlotte Mack, Paul Stewart, Kate Shourse, and Bill Perkins. 2014. "Climate Change Adaptation Cost in the US: What Do We Know?" *Climate Policy* 14(2): 242–282.
146 All *Congressional Record* data was derived from ProQuest Congressional's online database. In this instance, I searched for the term "climate change adaptation."
147 Committee hearing data was derived from ProQuest Congressional's online database. My initial search included the terms "climate change adaptation." Hearing data was derived by searching within all fields *except* full text. Once again, I manually sorted hearing data, removing hearings that were not germane to this topic.
148 Susanne C. Moser 2009. *Good Morning, America! The Explosive U.S. Awakening to the Need for Adaptation*. California Energy Commission and National Oceanic and Atmospheric Administration (NOAA): 8. Available at www.preventionweb.net/files/11374_MoserGoodMorningAmericaAdaptationin. pdf (accessed March 28, 2015).
149 E. Lisa F. Schipper. 2006. "Conceptual History of Adaptation in the UNFCC Process." *Reciel* 15(1): 82–92.
150 Intergovernmental Panel on Climate Change (IPCC). 2007. "6.3 Responses to Climate Change." Robust findings, Geneva: 56. Available at www.ipcc.ch/publications_and_data/ar4/syr/en/mains6-3.html (accessed March 28, 2015).
151 Roger Pielke, Jr., Gwyn Prins, Steve Rayner, and Daniel Sarewitz. 2007. "Lifting the Taboo on Adaptation." *Nature* 445(8): 597–598, 597.
152 Thomas R. Karl, Gerald A. Meehl, Christopher D. Miller, Susan J. Hassol, Anne M. Waple, and William L. Murray. 2008. *Weather and Climate Extremes in a Changing Climate*. U.S. Climate Change Program. June: VII. Available at http://downloads.globalchange.gov/sap/sap3–3/sap3–3-final-all.pdf (accessed March 28, 2015).
153 Quoted in the *Congressional Record*. 149 (April 2, 2003): H 643.
154 Quoted in the *Congressional Record*. 159(April 22, 2013): S282.
155 U.S. Congress. House. 2009. "Preparing for Climate Change: Adaptation Policies and Program." Subcommittee on Energy and Environment. March 25. 111th Cong., 1st sess. Washington, D.C.: U.S. Government Printing Office: 2.
156 U.S. Congress. House. 2009. "Planning for a Changing Climate—Smart Growth, Public Demand, and Private Opportunity." Select Committee on Energy Independence and Global Warming. June 18. 110th Cong., 2nd sess. Washington, D.C.: U.S. Government Printing Office.

172   *The Quest for Climate Change Policy*

157 U.S. Congress. Senate. 2008. "Climate Change Impact and Responses in Island Communities." Committee on Science Commerce, Science, and Transportation. March 19. 110th Cong., 2nd sess. Washington, D.C.: U.S. Government Printing Office.
158 U.S. Congress. Senate. 2009. "Drought, Flooding, and Refugees: Addressing the Impacts of Climate Change in the World's Most Vulnerable Nations." Subcommittee on International Development and Foreign Assistance, Economic Affairs, and International Environmental Protection. October 15. 111th Cong., 1st sess. Washington, D.C.: U.S. Government Printing Office.
159 U.S. Congress. House. 2008. "Planning for a Changing Climate and Its Impacts on Wildlife and Oceans: State and Federal Efforts and Needs." Subcommittee on Fisheries, Wildlife, and Oceans. June 24. 110th Cong., 2nd sess. Washington, D.C.: U.S. Government Printing Office.
160 The Climate Change Drinking Water Adaptation Act (S 2970) and, its sister bill in the House, the Climate Change Drinking Water Adaptation Act (H.R. 6297); Drinking Water Adaptation, Technology, Education, and Research Act (S. 1035); Climate Change Drinking Water Adaptation Research Act (H.R. 6297); Clean Coastal Environment and Public Health Act of 2009 (H.R. 2093); Climate Change Safeguards for Natural Resources Conservation Act (H.R. 2192); Water System Adaptation Partnerships Act of 2009 (H.R. 2969); Water Efficiency, Conservation, and Adaptation Act of 2009 (H.R. 3747); and the Coastal Zone Enhancement Reauthorization Act of 2007 (S. 1579).
161 The Coastal State Climate Change Planning Act (H.R. 5453—110th Congress) (H.R. 1905—111th Congress), Ocean and Coastal Adaptation Planning Act (S. 810); and the Chesapeake Bay Science, Education, and Ecosystem Enhancement Act of 2009 (S. 1224).
162 The Natural Resources Climate Adaptation Act (S. 1933), the Healthy Borderlands Act of 2009 (H.R. 3629), and the Global Warming Wildlife Survival Act (H.R. 2338).
163 The Public Health and Climate Change Adaptation (S. Res. 509) and the Climate Change Health Protection and Promotion Act (H.R. 2323).
164 Arctic Climate Adaptation Act (S. 1566).
165 Investing in Our Future Act of 2010 (H.R. 5783).
166 Native American Challenge Demonstration Project Act of 2009 (S. 980).
167 The Energy Independence and Security Act of 2007 (H.R. 6) required the Secretary of the Interior to develop an adaptation plan while the Omnibus Public Land Management Act of 2009 (H.R. 146) included a variety of provisions to track the implications of climate change for water systems.
168 The only two exceptions were the Save the Safeguarding America's Future and Environment (SAFE) Act (S. 1881), which established a federal program to protect natural resources from climate change, and the Climate Change Health Protection and Promotion Act (H.R. 3314), which developed a national strategic plan to help health professionals prepare for and respond to the spread of emerging diseases.
169 Ben Jervey. 2013. "Year After Sandy, Rebuilding for Storms and Rising Seas." *National Georgraphic.* Available at http://news.nationalgeographic.com/news/2013/10/131026-hurricane-sandy-anniversary-sea-level-rise-adaptation/ (accessed March 28, 2015).
170 Quoted in *Congressional Record.* 159 (March 14, 2013): H. 1401.
171 Examples include the Coastal Climate Change Planning Act (H.R. 764), Joint Ventures for Bird Habitat Conservation Act of 2013 (H.R. 1137), Northern Rockies Ecosystem Protection Act (H.R. 1187), Global Partnerships Act of 2013 (H.R. 1793), Climate Change Health Protection and Promotion Act (H.R. 2023), Coral Reef Conservation Amendments Act of 2013 (S. 839), Safeguard America's Future and Environment (SAFE) Act (S. 1202), and the Water Infrastructure Resiliency and Sustainability Act of 2013 (S. 1508).

172 See also: Environmental and Energy Study Institute. 2014. "Fact Sheet: Climate Adaptation at the Federal Level." January 16. Available at www.eesi.org/papers/view/fact-sheet-climate-adaptation-at-the-federal-level (accessed March 28, 2015).

173 Zack Colman 2014. "Democrats Shift from Climate 'Change' to 'Adaptation' to Woo Republicans." *The Washington Examiner.* April 11. Available at www.washingtonexaminer.com/democrats-shift-from-climate-change-to-adaptation-to-woo-republicans/article/2547109 (accessed March 28, 2015).

174 Steven Benen. 2014. "Pentagon: Climate Crisis an Immediate Security Threat." *MSNBC,* October 14. Available at www.msnbc.com/rachel-maddow-show/pentagon-climate-crisis-immediate-security-threat (accessed March 28, 2015).

175 Ibid.

176 Center for Climate and Energy Solutions. 2015. "State and Local Climate Adaptation." Available at www.c2es.org/us-states-regions/policy-maps/adaptation (accessed March 28, 2015).

177 Kristen L. Miller, Janet L. Kaminski Leduc, and Kevin E. McCarthy. 2012. "Sea-Level Rise Adaptation Policy in Various States." Connecticut Coastal Zone Management, October 26. Available at www.cga.ct.gov/2012/rpt/2012-R-0418.htm (accessed March 28, 2015); Hedia Adelsman and Joanna Ekrem. 2012. "Preparing for Climate Change: Washington State's Integrated Response Strategy." April, Georgetown Climate Center. Available at www.georgetownclimate.org/resources/preparing-for-a-changing-climate-washington-states-integrated-climate-response-strategy (accessed March 28, 2015); Sophie Yeo. 2014. "Hawaii Passes Climate Change Adaptation Law." *Responding to Climate Change,* June 12. Available at www.rtcc.org/2014/06/12/hawaii-passes-climate-change-adaptation-law/ (accessed March 28, 2015).

178 Maria Galluci. 2013. "6 of the World's Most Expensive Adaptation Plans." *Inside Climate News,* June 20. Available at http://insideclimatenews.org/news/20130620/6-worlds-most-extensive-climate-adaptation-plans (accessed March 28, 2015).

179 Joyce E. Coffee, Julia Parzen, Mark Wagstaff, and Richard S. Lewis. 2010. "Preparing for a Changing Climate: The Chicago Climate Action Plan's Adaptation Strategy." *Journal of Great Lakes Research* 36(2): 115–117; City of Chicago. Chicago Climate Action Plan. Available at www.chicagoclimateaction.org/pages/adaptation/11.php (accessed March 28, 2015).

180 Center for Climate and Energy Solutions 2015.

181 Pielke et al. 2007.

182 Environmental Protection Agency (EPA). 2012c. "Summary of the Clean Air Act." www.epa.gov/lawsregs/laws/caa.html (accessed March 28, 2015).

183 National Association of Clean Air Agencies. 2012. "Background and History of EPA Regulation of Greenhouse Gas (GHG) Emissions Under the Clean Air Act & National Association of Clean Air Agencies' Comments on EPA GHG Regulatory and Policy Proposals." February 17. Available at www.4cleanair.org/Documents/BackgroundandHistoryEPA RegulationGHGsFeb2012post.pdf (accessed March 28, 2015).

184 Environmental Protection Agency (EPA). 2010. "Endangerment and Cause or Contribute Findings for Greenhouse Gases under Section 202(a) of the Clean Air Act. EPA's Denial of Petitions for Reconsideration." www.epa.gov/climate-change/endangerment/petitions/decision.html (accessed March 28, 2015); Robert B. Moreno and Peter Zalzal. 2010. "Greenhouse Gas Dissonance: The History of EPA's Regulations and the Incongruity of Recent Legal Challenges." *UCLA Journal of Environmental Law and Policy* 30: 121–156; National Association of Clean Air Agencies. 2011. "Background and History of EPA Regulation of Greenhouse Gas (GHG) Emissions Under the Clean Air Act & National Association of Clean Air Agencies' Comments on EPA GHG Regulatory and Policy

Proposals." June 19. Available at www.nwcleanair.org/pdf/ClimateChange/Misc/History_%20EPA_Regulation_GHGs-July2011-post.pdf (accessed March 28, 2015).
185 National Association of Clean Air Agencies 2012.
186 Ibid. 5.
187 Moreno and Zalzal 2010.
188 Neela Banerjee. 2012. "U.S. Court Upholds EPA's Authority to Regulate Greenhouse Gases." *Los Angeles Times*, June 26. Available at http://articles.latimes.com/2012/jun/26/business/la-fi-epa-court20120627 (accessed March 28, 2015).
189 John M. Broder and Matthew L. Ward. 2010. "Clashes Loom, With E.P.A. Set To Limit Gases." *New York Times*, December 31, A01.
190 Ibid.
191 Ibid.
192 Russell Prugh. 2011. "Proposed Legislation Seeking to Block EPA Greenhouse Gas Regulation Picks Up Speed." Marten Law, February 17. Available at www.martenlaw.com/newsletter/20110217-epa-greenhouse-gas-regulation (accessed March 28, 2015); Alexa Jay. 2011. "House Republicans Prepare to Attack EPA Greenhouse Gas Rules." *Climate Science Watch*, January 10. Available at www.climatesciencewatch.org/2011/01/10/house-republicans-prepare-to-attack-epa-greenhouse-gas-rules/ (accessed March 28, 2015).
193 Billy Gribbin. 2011. "Energy Tax Prevention Act Passes House, Fails in Senate." Americans for Tax Reform, April 7. Available at http://atr.org/energy-tax-prevention-act-passes-house-a6032 (accessed March 28, 2015); Joe Kamalick. 2011. "US Congress Move to Block EPA Greenhouse Role." *ICIS.com*, February 3. Available at www.icis.com/Articles/2011/02/03/9431933/insight-us-congress-moves-to-block-epa-greenhouse-role.html (accessed March 28, 2015).
194 Gordon R. Alphonso, E. Brett Breitschwerdt, Neal J. Cabral, James Yancey Kerr, David L. Rieser, and Patricia F. Sharkey. 2010. "EPA Issues First-Ever Rule Regulating GHG Emissions from Light Duty Vehicles, Confirms GHG Stationary Source Permitting Begins January 2011." *Martindale.com*, April 14. www.martindale.com/environmental-law/article_McGuireWoods-LLP_979678.htm (accessed March 28, 2015); Moreno and Zalzal 2010; Brent D. Yacobucci and Robert Bamberger. 2010. *Automobile and Light Truck Fuel Economy: The CAFÉ Standards*. Congressional Research Service. April 23. Washington, D.C.: Library of Congress. Available at http://assets.opencrs.com/rpts/R40166_20100423.pdf (accessed March 28, 2015).
195 Yacobucci and Bamberger 2010.
196 Environmental Protection Agency. 2012b. "Region 9 Strategic Plan, 2011–14." Available at www.epa.gov/region9/strategicplan/air.html (accessed March 28, 2015).
197 Environmental Protection Agency. 2010b, "Prevention of Significant Deterioration and Title V Greenhouse Gas Tailoring Rule." *Federal Register* 75(106): 31513–31608.; Coalition for American Jobs. 2010. "Stationary Source GHG Regulation Under PSD." March 5. Available at http://coalitionforamericanjobs.com/2010/03/fact-sheet-stationary-source-ghg-regulation-under-psd/ (accessed March 28, 2015).
198 Linda M. Chappell 2010. *Regulatory Impact Analysis for the Final Prevention of Significant Deterioration and Title V Greenhouse Gas Tailoring Rule*. U.S. Environmental Protection Agency, Office of Air Quality Planning and Standards, Health and Environmental Impacts Division. Final Report, May. Available at www.epa.gov/ttn/ecas/regdata/RIAs/riatailoring.pdf (accessed March 28, 2015).
199 "Final Rule: Prevention of Significant Deterioration and Title V Greenhouse Gas Tailoring Rule." Fact Sheet. Available at www.epa.gov/nsr/documents/20100413fs.pdf (accessed March 28, 2015).

200 Robin Bravender. 2010. "EPA Issues Final 'Tailoring' Rule for Greenhouse Gas Emissions." *New York Times*, May 13. Available at www.nytimes.com/ gwire/2010/05/13/13greenwire-epa-issues-final-tailoring-rule-for-greenhouse-32021.html (accessed March 28, 2015).
201 Ibid.
202 Ibid.
203 Simon Lomax. 2010. "EPA Advisory Panel Splits Over U.S. Technology Rules for Carbon Dioxide." *Bloomberg*, September 28. Available at http://mobile. bloomberg.com/news/2010-09-28/epa-advisory-panel-splits-over-u-s-techno logy-rules-for-carbon-dioxide (accessed March 28, 2015).
204 Tracy E. Schelmetic. 2010. "EPA Releases Best Available Control Technology Guidelines for Polluters." *Green Technology World*, November 11. Available at http://green.tmcnet.com/topics/green/articles/116361-epa-releases-best-available-control-technology-guidelines- (accessed March 28, 2015).
205 Union of Concerned Scientists. 2014. "The Clean Power Plan: A Climate Game Changer." Available at www.ucsusa.org/our-work/global-warming/reduce-emissions/what-is-the-clean-power-plan#.VOFp5UtrWlJ (accessed March 28, 2015).
206 Environmental Protection Agency. 2014. *Carbon Pollution Emission Guide-lines for Existing Stationary Sources: Electric Utility Generating Units*, A Pro-posed Rule. Federal Register. June 18. Available at www.federalregister. gov/articles/2014/06/18/2014–13726/carbon-pollution-emission-guidelines-for-existing-stationary-sources-electric-utility-generating#h-9 (accessed March 28, 2015).
207 William Yeatman. 2015. "The Clean Power Plan's Dirty Secret." *U.S. News and World Report*, January 26. Available at www.usnews.com/opinion/ articles/2015/01/26/obamas-epa-clean-power-plan-has-a-dirty-secret (accessed March 28, 2015).
208 James Inhofe. 2014. "Letter to the Honorable Gina McCarthy." December 1. Available at www.inhofe.senate.gov/download/?id=be69bbb0-fe44–4542–98d 4–72cedca51897&download=1 (accessed March 28, 2015).
209 The White House. 2013. *The President's Climate Action Plan*. Executive Office of the President. June. Available at www.whitehouse.gov/sites/default/ files/image/president27sclimateactionplan.pdf (accessed March 28, 2015).
210 Ibid.
211 Jason Plautz. 2015. "Obama's Budget Goes All-In on Climate Change." *National Journal*, February 2. Available at www.nationaljournal.com/energy/ obama-budget-goes-all-in-on-climate-change-20150202 (accessed March 28, 2015).
212 The White House. 2015. *Investing in America's Future: The Budget for Fiscal Year 2016*: 21. Available at www.whitehouse.gov/sites/default/files/omb/ budget/fy2016/assets/investing.pdf (accessed March 28, 2015).
213 Center for Climate and Energy Solutions. 2012. *Climate Change Adaptation: What Federal Agencies Are Doing*. February. Available at www.c2es.org/doc-Uploads/federal-agencies-adaptation.pdf (accessed March 28, 2015).
214 Climate Change Adaptation Task Force. 2010. *Progress Report of the Inter-agency Climate Change Adaptation Task Force: Recommended Actions in Support of a National Climate Change Adaptation Strategy*. October 5, The White House Council on Environmental Quality. Available at www.whitehouse. gov/sites/default/files/microsites/ceq/Interagency-Climate-Change-Adaptation-Progress-Report.pdf (accessed March 28, 2015).
215 Climate Change Adaptation Task Force. 2011. *Federal Actions for a Climate Resilient Nation: Progress Report of the Interagency Climate Change Adaptation Task Force*. October 28. Available at www.whitehouse.gov/sites/default/files/ microsites/ceq/2011_adaptation_progress_report.pdf (accessed March 28, 2015).

216 A full list of agency plans is available at: www.performance.gov/content/
climate-change-federal-actions#supporting-info; Chris Mooney. 2014. "38
Federal Agencies Reveal Their Vulnerabilities to Climate Change—and
What They're Doing About It." *Washington Post*, Wonkblog, October 31.
Available   at   www.washingtonpost.com/blogs/wonkblog/wp/2014/10/31/38-
federal-agencies-reveal-their-vulnerabilities-to-climate-change-and-what-theyre-
doing-about-it/ (accessed March 28, 2015).
217 Anthony Adragna. 2015. "Rhetoric on EPA, Climate Show Divisions Among
Republican Presidential Hopefuls." *Bloomberg BNA*, January 27. Available at
www.bna.com/rhetoric-epa-climate-n17179922496/   (accessed   March   28,
2015).
218 Ibid.
219 Joel Achenbach and Juliet Eilperin. 2011. "Climate-change Science Makes for
Hot Politics." *Washington Post*, August 19. Available at www.washington
post.com/national/health-science/climate-change-science-makes-for-hot-politics/
2011/08/18/gIQA1eZJQJ_story.html?hpid=z1 (accessed March 28, 2015).

# 6 Anticipation
## A Distinctive Policy Type

This book identifies anticipatory problems as a distinctive type of policy issue, one that differs in substantial ways from other forms of government problem solving. The defining characteristic of the anticipatory policy problem is its temporal dimension. Specifically, anticipatory problems have yet to occur, although they are projected—anticipated—by some policymakers to emerge sometime in the future. This orientation toward the future means that anticipatory problems are marked by varied levels of uncertainty, as policymakers cannot definitively determine if, how, when, or to what extent they will materialize. Indeed, policymakers only truly "know" that these problems *could* occur sometime in the future and that, at least for the issues examined in this study, their occurrence would be extremely detrimental to human health, the economy, and the environment.

As evidenced in the nanotechnology, pandemic influenza, and climate change case studies, this temporal context and its associated uncertainty are overriding factors shaping the process of policy change. In each case, the anticipatory issue influenced policy development throughout and within the several stages of the cycle by which issues arise, programs are formulated, and new laws become implemented. Accordingly, this concluding chapter first examines and summarizes the features of this *anticipatory* policymaking process. This analysis specifically compares and contrasts the policy process engendered by anticipatory issues ("anticipatory policymaking") versus a more "reactive" or disaster-driven policy process.

After presenting these characteristic empirical features of anticipatory policymaking, this chapter offers some thoughts as to how government institutions might be modified to better promote anticipation as a routine endeavor. That government can and often does act in an anticipatory fashion has been demonstrated by this book. However, anticipatory policymaking is sporadic and, in some cases, can have deleterious results. This section argues that by institutionalizing anticipation within the U.S. Congress, improving the ability of federal agencies to generate and interpret problem indicators, and cultivating a larger culture of preparedness we can make tangible improvements in the capacity of the U.S. government— and its citizens—to engage in a proactive policymaking sequence.

This chapter closes by considering how this book fits into the larger body of literature on policy change. While notions of temporality factor prominently into research designs and underlying theoretical assumptions, relatively few works have isolated time as an important explanatory variable in its own right. This study shows that basic distinctions between the past, present, and, with respect to anticipatory policymaking, future are not only significant but can shape the policy process. This observation suggests a number of additional avenues of research, including the application of the anticipatory policy type to global and transnational policymaking contexts.

## The Dynamics of Anticipatory Policy Change

The anticipatory policy type is, in many respects, a conceptual counterweight to the work on policy change after disaster. This book asserts that proactive policymaking—policy change that occurs *before disaster*—differs in some important ways from reactive policymaking—policy change that occurs *after disaster*. As noted in Chapter 2, many observers believe a uniform and comprehensive "theory" of disaster-driven policymaking does not exist.[1] Indeed, even within the various "disaster" domains (e.g., the earthquake, hurricane, oil spill, nuclear power, and national security domains), one can find considerable variance with respect to the process of policy change.[2]

However, the absence of a uniform theory does not mean the burgeoning body of research on hazard and disaster policymaking is devoid of important and generalizeable findings. In fact, decades worth of research has provided an admirably comprehensive picture of "disaster-driven" policymaking (see Chapters 1 and 2). With an eye toward this work, the following section compares and contrasts policymaking in anticipation of an emergent hazard to policymaking after disaster. A summary of key differences and similarities is provided in Table 6.1.

### Pattern of Policy Change

In disaster domains, policy change typically occurs in the wake of a catastrophic focusing event. By aggregating death, economic loss, and physical destruction, disasters can come to represent powerful symbols of policy failure. However, policy change is rarely a forgone conclusion, even after a catastrophe. Organized interests must seize on these opportunities in the so-called "problem stream" and frame the incident as reason to question the status quo, injecting new ideas—or, more often than not, recycling old ones—and challenging the dominance of existing policy monopolies. Nor does this process necessarily emerge "out of the blue," so to speak. In many disaster domains, momentum will already be building and edging toward policy change before a focusing event accelerates these trends.[3] Focusing events can transform issues that were once only salient to expert communities into "public" problems worthy of citizen, media, and elected official attention.

Table 6.1 A Comparison of Disaster Policymaking and Anticipatory Policymaking

| | Disaster Policymaking | Anticipatory Policymaking |
|---|---|---|
| Pattern of Policy Change | Change occurs *after* a culminating event. Often lacks a public. | Change occurs *before* a culminating event. Often includes a public. |
| Discourse of Conflict | Discourse tries to interpret past. Disaster legitimizes claims of crisis. Considerable uncertainty, although disaster shapes certain aspects of policy discourse. | Discourse tries to imagine future. Analogy and implied comparison create a perception of crisis. Considerable uncertainty about future, which is strategically exploited by groups. |
| Venue Selection and Shopping | Focusing events facilitate venue shopping and change. Multilevel governance. | Absence of focusing event limits venue change, but accommodates venue adding. Multilevel governance. |
| Agenda Setting | Agenda setting is event driven. Focusing events catalyze issues onto the agenda. Issue attention is brief, but somewhat recurring. | Agenda setting is indicator driven. Trans-domain events "nudge" issues onto the agenda. Issue attention is sustained and iterative. |
| Policy Formulation and Design | Program design aims to recover and relieve. Variety of instruments utilized in policy design. | Program design aims to prepare and plan. Little use of coercive instruments, robust use of information, treasure, and market instruments. |
| Policy Implementation | Enacted policy aims to remedy programmatic deficiencies. Emphasis on measurable outcomes. Interest wanes across time. | Many programmatic deficiencies are discovered during implementation. Need to construct metrics of success. Interest sustained until problem appears. |

To some extent, the policymaking process surrounding anticipatory problems is very similar. Narratives, ideas, and institutions need to evolve before policy can change. Moreover, some anticipatory problems are truly novel whereas others exist in domains that are ripe for reform. Disasters and anticipatory problems share a number of more intrinsic similarities as well. Both threaten severe if not catastrophic harms. Both are highly complex and technical problems. Both are marked by tremendous uncertainty.

The two differ, however, with respect to the way in which they are revealed to policymakers, the media, the general public, and, in some cases, expert communities. Unlike most disasters, anticipatory problems provide a degree of forewarning. This distinction is not trivial. This book shows forewarning allows for policy change to *precede* the onset of a culminating and catastrophic event. Put differently, interest group mobilization is *proactive*, as opposed to being primarily reactive or event-driven.

This pattern of proactive policy change was evidenced in all three cases. The nanotechnology domain experienced radical change or "policy innovation" in that the federal government had very little sustained involvement in this area prior to the 21st Century Nanotechnology Research and Development Act. The Act vastly expanded the federal government's role in the stewardship of nanotechnology, mandating that it bankroll nanotechnology research and marketization, on the one hand, and requiring that it minimize the technology's risks, on the other. Similarly, in the years leading up to the 2009 swine flu outbreak, a series of national and state policies, including the Pandemic and All-Hazards Preparedness Act (PAHPA), the Public Readiness and Emergency Preparedness Act (PREPA), and the Model State Emergency Health Powers Act (MSEHPA), fundamentally reshaped the pandemic domain, introducing new policy instruments and altering the division of public health authority between state and national governments. And although the national government has been unable to craft uniform climate change mitigation or even adaptation policy, anticipatory policy change has occurred at the state level. In fact, recent executive actions, like the Climate Action Plan, aim to facilitate a nationwide, "bottom up" climate regime emanating from states and supported by the Environmental Protection Agency (EPA).

Another important distinction is worth highlighting. Birkland observed many disaster domains lack a "public," meaning only a small number of interest groups with close connections to the problem will mobilize during the policymaking process. Government officials and technical experts often dominate policymaking in disaster domains, although the public will demand action in the wake of catastrophe.[4] This was not the case in at least two out of the three domains examined in this book. Climate change is a highly competitive domain encompassing a diverse array of industry, environmental, and social justice groups. Climate change is an outlier, in some respects, as its baseline level of issue attention was much higher than the other cases. Indeed, it might be one of the most "hotly" contested political issues of the twenty-first century. And while nanotechnology is hardly a focal national issue, its distributive dimensions coupled with its potentially profound health, environmental, and social implications generated considerable mobilization from a variety of industry groups—spanning multiple sectors of the economy—as well as a bevy of activist groups.

Political mobilization was much less diffuse in the pandemic influenza case in the sense that government and public health experts led much of the policymaking process. Still it would be misleading to characterize the avian influenza case as being devoid of a public entirely. Controversial provisions, like vaccine liability and even state quarantine measures, mobilized a number of powerful lobbies, including vaccine manufacturers, civil rights groups, hospital associations, and trial attorneys. As noted in Chapter 4, avian flu seems to encompass elements of both the disaster and anticipatory domain types.

The presence of organized publics in the climate change, nanotechnology, and, to a lesser extent, pandemic domains *implies* certain areas of anticipatory policy are susceptible to widespread issue expansion. Whether this is an outgrowth of the anticipatory nature of these problems or a by-product of case selection is open to debate and certainly warrants further investigation (e.g., studies of other examples of anticipatory policymaking). In defense of the former, one could argue the anticipatory problems examined in this book threaten a much larger subsection of society than most disasters and that this feature, in turn, promotes issue expansion. All of these problems were "global"—not regional, not national—threats. Moreover, as described in the next section, because they are marked by such inherent uncertainty, it is easy for stakeholders to "imagine" themselves suffering at the expense of anticipatory problems. Relative to anticipatory problems, some, but certainly not all, disasters narrow our perception of a problem population by visibly highlighting the suffering of a specific class of victims—individuals working in a certain occupation, living in a certain region, participating in a certain activity. In other words, fear of the unknown, it seems, is very powerful and can provoke widespread concern about anticipatory problems.

## Discourse of Conflict

While disasters are real, focusing events are politically constructed. A disaster needs to be defined as a symbol of policy failure before it can promote radical policy change. As such, the discourse that emerges in the wake of disaster at least in part works to interpret and define that particular event. What caused the disaster? Was government responsible and, if so, how? Who were its victims? Were they complicit in the damages? How severe was the disaster? This assignment of meaning and blame allows organized interests to derive broader "lessons" for government that can later serve as the basis for policy change. As indicated above, these lessons need not be novel. Often times, they percolate in policy communities for years before an event elevates their relevancy. Because these same narratives can shape future policy outputs, competing groups will struggle to ensure their preferred definition sticks. Problem definition in the wake of disaster therefore builds from an understanding of a *past* event, which in turn informs the type of solutions that will prevent or mitigate a similar occurence in the *future*.[5]

The absence of an actual event alters this causal chain in the case of anticipatory problems.[6] In all three cases, problem definition hinged not on making sense of a past event, but on imagining what the "future" might bestow. Nanotechnology was obviously the most extreme example because many of its risks were not known at the time of policy action. Pandemic influenza was also seen as a looming or future threat, particularly prior to the actual outbreak of swine influenza in 2009. Although analogies could be drawn between bird flu and previous outbreaks, the pandemic issue was marked by great uncertainty and was seen as an emerging issue. Certainly by 2007, climate change already represented a very real and present danger, especially for those living in coastal communities. However, policy discourse remained focused on the future and many feared the most serious consequences—a climate change tipping point—were decades away.

The implications of this distinctive temporal orientation are profound. In disaster domains, problem definers usually point to tangible examples of death and destruction—actual disasters—as proof of crisis and government failure.[7] This is not possible in the case of anticipatory problems so perceptions of crisis must be creatively engineered. This book shows analogy and other forms of implied comparison were readily deployed to create a perception of crisis and minimize uncertainty about the future.[8] This strategy was used by both proponents and opponents of policy change. Climate change was analogized to World War II as well as a Medieval Warm Period. Avian influenza was analogized to the 1918 Spanish flu and the 1976 swine flu debacle. Nanotechnology was analogized to the Industrial Revolution, asbestos, and Genetically Modified Organisms (GMOs). Analogies communicated a cautionary story about the dangers of government action or inaction, directing attention away from the fact that the future is always marked by great uncertainty.

Analogy is not unique to anticipatory policy problems. Brändström, Bynander, and 't Hart show historical analogy is often used to make sense of unfolding or existing crisis situations.[9] Examples from the past can guide decision making, helping to reduce the complexities of these highly stressful situations. In the case of anticipatory problems, however, analogy serves a much more fundamental function in that advocates of policy change do not use it to make sense of an existing or developing crisis, but to *create the perception of looming crisis*. In other words, analogy is a precursor to lesson learning. Before government can even begin to formulate anticipatory policy, it must first acknowledge that an actual problem—a potential crisis—exists. Analogy serves this function.[10]

The significant uncertainty associated with imagining the future might have its advantages as well. Rose argues that defining problems that have yet to occur allows policymakers to skirt certain evidentiary expectations and strategically "link" their desired policy solutions to the issue in question.[11] He writes:

Although policymakers cannot draw lessons from events that have yet to occur, they can try to anticipate events. In doing so, they may treat the future as an extension of the present in order to bound speculation by existing knowledge. Theorists can claim future success for their prescriptions on the grounds that predictions follow logically from premises, whether or not their premises are plausible. Politicians can exploit uncertainty about the future by willfully asserting faith in their proposals, which have yet to be proven wrong.[12]

In all three of my cases, interest groups readily exploited this uncertainty about the future in order to reframe entire policy debates and, in many cases, better advocate for desired policy outcomes. In the climate change case, state policymakers and even President Obama worked to recast mitigation policy as an economic opportunity, not a zero-sum trade off between risk reduction and profit. Mitigation policy, they argued, would birth a new green economy. In the avian influenza case, global health advocates seized on concerns about a looming pandemic and argued greater investment in foreign public health infrastructures would reduce the likelihood of H5N1's spread. Of course, this solution was hardly new and many entrepreneurs had been touting the importance of a "one world" approach to health for decades. This trend was also evidenced in the nanotechnology case. In this instance, however, nanotechnology was presented as a solution to a range of policy problems. Industries from an array of fields argued that, with the help of nanotechnology R&D money, they could vastly improve society, providing more durable clothing, better drug delivery devices, and advanced weaponry.

Disasters are also marked by tremendous uncertainty. Confusion will inevitably emerge as to cause and broader implications of such tragic events. Moreover, because focusing events are ultimately political constructs, interest groups are free to frame and cast the problem as they see fit, provided their desired audience will listen. Yet the presence of an event likely lends a little more structure to this process. Indeed, disasters inevitably highlight certain problem features (e.g., a specific class of victims, connections to a certain industry, tangible measures of human and economic loss). Such empirical evidence simply does not exist within the context of anticipatory problems. While proponents of policy change likely benefit from this tangible proof of crisis, they are much more constrained in terms of their ability to speculate about the future. Anticipatory policymaking, therefore, is typically more "conjectural" than disaster policymaking.

### Venue Selection and Shopping

Crises can come to represent an indictment on existing policy monopolies, including the entrenched government institutions or "venues" supporting their dominance. This challenge to the status quo allows organized interests to appeal to new venues that offer a fresh approach to disaster policymaking and are not beholden to subsystem elites. Crisis, in other words, can provide access to new venues, although the onus is ultimately on organized interests to capitalize on this opportunity.[13]

Interestingly, very little overt venue change was reported in the avian influenza, climate change, and nanotechnology cases. A core set of venues steered much of the anticipatory policymaking process, often from start-to-finish. Public health committees and agencies, like the Senate Committee on Health, Education, Labor, and Pensions (HELP) and the Department of Health and Human Services (HHS), dictated most pandemic policymaking. Similarly, the House and Senate Science committees devised the 21st Century Nanotechnology Research and Development Act while the EPA guided most nanotechnology regulation. And, of course, the EPA has always been at the forefront of the climate change debate, as have the House Committee on Energy and Commerce and the Senate Committee on Environment and Public Works.

Without a focusing event to demonstrate a clear example of policy failure, dislodging issues from these venues was extremely difficult. Organized interests did, however, opt for a strategy of "venue adding," meaning they did not abandon their work in dominant venues but rather added additional venues.[14] For example, proponents of an all-hazards approach to pandemic preparedness succeeded in gaining the support of committees with jurisdiction over security issues, like the House Committee on Homeland Security, but remained committed to fighting for policy change in more traditional "public health" settings. Similarly, advocates of stronger public health statutes worked to promote policy change at both the national and, in some cases, state levels. While the EPA remained ground zero in the fight to regulate nanotechnology, a number of other agencies were engaged as potential rule-makers, including the Food and Drug Administration (FDA) and the National Institute for Occupation Safety and Health (NIOSH).[15] State and local governments were also involved in the regulatory process, including Massachusetts, Maine, and Berkeley, California. Venue adding was rampant in the climate change case, but the addition states did not come at the expense of national efforts. In fact, state policy has supported the movement for national policy change, as evidenced in the Climate Action Plan's desire to build a "bottom up" emissions mitigation regime.

Finally, anticipatory policymaking and disaster-driven policymaking share the same propensity for multilevel governance. This is in large part a testament to the fact that states retain a great deal of authority over disaster preparedness, environmental safety, and public health.[16] As indicated in the previous paragraph, this was boldly illustrated in all three cases, which saw states play an instrumental role in crafting and enacting anticipatory policy change.

## Agenda Setting

Problems in disaster domains reach the institutional agenda in the wake of focusing events, which come to symbolize government failure.[17] Problem indicators, by contrast, are the primary agenda catalysts for anticipatory problems. This dynamic was perhaps most clearly illustrated in the pandemic case where policymaker attention closely coincided with the accumulation of avian influenza cases and deaths abroad. Both "rounds" of nanotechnology policymaking were indicator-driven as well. In the first round, indicators comparing the number of patents and peer-reviewed publications in the U.S. versus other countries fueled concerns that America was lagging in this emerging area of technology and that greater government investment and involvement was needed. In the second round, proponents of expanded funding for risk research used the upward trend in peer-reviewed publications documenting nanotechnology's health and environmental dangers to justify their claims that nanotechnology was cause for concern. And, of course, records of annual average temperatures have long been seen as one of the most important indicators of climate change.

Like focusing events, the power of indicators is derived not from the measures themselves but from how they are interpreted and integrated into larger narratives. According to Stone, the use of quantitative indicators allows policymakers to "project trends into the future to demonstrate decline is just around the bend."[18] The ability to project into the future is critical to mobilizing support for anticipatory policy problems, which are, by nature, noticeably abstract. In the nanotechnology and avian influenza cases, it was nearly impossible to cite a single instance where an American citizen was actually harmed at the time of policymaking. As such, measures depicting a growing number of avian influenza cases and deaths abroad as well as the number of articles documenting the potential risks of nanotechnology supported the proposition that crisis was forthcoming even if present harms were minimal. And although many believed climate change already constituted a crisis, projections of future warming trends were used to show the dangers of government inaction, supporting claims that a tipping point needed to be avoided at all costs.

While indicators played an integral role in all three cases, evidence suggests the agenda setting process was helped along by disaster events. Hurricane Katrina heightened concerns with avian influenza, as policymakers were simply unwilling to be "caught off guard" by another catastrophe. Global warming was also influenced by an exogenous event, the *Deepwater Horizon* oil spill, which effectively halted Senate debate of the Clean Energy Jobs and American Power Act. That no comparable pattern was observed in the nanotechnology case may simply be testament to this field's newness (and the possibility that the right triggering event has not yet occurred).

It seems the fluid and unsettled nature of policy analysis regarding anticipatory problems creates permeable boundaries of interpretation. Under these conditions, events occurring in domains that may or may not be logically "related" to a given anticipatory problem can help "nudge" the problem onto the agenda, especially when indicators of an emerging threat have already captured policymaker attention. While scholars have noted the capacity of sufficiently large events (e.g., the September 11, 2001 terrorist attacks) to provoke change across multiple subsystems, agenda setting in the wake of disaster tends to be fairly catalytic, propelling one or a handful of domains, most of which were directly impacted by the disaster, to the top of the institutional agenda.[19] In the case of anticipatory problems, these events, which were only tangentially associated to the anticipatory problem in question, did not necessarily catalyze issues onto the agenda but more reinforced patterns that were already underway.

Issue attention in disaster domains tends to be fairly brief. Disaster provides an impetus to "do something fast" thus prompting a spike in attention and activity. Once the immediate trauma of the event fades, issue attention will inevitably wane. This cycle is, however, fairly persistent, as many of these "routine hazards" and disasters recur with relative frequency.[20]

The three cases examined in this book are demonstrative of a slightly different pattern. Specifically, issue attention was not catalytic but fairly protracted and iterative, encompassing two distinct but at times overlapping "stages" of policy activity. Virtually all of the figures documenting congressional and media attention portray two noticeable "bumps" in activity, usually occurring in relatively quick succession of one another. In the nanotechnology case, the first round of activity resulted in the establishment of the 21st Century Nanotechnology Research and Development Act, while the second was akin to a period of policy maintenance, as organized interests worked to reassert the importance of risk research. A very similar pattern was evidenced in the avian influenza case. In this instance, the first round of anticipatory policy activity resulted in a number of important pandemic preparedness and public health statutes, including PAHPA, PREPA, and MSEHPA. A second round of policy maintenance was observed in this case as well, although this activity was more the product of changes in the actual problem rather than the emergence of a coherent counter-mobilization effort. Specifically, the outbreak of swine influenza in 2009 prompted Congress to revisit many of the programs established in previous sessions. Whether this second iteration would have occurred in the absence of the H1N1 outbreak is uncertain. The climate change case also included two overlapping periods of attention, the first occurring between roughly 2006 and 2007 and the second between 2009 and 2010. In this instance, changes in party control of Congress and, later, the White House precipitated the two stages of agenda activity.

With respect to anticipatory problems, the absence of an actual event removes the post-disaster pressure to "do something fast." This is not to say that there is a lack of urgency per se, only that anticipatory problems have a slightly longer shelf life on the institutional agenda. Indicators of an emerging threat slowly reveal themselves over a period of months, if not years, gradually stoking fears and maintaining issue attention. Anticipatory policy is very susceptible to tinkering and alteration, especially when the anticipated problem has yet to come to fruition. This does not mean, however, that the issue attention cycle for anticipatory problems is ageless. Indeed, as evidenced in the stark decline in congressional and media attention documented at the end of all three cases, diminished novelty almost always amounts to waning interest.

## Policy Formulation and Design

Broadly speaking, the processes of *recovery*, on the one hand, and *preparedness*, on the other hand, shift our temporal lens from the past to the future. While the disaster management cycle advocates for balance, elected officials overwhelmingly emphasize recovery over planning.[21] Although disaster relief spending is important to helping communities recover, this type of "distributive" policymaking also pays electoral dividends. Relief spending is rewarded with a far greater "vote share" than preparedness spending.[22] Of course, recovery, preparedness, and planning can occur simultaneously. As evidenced in the Disaster Relief Appropriation Act of 2013 or the "Hurricane Sandy Relief" bill, recovery packages can include provisions requiring that funding recipients alter their living patterns in order to mitigate the likelihood of future harm.

This said, anticipatory policies take the characteristic form of strategies and preparedness plans. The three cases examined herein are littered with examples of government planning, including the *National Strategy for Pandemic Influenza*, the various influenza plans developed by federal agencies, and the Climate Action Plan. Planning was also central to the 21st Century Nanotechnology Research and Development Act, which set out to proactively identify and manage nanotechnology's risks, a notable departure from the "wait-and-see" regulatory approach typically associated with new technologies.

Plans are a useful strategy for readying for an emerging event and can help structure the anticipatory policymaking process. For example, the various "pillars" of the *National Strategy* not only established important targets for government agencies, but narrowed the types of issues Congress focused on in the years leading up to the 2009 swine influenza outbreak. Similarly, the three goals of R&D, marketization, and risk reduction, all of which were codified in the 21st Century Nanotechnology Research and Development Act, continue to guide the federal government's stewardship of nanotechnology to this day. The Climate Action Plan, which establishes long term emissions targets and presents various paths to achieving these goals, is another fine example of planning policy. Plans allow government to work toward policy outputs that might not confer any tangible benefits until years in the future, a departure from the remedial designs that "fix" existing social ills.

Planning serves an important political function as well. Citizens and other stakeholders can demand proactive policy when they feel threatened, in turn necessitating that policymakers take action on risks that do not currently exist.[23] Although policymakers can rarely *prevent* anticipatory problems, they can certainly take steps to prepare and plan for them. This was the case with respect to avian influenza, which, for a time, was widely perceived as a significant threat to America's safety. Through the *National Strategy*, President Bush could demonstrate tangible action on an essentially intangible problem, thereby satiating stakeholder appetite for a government response. The Climate Action Plan served a similar function, although President Obama was likely trying to assuage a much more targeted Democratic base, many of whom were disillusioned by the government's inability to pass substantive legislation in 2010. And although public concern with nanotechnology never approached the levels observed in the other cases, the lessons of the GMO and biotechnology debates implied proactive risk governance and planning might pay political dividends in the long run.

The distinctions between anticipatory and disaster policy become less clear when we look at the actual instruments utilized in these types of policy designs. From direct regulation to information sharing, a variety of policy instruments can be integrated into post-disaster policies.[24] Virtually all options are on the table. The same holds true for anticipatory policymaking, although this book shows policymakers put a special emphasis on information (nodality) instruments.[25] Specifically, in all three cases the government either directly provided information in hopes of changing behavior or, more commonly, funded private initiatives to generate and disseminate information. One of the primary goals of the National Nanotechnology Initiative (NNI) was to produce information and research on nanotechnology's health and environmental risks.[26] Most importantly, this information fed back into the political system and came to represent an important indicator of nanotechnology's risk, in effect kick starting the second round of policy activity. In the avian influenza case, information sharing between various public and private actors was an important component of preparedness planning, as was disease surveillance and data documenting the progress of vaccine development. And, of course, the government generates an abundance of information on the risks associated with climate change. For example, the International Panel on Climate Change (IPCC) and the National Oceanic and Atmospheric Administration (NOAA) have long produced comprehensive records of annual average temperatures, which are arguably the most important indicators of warming trends. When the government finally took to the task of adapting to climate change, a number of proposed policies as well as the Climate Action Plan mandated an extensive period of vulnerability assessment in order to produce information on organizational needs and regional risks.

Because they provide a degree of foreshadowing, anticipatory problems lend themselves to information gathering. Many natural disasters, by contrast, are unexpected events.[27] Moreover, given their inherent uncertainty, information gathering is critical to helping officials track and understand anticipatory problems as they unfold across time. At the time of policy deliberation and action, tremendous uncertainty will abound as to the actual features of these problems. Information instruments help minimize that uncertainty.

Despite the fact that private actors and industries were deeply involved in the avian influenza, nanotechnology, and climate change cases, anticipatory policies put very little emphasis on "command and control" regulatory instruments. Instead, policymakers usually opted to use market or treasure instruments, which strived to incentivize private actors to perform certain functions. This was best evidenced in the "cap-and-trade" policy designs, which were the centerpiece of most climate change mitigation policy. Cap-and-trade created an artificial marketplace for carbon and struck a balance between industry's concern with the "bottom line" and government's desire to reduce emissions. Nanotechnology producers have been very resistant to "command and control" regulations, especially the prospect of Toxic Substances Control Act (TSCA) reform. In fact, most of the government's information gathering programs, which are arguably the least intrusive regulatory options in this context, have been voluntary and "sweetened" with grants and other forms of financial inducement. Grants have also been the most important mechanism for promoting nanotechnology R&D. Of course, this domain is very nascent and more prescriptive regulation might be forthcoming. A pandemic outbreak *could* trigger more stringent command-and-control regulations, like forced quarantines and vaccinations. Yet, even in the avian influenza case, policymakers were committed to creating market arrangements that were extremely favorable to vaccine manufacturers, including limited liability laws and buyback programs.

Command-and-control regulation is always a tough sell, even after disaster. The absence of an actual event and the speculative nature of anticipatory problems certainly do not make this "sell" any easier, making market mechanisms and treasure instruments particularly enticing options for inducing compliance. The actual process of policy formulation in all three cases was noticeably consensual—some might argue too consensual—and policymakers took great pains to assuage the concerns of industry.[28] This dynamic was especially prominent in the avian influenza and nanotechnology cases because, at the time of policy formulation, there was no evidence that industry had done anything wrong or was somehow responsible for policy failure. Indeed, nanotechnology producers frequently noted that it would be unfair and even stigmatizing to regulate them when their products had not been shown to cause harm. And while there are certainly factions within the national government that would gladly impose stringent regulations on greenhouse gas emitters, this approach has not been politically feasible.

*Policy Implementation*

Event-driven implementation works to address the systemic failures revealed by disaster by executing a new policy or more vigorously enforcing an existing law. Part of this process will include the distribution of relief funds and the initiation of recovery efforts. However, post-disaster implementation also includes actions taken to reduce vulnerabilities and, ideally, minimize future harms.[29]

This same fundamental desire to address failure is paramount to anticipatory policy as well. However, anticipatory policymaking is marked by a number of distinguishing characteristics. For one, the absence of a disaster event means many of the "failures" that disasters reveal prior to the policy formulation stage may not be discovered until the implementation stage. Because anticipatory policymaking is an inherently prospective process and because the empirical features of anticipatory issues often take years to fully blossom, many important revelations about programmatic deficiencies or "alarmed discoveries" do not occur until the implementation process is well underway.[30]

Nanotechnology is perhaps the most extreme example of this dynamic. At the time of the 21st Century Nanotechnology Research and Development Act's passage, the technology's risks were extraordinarily abstract, undefined, and even unknown. In turn, most of the implementation process was devoted to determining which statutes, if any, could even be applied to this radically novel technology—let alone regulating manufacturers. The sort of rigorous enforcement observed in the wake of disaster still seems years away. The Climate Action Plan's adaptation programs also called for an extended period of planning and risk assessment, particularly within federal agencies and at the regional level, a testament to the fact that there remains tremendous uncertainty about the actual sources of climate vulnerability. This does not mean the Climate Action Plan is devoid of actionable items, but the immediate impulse to "fix" something seems much less pronounced without a catastrophic event. To some extent, the fixing process cannot commence until bureaucrats discover where, exactly, the system is broken. If anything, the pandemic case, which offered a rare opportunity to examine anticipatory policy implementation both before and after disaster, bears witness to the challenges and uncertainties associated with proactive implementation. Despite best efforts to prepare, a slew of unanticipated problems emerged during the swine flu pandemic, including difficulties acquiring and distributing a vaccine. Subsequent preparedness plans will likely draw from these lessons and remedy these flaws, a luxury that was not afforded during the initial—anticipatory—implementation process.

This dynamic is not distinctive to anticipatory problems and implementation is always rife with challenges and uncertainty. However, this study shows that these difficulties are exacerbated by the speculative nature of anticipatory policymaking, which puts a great deal of pressure on implementers to "learn on the fly." Thus, whereas post-disaster implementation is typically marked by renewed vigor and commitment to enforcing rules and regulations, anticipatory problems necessitate a period of active reflection while bureaucrats try to determine the actual risks associated with these problems as well as the applicability of existing programs. This phenomenon also helps explains the iterative agenda setting process described above. Feedback generated during the implementation stage revealed programmatic flaws that needed to be addressed legislatively, including gaps in funding for both nanotechnology risk research and pandemic preparedness programs. In other words, even after punctuations or policy change, the policy system remained slightly destabilized, in turn inducing subsequent action and policy maintenance.[31] The observed instability was partially an outgrowth of the relative newness of the policies themselves, which made them inherently unstable, and partially an outgrowth of the enormous uncertainty surrounding these anticipatory problems, which meant implementers were the first to discover many programmatic shortcomings.

The sustainability of any reform will partially hinge on agencies' ability to demonstrate success, both in terms of meeting programmatic goals and in improving actual social conditions.[32] Bureaucrats working in disaster fields benefit from being able to distribute large sums of relief money and other resources to disaster areas, a fairly measurable policy output. What is more, agencies often have a propensity to target "low hanging" (easily achievable) items first because doing so allows them to demonstrate progress.[33] Still, even in disaster domains, implementers will eventually encounter many of the same challenges associated with anticipatory policy implementation. Disaster preparedness inevitably requires that bureaucrats ready for problems that have yet to occur—future disasters. Preparedness is both administratively difficult, because it is challenging to measure progress in the absence of a manifest problem, and politically contested, because proactive measures are rarely rewarded.

These challenges are vastly magnified with respect to anticipatory problems. For one, bureaucrats cannot point to relief or recovery programs as evidence of a policy's worth. Nor can they point to these programs as evidence that they are making progress toward achieving a certain goal. Moreover, it is impossible to demonstrate actual improvement in a problem that simply does not exist. Indeed, in anticipatory contexts, bureaucrats cannot even make the case that their work is somehow an extension of the lessons revealed by some past event or example of policy failure. The demonstration of programmatic success, in other words, is extraordinarily speculative.

Given this precarious arrangement, policymakers in all three cases took great pains to construct proxy benchmarks of success. Soon after establishing the *National Strategy*, Bush Administration officials developed a series of "scorecards," which tracked the progress of various preparedness goals. Similarly, updates and progress reports documenting movement toward adaptation goals were widely circulated soon after the establishment of Climate Change Adaptation Task Force and, later, the Climate Action Plan. And in the nanotechnology case, market forecasters provided various projections of the potential economic impact of nanotechnology R&D well after the passage of the 21st Century Nanotechnology Research and Development Act.

Constructed measures are, in large part, more political than practical. The agencies and individuals responsible for implementing anticipatory programs are severely limited in terms of their capacity to demonstrate tangible reductions in an existing problem. At best, they can resort to "bean counting" or a tabulation of the amount of new programs developed or grants allocated, but even here resources are rarely being siphoned toward an injured clientele that can attest to the program's importance. As such, the scorecard plays an important role in convincing others that we are in fact "prepared" and that costly programs are worthy of sustained government investment, despite the fact that it might take months, if not years, for these efforts to reveal their actual value.

All of which begs the questions: How long will policymakers continue to support programs in the absence of an actual problematic condition? Will policymakers remain committed to programs with very few immediate benefits? The literature on policy change after disasters documents a waning in interest in the years following a focal event, particularly after memories of the disaster begin to fade.[34] This book did not observe any obvious waning in bureaucratic attention in the aftermath of anticipatory policy change. In all three cases, implementation was and remained an active and important stage in the policymaking process. Debate over the regulation of nanomaterials continues to grip the EPA and many observers believe a wholesale reform of TSCA is forthcoming. Moreover, the Climate Action Plan virtually ensures the EPA will remain actively involved in climate change policy, at least for the duration of President Obama's term. Although the period between policy enactment and the actual outbreak of a pandemic was fairly short, bureaucrats also remained diligent in their preparedness efforts in the run up to the 2009 swine flu outbreak.

Fear of the unknown is very powerful and can stoke the urge to prepare, which partially explains the sustained attention described above. Moreover, given their immense technical complexity and uncertainty, anticipatory problems inevitably vest a great deal of responsibility in the hands of bureaucrats. In both the nanotechnology and climate change cases, this responsibility effectively made the EPA a prominent policy *making* venue in its own right. Such an elevated role in the policymaking process negates the possibility that agencies will fade into the shadows. Of course, this book did observe an eventual waning in congressional interest, so it is possible that a similar decline will occur at the agency level in coming years as well. Having described the basic features of the anticipatory policy process, the next section considers how these findings might be used to improve government performance.

## Anticipation: A Road to Better Government?

The three cases explored in this book present anticipation as a laudable and important policy goal. Yet, a handful of recent examples suggest that anticipation may not always be such a good thing. The 2003 invasion of Iraq is a case and point. Deemed a "preemptive war," the Bush Administration justified the invasion on the grounds that it would forestall Iraq's use of weapons of mass destruction (WMDs), which were allegedly hidden in clandestine laboratories scattered throughout the Middle Eastern country. Yet these WMDs were never found, and to its critics the Bush Administration's preemptive strategy amounted to arguably the biggest foreign policy debacle of the twenty-first century, killing or wounding thousands of American soldiers and costing the country trillions of dollars.[35] Many believe the war also permanently undermined the United Nations (UN) by demonstrating the ability of a single nation to act unilaterally and defy established norms of international engagement.[36]

The response of a handful of U.S. state governments to the 2014 Ebola outbreak offers yet another illustration of anticipation gone wrong. First discovered in West Africa in March 2014, Ebola is suspected to have infected roughly 25,000 individuals worldwide, killing more than 10,000 of them.[37] An opportunistic disease that can only spread via direct contact, Ebola tends to exploit weaker public health systems that are unable to properly identify, quarantine, and treat the infected. Thus, when Ebola first appeared in the U.S. in September 2014, CDC officials were alarmed, but confident that the country's strong health care system could contain the outbreak. The agency issued guidelines recommending voluntary at-home monitoring for high-risk travelers returning from West Africa. The guidelines also recommended monitoring without isolation for returning health care workers.[38]

In late October, the identification of three additional cases (two in Texas and another in New York), prompted governors in New York, New Jersey, and Illinois to break from CDC guidelines and issue their own public health emergency protocols, mandating home quarantine for *all* individuals returning from West Africa. Another seven states followed suit shortly thereafter and imposed stringent monitoring requirements.[39] Public health experts and the Obama Administration chided the measures for being overly protective and not grounded in sound science. In fact, many feared enhanced quarantine measures would discourage aid workers from travelling to West Africa and, therefore, undermine international efforts to control the outbreak at its source.[40] The mandatory quarantines, did, in fact, prove to be an overreaction, and the much-feared U.S. Ebola crisis never came to fruition.

Misguided application of the anticipatory approach can also be seen when the practice of preventative medicine, a staple of Democratic health care reform proposals, is carried too far.[41] Preventative medicine tries to identify and treat health ailments during their early stages, as opposed to waiting for smaller problems to evolve into potentially life threatening diseases. Beyond saving lives, proponents are quick to note that late-intervention "emergency care" constitutes a large portion of wasteful health expenditures.[42]

While prevention is critical to ensuring good health outcomes, many critics make the case that the medical system overemphasizes prevention and has even opted for a more "defensive" medical approach. Welch, Schwartz, and Woloshin, for example, argue that modern medicine's heightened ability to detect and identify abnormalities has prompted an epidemic of overdiagnosis with many patients receiving treatments and drug prescriptions doing them more harm than good. These authors cite a wealth of instances of overdiagnosed diseases, including diabetes, hypertension, and even a number of cancers. One study cited by the authors found that overdiagnoses of breast cancer are, in fact, more prevalent than actual cancer diagnoses.[43] According to the president of the Royal College of General Practitioners, Iona Heath:

> The evidence review suggests that for every 2,000 women invited to screening for 10 years one death from breast cancer will be avoided but that 10 healthy women will be overdiagnosed with cancer. This overdiagnosis is estimated to result in six extra tumorectomies and four extra mastectomies and in 200 women risking significant psychological harm relating to the anxiety triggered by the further investigation of mammographic abnormalities.[44]

In other words, excessive preventative medicine is not only failing to make us healthier, but is likely making us sicker under some circumstances.

Anticipation, then, seems both a potential blessing and curse. With these challenges in mind, the following section considers a number of practical suggestions that might improve the capacity of government—indeed society—to anticipate. It is not supposed these reforms would transform government into some oracular entity, devoid of political conflict. Instead, they simply try to position government to better deal with emergent problems, in particular hazards and risks. For better or worse, how decision makers behave in the face of looming crisis will always remain a question of politics.

### Expand the Information Gathering and Processing Capabilities of Federal Agencies

Spurred by the growing number of economists working in national government and policy institutes, interest in developing a comprehensive set of "social indicators" surged in the 1970s. This so-called social indicator movement strived to develop robust statistical measurements of social problems, which could then be used to evaluate the effectiveness of social welfare programs and other government initiatives. Interest in social indicators waned throughout the 1980s, as scholars and policymakers alike began to question the goal of quantifying all social problems in an accurate and timely way. Despite a minor revival in the 1990s, the social indicators movement has yet to reclaim the enthusiasm with which it was once associated.[45]

This present study does not support renewal of the social indicator movement, nor any notion that all policy problems can be reduced to statistics. At the same time, however, this book has documented that sound policy indicators are vital to identifying and responding to anticipatory issues. With this in mind, federal agencies, which are some of the single most important producers of this information, stand to gain enormously from launching an initiative to improve their ability to identify, analyze, and evaluate problem indicators. This initiative should encompass the entirety of the federal bureaucracy—all agencies, offices, and commissions—and solicit input from inside and outside the public sector. The goal of this project is not to question the validity of specific indicators, although that may well be one outcome of the process, but rather to review and come up with suggestions for strengthening the use of indicators in general within the federal government, be it through data collection techniques, fieldwork, or any other activity leading to the identification of new problems.

The details of such an unprecedented and far-reaching review are obviously beyond the bounds of this study. A logical first step might include requiring that a multi-agency task force establish some basic goals and perhaps even a framework for evaluating the information gathering and processing capabilities of individual agencies. Moreover, this initiative should work to challenge the propensity of government officials to fixate on only a handful of metrics. In other words, the initiative should encourage officials to think critically about the various trade-offs associated with selecting one problem measure over another. Unlike the social indicators movement, this process should not be dogmatic in terms of its commitment to quantitative metrics and should also consider a range of more contextual or qualitative metrics that might improve problem identification and program evaluation.

Building on the findings of this review, the task force should develop a series of best practices and recommendations that can be applied by all executive branch agencies. While such a process is unlikely to stop the strategic use of indicators for political gain, greater "indicator literacy" could help prevent wanton manipulation of these measures. This type of critical reflection will become even more important as we continue to move toward greater cross-departmental collaboration on important issues, like homeland security. Indeed, as May, Jochim, and Sapotichne note, subsystem actors, including bureaucrats, have wildly different perceptions of the meaning of preparedness, despite the fact that they technically work under the same "all-hazards" umbrella.[46] One can imagine this variance becomes even more extreme when we drill deeper and look at the types of metrics used to create and evaluate preparedness policies within different federal agencies. While uniformity is not the goal of this initiative, a renewed commitment and understanding of the information production process could very well improve agency performance and collaboration.

### Bolster the Capacity of Congress to Anticipate

Anticipation was once embedded in the institutional fabric of the U.S. Congress. Take the now-defunct federal Office of Technology Assessment (OTA) for example. OTA was created through the Technology Assessment Act of 1972 (P.L. 92–484) to provide Congress with objective analysis of scientific and technological change. As part of its mission, OTA was responsible for providing foresight, or early alerts, regarding new developments that could impact federal policy. OTA was a means for the institution to forecast changes and developments, especially emerging technologies, and, at least in theory, allowed Congress to enact policy change in anticipation of these events. OTA published more than 750 studies, covering topics from health care to global climate change, polygraphs to acid rain.[47]

An even more compelling example is the Congressional Clearinghouse for the Future, established in 1976 by members of Congress who wanted to integrate greater foresight into the policymaking process. The Congressional Clearinghouse for the Future was a Legislative Service Organization (LSO), which meant that interested members had to pool resources and time in order to support the entity. At one time, the Clearinghouse enjoyed a bipartisan membership of more than 100 members from both the House and Senate. The Clearinghouse actually proposed three separate pieces of legislation, including the Critical Trends Assessment Act (S. 1345), which sought to establish a permanent agency within the Executive Branch responsible for long term planning; the Global Resources, Environment and Population Act (S. 1025/H.R. 2491), which called for an interagency council to coordinate national planning for population and demographic change; and the Cost-Conscious Congress initiative, which asked executive agencies to conform to the Carter Administration's "Global 2000 Report," a multi-year study of the future implications of changing population, natural resource, and environmental trends.[48] None of these initiatives were able to mobilize the support needed for passage and all three died in the committee setting. Moreover, both OTA and the Clearinghouse were created during the 1970s, a period marked by noticeable interest in forecasting.[49] This interest in forecasting eventually waned and neither of these entities survived to the new century.

Based on the research in this book, it seems implausible that establishing a separate legislative entity for the purpose of anticipation could promote better foresight. The case studies show it is experts and specialists working in relatively narrow policy domains who often initiate and spearhead the anticipatory policy process. These individuals and groups have the competence to analyze complex indicators pertaining to emerging trends in their respective fields and, equally important, their professional, industrial, and corporate standing gives them legitimacy as policy participants. Their role, then, is just the opposite of generalists who would be broadly tasked with anticipating emerging issues in a multitude of fields under simultaneous observation.

With this fact of specialization in mind, my recommendation is to embed relevant experts within the structure of congressional committees and to charge them with the primary task of anticipating emerging trends and issues specific to the business of the committee to which they are attached. This notion of embedded expertise has been applied in other anticipatory governance models, the most notable example being Guston and Sarewitz's "real-time technology assessment" approach.[50] Real-time technology assessment places social scientists and philosophers within laboratory settings and tasks them with fostering discourse regarding the likely social and ethical implications of the products and technologies being developed. The basic premise underlying this model is that by fostering these discussions "upstream," critical problems might be identified during the R&D process, as opposed to waiting for unpleasant surprises after a product has already gone to market.[51]

Congressional anticipation specialists, an apt if awkward designation, should be hired as full-time committee staff and made responsible for identifying anticipatory problems, gathering empirical information relating to emerging problems, aiding in the analysis and interpretation of information provided by external stakeholders, and helping to forecast the long-term social implications of policy proposals. This information should be relayed, in turn, to committee members, who will thereby be better equipped to legislate in regard to anticipatory problems and to develop an orientation to the future as a fundamental aspect of their institutional role.

## Create a Culture of Preparedness and Anticipation

The gap between relief and preparedness—reactive and proactive—spending in this country is astronomical. One estimate holds that between 1985 and 2008 the federal government spent more than $82 billion on disaster relief programs but only $7.5 billion on preparedness.[52] Harkening back to the nanotechnology case, less than 8 percent of all National Nanotechnology Initiative funding in FY2015 was allocated for risk research. Moreover, medical treatment trumps prevention by a $1:99$ cents ratio.[53]

It stands to reason that closing these gaps would go a long way toward improving government's capacity to anticipate. Indeed, sustained and adequate funding is in many respects a baseline requirement for successful preparedness, prevention, and anticipatory programs. These activities are extraordinarily complex. But without financial support, they are nearly impossible.

This study implies achieving this end will require far more than simply adjusting our budgetary priorities. Elected officials underfund preparedness because, quite frankly, voters allow them to. In fact, voters *reward* them for doing so.[54] Of course, at a time when many Americans are struggling to manage their own futures, the prospect of using scarce tax dollars to fight problems that are years, if not decades, away is likely an unappealing option. And even among those Americans who do support these types of programs, many will likely harbor a degree of skepticism about the ability of government to fund programs that will, in fact, improve our readiness, resilience, and ability to anticipate. Foresight and strategic advantage, many believe, are best left to "nimble" private sector firms.

While changing these perceptions is no doubt a gargantuan task, a number of small steps could help remedy this overt indifference towards anticipation. For one, the American public needs to be reminded of the value of prevention, preparedness, and other anticipatory activities. Preparedness and prevention can save both lives and money. History implies relaying this information to the general public can go a long way toward changing opinion and behavior. From anti-smoking campaigns to discouraging drunk driving, the safety and disease domains—the very domains that stand to benefit most from this proposal—have historically been very effective in terms of their ability to use information instruments to change cultural norms.

Moreover, advertising and information campaigns should encourage citizens to become active participants in the preparedness process, not simply passive consumers of government information. In many respects, this movement is already underway. Established in 2003, the "Ready" program, a public advertising campaign run by the national government, aims to make the American public more proficient at and committed to preparing for hazards and disasters. The campaign is largely disseminated through online advertising and social media, and employs a distinctly all-hazards approach to preparedness, encouraging citizens to plan for a number of emerging threats, including terrorism, pandemics, natural disasters, and technological hazards.[55] The campaign has also partnered with entertainment companies to run a series of advertisements describing the importance of preparedness. In a recent campaign, which is targeted toward children, the popular Disney character "Big Hero" describes the key elements of emergency readiness.[56]

Creating a culture of preparedness and promoting individual preparedness from the "bottom up" will also improve communities' capacity to weather emergencies, as evidenced in the growing body work examining the relationship between social capital and community resilience.[57] The CDC's "Preparedness 101: Zombie Apocalypse" provides a useful model. Created in 2011, "Preparedness 101: Zombie Apocalypse" consists of a series of blogs and videos appearing on the CDC's website that use a hypothetical zombie apocalypse scenario to promote individual and household preparedness. As part of the program, viewers are asked to upload their own videos describing how they would prepare for a zombie apocalypse.[58] The program has grown into somewhat of a cultural phenomenon and, less than one week after launching "Zombie Apocalypse," the CDC's website crashed after receiving close to 3,000 hits. In fact, hours after the initial posting, the term "zombie apocalypse" was trending on Twitter.[59]

Cultural change is slow, but this book believes the more citizens value anticipation in their own lives, the more they will come to see it as an important government function. Increasing funding for preparedness and expanding federal programs, like the "Ready" campaign and the CDC's "zombie apocalypse," are great first steps toward cultivating a culture of preparedness and anticipation. However, these campaigns cannot rely on the national government alone. Additional resources need to be directed toward state and local governments, allowing them to craft their own outreach campaigns. States and localities are best equipped to contour programs to meet the unique cultural needs of their citizens. Most importantly, by empowering citizens to become active participants in the preparedness process, governments at all levels will enjoy greater community resilience and, ideally, increased voter demand for anticipatory policies.

## The Future of Anticipation

The assumption that human behavior is embedded within a larger temporal context is central to the study of public policy. Consider, for example, the study of policy change. Many scholars believe that, in order to explain how and why change occurs, one must examine domains over a period of ten to 50 years—if not longer.[60] This methodological assumption is derived from the understanding that actors, institutions, ideas, opinions, cultures and other causal variables evolve *across time*. As such, fixating on one particular moment cannot adequately capture this dynamism and is bound to provide a shallow description of the process of policy change.

The literature on lesson drawing, which is one of the few streams of research to directly underscore the importance of temporal variables, is also demonstrative of the importance of time. Policymakers readily draw "lessons from the past" in order to identify, frame, and advocate for preferred policy solutions.[61] As this study shows, lessons often take the form of analogies, which provide convenient anecdotes for drawing parallels between a present moment and a past experience. Similarly, the concept of policy learning assumes that the uptake of different ideas and lessons will vary across time. Birkland demonstrates the extent to which focusing events can induce learning. He ultimately concludes that very few ideas are original and most policies are recycled and hybridized, again, *across time*.[62]

Time, in other words, matters a great deal to the study of public policy. Yet, despite its importance, a number of scholars have documented a systematic devaluing of temporal variables, which are often ignored or reduced to mere methodological sidelights. Indeed, it is exceedingly rare for studies of policymaking and policy change to explicitly acknowledge and discuss the importance time. Pierson refers to this phenomenon as the temporal "decontextualization" of the social sciences, which tend to fixate on rapid evolving processes as opposed to those that play out over a long period of time. Writes Pierson:

> Identifying and exploring distinct mechanisms that generate striking temporal relationships—whether these involve positive feedback, sequencing, or slow-moving causes and outcomes—can make important contributions to a wide range of theoretical traditions. There are considerable insights to be derived from a focus on the temporal dimensions of social processes, irrespective of the specific claims an analyst may want to advance about the "major" drivers of social outcomes.[63]

Pollitt concurred with Pierson, arguing "strong temporal patterns are present in a great deal of policymaking and public management."[64] Pollitt advocates for both policy scholars *and* practitioners to "actively learn to live with the past" and acknowledge that many phenomenon take a great deal of time to unfold.[65]

This book is, first and foremost, intended to advance our understanding of policy change, particularly as it relates to the study of risks, hazards, and crises. However, in some respects, it also echoes the important themes introduced by Pollitt and Pierson, who boldly challenged theoretical orthodoxy and reasserted the importance of temporal variables. Not only does this book argue time matters, but the concept of an anticipatory policy type is founded on the assumption that time can constitute an explanatory variable in its own right. However, whereas many of my predecessors have looked to the past for explanations, this work looks to the future and considers its implications on policymaking and political behavior.

This book shows that the acute uncertainty associated with anticipatory problems can induce a policymaking process that differs in some very important ways from, for lack of a better term, a more reactive process. As described above, the distinctive temporal dimensions of anticipatory issues impact virtually every stage of the policymaking process, from the identification and definition of problems to the implementation of anticipatory policies and programs.

In light of these important findings, a number of future research directions can be identified. Perhaps most obviously, it only makes sense that a conceptual device founded on the assumption that time matters be applied in a variety of spatial contexts. Specifically, the anticipatory policymaking concept should be applied to political settings outside of the U.S. Rose's seminal work on lesson learning across time and space obviously exemplifies the importance of such a multidimensional approach.[66] It is notable that all three of the problems examined in this book were global in nature and transcended national boundaries. This is obviously a departure from much of the literature on disaster policymaking, which often focuses on fairly localized events (e.g., hurricanes, tornadoes, earthquakes, floods).

A blossoming stream of policy research has identified global and even transnational policy as the proverbial "next frontier" for theory building, allowing scholars to test the explanatory power of our concepts outside of the rigid confines of the "state."[67] Nor is this research purely conceptual. A great deal of policymaking is, in fact, occurring in supranational or transnational settings, often without any sustained involvement from government actors. For example, a laundry list of private companies and non-profit organizations have voluntary entered into binding climate change mitigation programs, including emissions capitation agreements.[68] This dynamic is obviously a departure from the gridlocked policymaking arrangement currently plaguing U.S. national institutions.

Moreover, international institutions like the UN and World Health Organization (WHO) play an important role in governing many of the problems examined in this book and therefore represent obvious case material for advancing a theory of anticipatory policymaking. A number of recent works have already broken this conceptual ground. For example, Jutta Joachim's 2007 book, *Agenda Setting, The UN, and NGOs: Gender Violence and Reproductive Rights*, uses social movement theory to elucidate patterns of agenda setting in the UN. To what extent can global and transnational actors and institutions create policies in anticipation of an emerging event? What sorts of instruments are available to risk regulators in these settings? Does the same aversion toward preparedness and proactive policy exist in global and transnational settings?

Moreover, the three cases examined in this book are but a sampling of the types of risks and hazards that might be construed as "anticipatory." In the public health domain alone, there are scores of emerging diseases that require planning and preparedness policymaking, including chikungunya, enterovirus D-68, West Nile Virus, various potential pandemic influenza strains, and, most recently, measles. Similarly, a number of "near miss" asteroid events imply some of the greatest threats to humankind might be lurking outside of our atmosphere. Much like climate change, identifying and diverting these "near-Earth objects" will require unprecedented global cooperation and investment. And in the U.S. the increasing threat of cyber attacks has stirred discussions about how government and private companies can anticipate and prepare for these incredibly costly and potentially dangerous data breaches.

Additional research might also test the theoretical export of the anticipatory policy type to topics beyond risks and hazards. The expansive field of social policy, for example, is facing a number of emerging crises that will likely necessitate anticipatory policy change. The threat of social security insolvency looms large in many industrialized nations coping with aging populations. Similar challenges will likely emerge in the health care domain or any other policy area with programs that use intergenerational distributions. In sum, it is plausible that the concept of anticipatory policymaking could inform these discussions as well.

In Ignacio Palacios-Huerta's 2014 edited volume, *In 100 Years: Leading Economists Predict the Future*, a group of world renowned economists set out to predict what the world will look like in 2114.[69] Not surprisingly, the esteemed group, which included a number of Nobel Prize winning scholars, predicted that many of the problems examined in this book will continue to plague humankind for decades to come. Unforeseen diseases will emerge and spread. Global warming will grow in severity. Technological innovations will proliferate.[70] Anticipatory problems, it seems, will abound well into the future.

Of course, Palacios-Huerta's book is the latest in a long line of attempts to forecast and understand the future, a tradition the editor traces to John Maynard Keynes' famous 1930 lecture on the future of the world in 2030.[71] But the desire to understand and more importantly shape our future transcends any one field of study and is in fact inherent to the human condition. This book shows that this same desire can even stimulate policy change and innovation.

Ironically, the same institutions that were designed to help us shape our collective destiny—our collective future—militate against virtually all forms of proactive social reform. Government is, by its very nature, reactive. Yet, as technological and social innovations continue to improve humankind's ability to foresee and forecast emerging threats, policymakers will no doubt face increased pressure to proactively govern. Only *time*, it seems, will tell whether government embraces an anticipatory future or remains wedded to a reactionary past.

## Notes

1 Paul 't Hart and Arjen Boin. 2001. "Between Crisis and Normalcy: The Long Shadow of Post-Crisis Politics." In *Managing Crises: Threats, Dilemmas, Opportunities*, eds. Uriel Rosenthal, Arjen Boin, and Louise Comfort. Springfield: Charles C. Thomas: 28–46.

2 Thomas A. Birkland. 2006. *Lessons of Disaster: Policy Change After Catastrophic Events*. Washington, D.C.: Georgetown University Press; Thomas A. Birkland. 1997. *After Disaster: Agenda Setting, Public Policy, and Focusing Events*. Washington, D.C.: Georgetown University Press.

3 Birkland 1997; Birkland 2006; John W. Kingdon. 2003. *Agendas, Alternatives, and Public Policies, 2nd Ed.* New York: Addison-Wesley Educational Publishers; Frank R. Baumgartner and Bryan D. Jones. 1993. *Agendas and Instability in American Politics*. Chicago: University of Chicago Press.

4 Birkland 2006, 165.

5 Birkland 1997; Birkland 2006; Daniel Nohrstedt. 2008. "The Politics of Crisis Policymaking: Chernobyl and Swedish Nuclear Energy Policy." *Policy Studies Journal* 36(2): 257–278.

6 Scholars of disaster policymaking concede this fact. For example, as indicated in Chapter 2, Birkland (1997) described the policymaking process surrounding the Three Mile Island nuclear disaster as being conjectural, a testament to the fact that no one was actually hurt by this near miss event. As such, policymakers had to speculate and conjecture what an actual meltdown might look like in terms of its human and economic tool. This dynamic is obviously magnified greatly in the case of anticipatory policymaking, which is largely devoid of any event, including a "near miss."

7 Birkland 1997, 2006; Kingdon 2003; Nohrstedt 2008.

8 Indicators are critical to this process, as described in greater detail below, in the agenda setting section.

9 Annika Brändström, Fredrik Bynander, and Paul 't Hart. 2004. "Governing By Looking Back: Historical Analogy and Crisis Management." *Public Administration* 82(1): 191–219.

10 Interestingly, Brändström, Bynander, and 't Hart (2004) provide an example of anticipatory policymaking, although they do not define it as such. They note that following the 1999 Austrian parliamentary elections, many feared an extreme right wing party, the Freiheitliches Partei Osterreich (FPO), would enter into a governing coalition with the Christian-democratic party. In anticipation of what some saw as a potential political crisis, many European policymakers analogized the FPO to the Nazi party and their leader to Adolf Hitler. The authors note that problem definers hoped to project "a troublesome future from the analogy with a catastrophic past" (197).

11 Richard Rose. 1993. *Lesson-Drawing in Public Policy: A Guide to Learning Across Time and Space*. Chatham: Chatham House Publishers.
12 Ibid., 91.
13 Daniel Nohrstedt and Christopher M. Weible. 2010. "The Logic of Policy Change After Crisis: Proximity and Subsystem Interaction." *Risk, Hazards, & Crisis in Public Policy* 1(2): 1–32; Sarah B. Pralle. 2006. *Branching Out, Digging In: Environmental Advocacy and Agenda Setting*. Washington, D.C.: Georgetown University Press.
14 Sarah Pralle. 2006. "The 'Mouse That Roared': Agenda Setting in Canadian Pesticides Politics." *Policy Studies Journal* 34(2): 171–194.
15 As a point of fact, the National Aeronautics and Space Administration (NASA), Department of Defense (DOD), and Department of Energy (DOE) have considered regulations as well.
16 Thomas A. Birkland. 2010. "Federal Disaster Policy: Learning, Priorities, and Prospects for Resilience." In *Designing Resilience: Preparing for Extreme Events*, eds. Louise K. Comfort, Arjen Boin, and Chris C. Demchak. Pittsburgh: University of Pittsburgh Press: 106–128; Rebecca Katz. 2011. *Essentials of Public Health Preparedness*. Sudbury: Jones & Bartlett Learning.
17 Birkland 1997.
18 Deborah Stone. 2002. *Policy Paradox: The Art of Political Decision Making*. New York: W.W. Norton: 191.
19 Birkland 1997; Peter J. May, Ashley E. Jochim, and Joshua Sapotichne. 2011. "Constructing Homeland Security: An Anemic Policy Regime." *Policy Studies Journal* 39(2): 285–307.
20 Birkland 1997; Thomas A. Birkland. 2004. "Learning and Policy Improvement After Disaster." *American Behavioral Scientist* 48(3): 341–364.
21 Jaime Sainz-Santamaria and Sarah Anderson. 2013. "The Electoral Politics of Disaster Preparedness. *Risk, Hazards, & Crisis in Public Policy* 4(4): 234–249.
22 Andrew Healy and Neil Malhotra. 2009. "Myopic Voters and Natural Disaster Policy." *American Political Science Review* 103(3): 387–406.
23 Brian J. Gerber and Grant W. Neeley. 2005. "Perceived Risk and Citizen Preferences for Governmental Management of Routine Hazards." *Policy Studies Journal* 33(3): 395–418.
24 David A. Moss. 2002. *When All Else Fails: Government as the Ultimate Risk Manager*. Cambridge: Harvard University Press; Thomas A. Birkland. 2009. "Disasters, Catastrophes, and Policy Failure in the Homeland Security Era." *Review of Policy Research* 26(4): 423–438; Thomas A. Birkland and Sarah Waterman. 2008. "Is Federalism the Reason for Policy Failure in Hurricane Katrina?" *Publius* 38(4): 692–714.
25 Michael Howlett and M. Ramesh. 2003. *Studying Public Policy: Policy Cycles and Policy Subsystems, 2nd Ed*. London: Oxford University Press.
26 Of course, this research was supported by federal grant money, which is a treasure instrument.
27 Peter J. May and Chris Koski. 2013. "Addressing Public Risks: Extreme Events and Critical Infrastructures." *Review of Policy Research* 30(2): 139–159, 139.
28 Jeremy J. Richardson, Gunnel Gustafsson, and Grant A. Jordan. 1982. "The Concept of Policy Style." In *Policy Styles in Western Europe*, ed. Jeremy Richardson, Herts: George & Unwin: 1–16.
29 Birkland 2006.
30 Thomas A. Birkland. 2004. "The World Changed Today: Agenda-Setting and Policy Change in the Wake of the September 11 Terrorist Attacks." *Review of Policy Research* 21(2): 179–200.
31 Baumgartner and Jones 1993.
32 Eric M. Patashnik. 2008. *Reforms at Risk: What Happens After Major Policy Changes Are Enacted*. Princeton: Princeton University Press.

33 Birkland 1997, 2006.
34 Birkland 2006.
35 Demetrious James Caraley. 2004. *American Hegemony: Preventing War, Iraq, and Imposing Democracy.* New York: American Academy of Political Science; Eli Clifton. 2012. "Nine Years Since the Beginning of the Iraq War." *Think Progress*, March 19. Available at: http://library.williams.edu/citing/styles/chicago2.php (accessed March 28, 2015).
36 "Chirac: Iraq War Undermined U.N." *CNN*, September 23. Available at http://articles.cnn.com/2003–09–23/us/sprj.irq.annan_1_iraq-policy-iraq-war-invasion?_s=PM:US (accessed March 28, 2015).
37 Centers for Disease Control and Prevention. 2015. "Countries With Widespread Transmission." March 14. Available at: www.cdc.gov/vhf/ebola/outbreaks/2014-west-africa/case-counts.html (accessed March 28, 2015).
38 Centers for Disease Control and Prevention. 2014. "Interim U.S. Guidance for Monitoring and Movement of Persons with Potential Ebola Virus Exposure." December 24. Available at www.cdc.gov/vhf/ebola/exposure/monitoring-and-movement-of-persons-with-exposure.html (accessed March 28, 2015).
39 Holly Yan and Greg Botelho. 2014. "Ebola: Some U.S. States Announce Mandatory Quarantines—Now What?" *CNN* October 27. Available at www.cnn.com/2014/10/27/health/ebola-us-quarantine-controversy/ (accessed March 28, 2015).
40 Dan Friedman. 2014. "CDC Issues New Guidelines for Medical Workers Who Treated Ebola Patients in West Africa." *New York Daily News*, October 27. Available at www.nydailynews.com/life-style/health/cdc-issues-new-guidelines-medical-workers-treated-ebola-patients-article-1.1989314 (accessed March 28, 2015).
41 Charles Krauthammer. 2009. "Preventative Care Isn't the Magic Bullet for Health Care Costs." *Washington Post*, August 14. Available at www.washingtonpost.com/wp dyn/content/article/2009/08/13/AR2009081302898.html (accessed March 28, 2015).
42 Kris Coyner. 2011. "Preventative Care Can Cut Costly Emergency Room Visits." *The Olympian*, January 28. Available at www.theolympian.com/2011/01/28/1522894/preventative-care-can-cut-costly.html (accessed March 28, 2015).
43 Gilbert H. Welch, Lisa Schwartzl, and Steve Woloshin. 2011. *Over-Diagnosed: Making People Sick in the Pursuit of Health.* Boston: Beacon Press.
44 Ibid., 88.
45 Clifford W. Cobb and Craig Rixford. 1998. *Lessons Learned from the History of Social Indicators.* San Francisco: Redefining Progress.
46 May et al. 2011.
47 Gerald L. Epstein. 2009. "Restart the Congressional Office of Technology Assessment." *Science Progress*, March 31. Available at www.fas.org/ota/2009/03/31/restart-the-congressional-office-of-technology-assessment/ (accessed March 28, 2015); "The OTA Legacy." 2012. Princeton University Archives of the Office of Technology Assessment. Available at www.princeton.edu/~ota/ (accessed March 28, 2015).
48 Woodrow Wilson International Center for Scholars. 2002. "Congressional Clearinghouse on the Future." October 1. Available at www.wilsoncenter.org/article/congressional-clearinghouse-the-future (accessed March 28, 2015).
49 Clement Bezold. 2006. "Anticipatory Democracy Revisited." Paper Prepared for the Finnish Parliament's 100 Year Anniversary: 38–51.
50 David H. Guston and Daniel Sarewitz. 2002. "Real-Time Technology Assessment." *Technology in Society* 23(4): 93–109.
51 Ibid.
52 Sainz-Santamaria and Anderson 2013.

53 Halley S. Faust. 2005. "Prevention vs. Cure: Which Takes Precedence?" *Medscape Public Health & Prevention*, 3 (May).

54 Healy and Neil Malhotra 2009.

55 See: www.ready.gov.

56 See: www.youtube.com/watch?v=nK1XcIp6hrk.

57 Daniel P. Aldrich. 2012. *Building Resilience: Social Capital in Post-Disaster Recovery*. Chicago: University of Chicago Press.

58 Centers for Disease Control and Prevention (CDC). 2011. "Preparedness 101: Zombie Apocalypse." May 16. Available at http://blogs.cdc.gov/publichealth matters/2011/05/preparedness-101-zombie-apocalypse/ (accessed March 28, 2015).

59 Mike Stobbe. 2011. "CDC's Zombie Apocalypse Advice An Internet Hit." *Huffington Post*. May 20. Available at www.huffingtonpost.com/2011/05/21/ zombie-apocalypse-advice-cdc_n_865078.html (accessed March 28, 2015).

60 Birkland 1997, 2006.

61 Rose 1993.

62 Birkland 2006.

63 Paul Pierson. 2004. *Politics in Time: History, Institutions, and Social Analysis*. Princeton: Princeton University Press: 176.

64 Christopher Pollitt. 2008. *Time, Policy, Management: Governing With the Past*. Oxford: Oxford University Press: 181.

65 Ibid.

66 Rose 1993.

67 Diane Stone. 2008. "Global Public Policy, Transnational Policy Communities, and their Networks." *Policy Studies Journal* 36(1): 19–38.

68 Liliana B. Andonova, Michele M. Betsill, and Harriet Bulkely. 2009. "Transnational Climate Governance." *Global Environmental Politics* 9(2): 52–73.

69 Ignacio Palacios-Huerta. 2014. *In 100 Years: Leading Economists Predict the Future*. Boston: The MIT Press.

70 Ibid.

71 Ibid.

# References

Achenbach, Joel and Juliet Eilperin. 2011. "Climate-change Science Makes for Hot Politics." *Washington Post*, August 19. Available at www.washingtonpost.com/national/health-science/climate-change-science-makes-for-hot politics/2011/08/18/gIQA1eZJQJ_story.html?hpid=z1 (accessed March 28, 2015).

Adelsman, Hedia and Joanna Ekrem. 2012. "Preparing for Climate Change: Washington State's Integrated Response Strategy." April, Georgetown Climate Center. Available at www.georgetownclimate.org/resources/preparing-for-a-changing-climate-washington-states-integrated-climate-response-strategy (accessed March 28, 2015).

Adragna, Anthony. 2015. "Rhetoric on EPA, Climate Show Divisions Among Republican Presidential Hopefuls." *Bloomberg BNA*, January 27. Available at www.bna.com/rhetoric-epa-climate-n17179922496/ (accessed March 28, 2015).

Aldred, Jessica. 2007. "Timeline: Al Gore." *Guardian*, October 12. Available at www.guardian.co.uk/environment/2007/oct/12/climatechange1 (accessed March 28, 2015).

Aldrich, Daniel P. 2012. *Building Resilience: Social Capital in Post-Disaster Recovery*. Chicago: University of Chicago Press.

Allen, Mike. 2005. "Bush v. Bird Flu." *Time*, November 1. Available at www.time.com/time/nation/article/0,8599,1125104,00.html (accessed March 29, 2015).

Alonso-Zaldivar, Ricardo. 2005. "Bush's Flu Plan Stresses Vaccine." *Los Angeles Times*, November 2, 2.

Alphonso, Gordon R. E. Brett Breitschwerdt, Neal J. Cabral, James Yancey Kerr, David L. Rieser, and Patricia F. Sharkey. 2010. "EPA Issues First-Ever Rule Regulating GHG Emissions from Light Duty Vehicles, Confirms GHG Stationary Source Permitting Begins January 2011." *Martindale.com*, April 14. Available at www.martindale.com/environmental-law/article_McGuireWoods-LLP_979678.htm (accessed March 28, 2015).

Altman, Lawrence. 1997. "When a Novel Flu is Involved, Health Officials Get Jumpy." *New York Times*, December 30, F7.

American Bar Association. 2006. "Regulation of Nanoscale Materials under the Toxic Substances Control Act." Section of Environment, Energy, and Resources. June. Available at www.americanbar.org/content/dam/aba/migrated/environ/nanotech/pdf/TSCA.authcheckdam.pdf (accessed March 27, 2015).

American Civil Liberties Union. 2002. "Model State Health Emergency Powers Act." January 1. Available at www.aclu.org/technology-and-liberty/model-state-emergency-health-powers-act (accessed March 28, 2015).

American Civil Liberties Union. 2009. "Maintaining Civil Liberties Protections in Response to the H1N1 Flu." An ACLU White Paper. November: 7. Available at www.aclu.org/files/assets/H1N1_Report_FINAL.pdf (accessed March 28, 2015).

Anderson, Glen and David Sullivan. 2009. "Reducing Greenhouse Gas Emissions: Carbon Cap and Trade and the Carbon Tax." National Conference of State Legislatures, July. Available at www.ncsl.org/documents/environ/captrade.pdf (accessed March 28, 2015).

Anderson, James E. 2003. *Public Policymaking, 5th Edition.* Boston: Houghton Mifflin Company.

Andonova, Liliana B., Michele M. Betsill, and Harriet Bulkely. 2009. "Transnational Climate Governance." *Global Environmental Politics* 9(2): 52–73.

Anson, Robert Sam. 1999. "The Y2K Nightmare." *Vanity Fair* 80.

Antonelli, Angela and Bett D. Schaefer. 1998. "Why Kyoto Signing Signals Disregard For Congress." The Heritage Foundation, November 23. Available at www.heritage.org/research/reports/1998/11/why-the-kyoto-signing-signals-disregard-for-congress (accessed March 28, 2015).

Arkansas Department of Health. 2011. "Epidemiology." Available at www.healthy.arkansas.gov/programsServices/epidemiology/Pages/default.aspx (accessed March 28, 2015).

Atkin, Emily. 2014. "What Climate Scientists Have to Say About Obama's Deal With China." *Climate Progress*, November 12. Available at: http://thinkprogress.org/climate/2014/11/12/3591341/china-climate-deal-scientists-reaction/ (accessed March 28, 2015).

"Avian flu, Bioterror and Washington." 2005. *Washington Times*, September 29, A22.

Banerjee, Neela. 2012. "U.S. Court Upholds EPA's Authority to Regulate Greenhouse Gases." *Los Angeles Times*, June 26. Available at http://articles.latimes.com/2012/jun/26/business/la-fi-epa-court20120627 (accessed March 28, 2015).

Bardach, Eugene. 1977. *The Implementation Game: What Happens After a Bill Becomes a Law.* Cambridge: MIT Press.

Barringer, Felicity. 2006. "Officials Reach California Deal to Cut Emissions." *New York Times*, August 30. Available at www.nytimes.com/2006/08/31/washington/31warming.html?pagewanted=print (accessed March 28, 2015).

Barry, John M. 2004. *The Great Influenza: The Story of the Deadliest Pandemic in History.* New York: Viking Press.

Baum, Rudy. 2003. "Nanotechnology: Drexler and Smalley Make the Case For and Against 'Molecular Assemblers.'" *Chemical & Engineering News* 81(48): 37–42.

Baumgartner, Frank R. and Bryan D. Jones. 1993. *Agendas and Instability in American Politics.* Chicago: University of Chicago Press.

Baumgartner, Frank R. and Bryan D. Jones. 2002. "Positive and Negative Feedback in Politics." In *Policy Dynamics*, eds. Frank R. Baumgartner and Bryan D. Jones. Chicago: University of Chicago Press: 3–28.

Baumgartner, Frank R., Bryan D. Jones, and John D. Wilkerson. 2002. "Studying Policy Dynamics." In *Policy Dynamics*, eds. Frank R. Baumgartner and Bryan D. Jones. Chicago: University of Chicago Press: 29–46.

BCC Research. 2010. "Nanotechnology: A Realistic Market Assessment." July. Available at www.bccresearch.com/market-research/nanotechnology/nanotechnology-realistic-market-assessment-nan031d.html (accessed March 27, 2015).

Beillo, David. 2007. "Climate Change's Uncertainty Principle." *Scientific American*, November 29. Available at www.zcommunications.org/climate-changes-uncertainty-principle-by-david-biello (accessed March 28, 2015).

Benen, Steven. 2014. "Pentagon: Climate Crisis An Immediate Security Threat." *MSNBC*, October 14. Available at www.msnbc.com/rachel-maddow-show/ pentagon-climate-crisis-immediate-security-threat (accessed March 28, 2015).

Bennet, Colin J. 1991. "What Is Policy Convergence and What Causes It?" *British Journal of Political Science* 21(2): 215–233.

Berger, Michael. 2007. "Regulating Nanotechnology: Incremental Approach or New Regulatory Framework?" *Nanowerk*, June 5. Available at www.nanowerk. com/spotlight/spotid=2027.php (accessed March 27, 2015).

Bergeson, Lynn L. 2009. "EPA Publishes NMSP Interim Report." Nano and Other Emerging Chemical Technologies Blog, Bergeson & Campbell, PC, January 14. Available at http://nanotech.lawbc.com/2009/01/articles/united-states/federal/ epa-publishes-nmsp-interim-report/ (accessed March 27, 2015).

Bergeson, Lynn L. 2014. "EPA Withdraws Direct Final SNUR for Functionalized Carbon Nanotubes." Nano and Other Emerging Chemical Technologies Blog. Bergeson & Campbell, PC, December 22. Available at http://nanotech.lawbc. com/2014/12/articles/united-states/federal/epa-withdraws-direct-final-snur-for-functionalized-carbon-nanotubes-generic/ (accessed March 27, 2015).

Bergeson, Lynn L. 2014b. "TSCA Reform a Viable Contender For Serious Legislative Attention Next Year." *JD Supra Business Advisor*. December 4. Available at www.jdsupra.com/legalnews/tsca-reform-a-viable-contender-for-serio-04794/ (accessed March 27, 2015).

Berke, Philip R. and Timothy Beatley. 1992. *Planning for Earthquakes: Risks, Politics, and Policy*. Baltimore: The Johns Hopkins University Press.

"Berkeley to be First City to Regulate Nanotechnology." 2006. *New York Times*, November 12. Available at www.nytimes.com/2006/12/12/technology/12iht-nano. 3870331.html?_r=0 (accessed March 27, 2015).

Berube, David M. 2005. *Nano-Hype: The Truth Behind the Nanotechnology Buzz*. Amherst: Prometheus Books.

Besel, Richard D. 2013. "Accommodating Climate Change Science: James Hansen and the Rhetorical/Political Emergence of Global Warming." *Science in Context* 26(1): 135–152.

Bevins, Sue. 2002. "The Model State Health Emergency Powers Act: An Assault on Civil Liberties in the Name of Homeland Security." The Heritage Foundation. June 10. Available at www.heritage.org/research/lecture/the-model-state-emergency-health-powers-act (accessed March 28, 2015).

Beyond Pesticides. 2014. "Lawsuit Challenges EPA's Failure to Regulate Nano-material Pesticides." Daily News Blog. Available at www.beyondpesticides.org/ dailynewsblog/?p=14685 (accessed March 27, 2015).

Bezold, Clement. 2006. "Anticipatory Democracy Revisited." Paper Prepared for the Finnish Parliament's 100 Year Anniversary: 38–51.

Birnbaum, Jeffrey H. 2005. "Vaccine Funding Tied to Liability." *Washington Post*, November 17. Available at www.washingtonpost.com/wp-dyn/content/article/ 2005/11/16/AR2005111602238.html (accessed March 28, 2015).

Birkland, Thomas A. 1997. *After Disaster: Agenda Setting, Public Policy, and Focusing Events*. Washington, D.C.: Georgetown University Press.

Birkland, Thomas A. 2004. "Learning and Policy Improvement After Disaster." *American Behavioral Scientist* 48(3): 341–364.

Birkland, Thomas A. 2004b. "The World Changed Today: Agenda-Setting and Policy Change in the Wake of the September 11 Terrorist Attacks." *Review of Policy Research* 21(2): 179–200.

Birkland, Thomas A. 2006. *Lessons of Disaster: Policy Change After Catastrophic Events*. Washington, D.C.: Georgetown University Press.

Birkland, Thomas A. 2009. "Disasters, Catastrophes, and Policy Failure in the Homeland Security Era." *Review of Policy Research* 26(4): 423–438.

Birkland, Thomas A. 2010. "Federal Disaster Policy: Learning, Priorities, and Prospects for Resilience." In *Designing Resilience: Preparing for Extreme Events*, eds. Louise K. Comfort, Arjen Boin, and Chris C. Demchak. Pittsburgh: University of Pittsburgh Press: 106–128.

Birkland, Thomas A. and Sarah Waterman. 2008. "Is Federalism the Reason for Policy Failure in Hurricane Katrina?" *Publius* 38(4): 692–714.

Boin, Arjen, Louise K. Comfort, and Chris C. Demchak. 2010. "The Rise of Resilience." In *Designing Resilience: Preparing for Extreme Events*, eds. Louise K. Comfort, Arjen Boin, and Chris C. Demchak. Pittsburgh: University of Pittsburgh Press: 1–12.

Boin, Arjen and Paul 't Hart. 2003. "Public Leadership in Times of Crisis: Mission Impossible?" *Public Administration Review* 63(5): 544–553.

Boin, Arjen, Paul 't Hart, Eric Stern, and Bengt Sundelius. 2005. *The Politics of Crisis Management: Public Leadership under Pressure*. Cambridge: Cambridge University Press.

Bonan, Gordon B. 2008. "Forests and Climate Change: Forcing, Feedbacks, and the Climate Benefits of Forests." *Science* 320 (5882): 1444–1449.

Bosso, Christopher J. 1987. *Pesticides and Politics: The Life Cycle of a Public Issue*. Pittsburgh: University of Pittsburgh Press.

Brändström, Annika, Fredrik Bynander, and Paul 't Hart. 2004. "Governing By Looking Back: Historical Analogy and Crisis Management." *Public Administration* 82(1): 191–219.

Bravender, Robin. 2010. "EPA Issues Final 'Tailoring' Rule for Greenhouse Gas Emissions." *New York Times*, May 13. Available at www.nytimes.com/gwire/2010/05/13/13greenwire-epa-issues-final-tailoring-rule-for-greenhouse-32021.html (accessed March 28, 2015).

Breiman, Robert F., Abdulsalami Nasidi, Mark A. Katz, and John Vertefeuille. 2007. "Preparedness for Highly Pathogenic Avian Influenza Pandemic in Africa." *Emerging Infectious Diseases* 13(10): 1453–1458.

Bristol, Nellie. 2005. "US President Releases Long-awaited Preparedness Plan." *The Lancet* 366(9498): 1683.

Broder, John M. and Matthew L. Ward. 2010. "Clashes Loom, With E.P.A. Set To Limit Gases." *New York Times*, December 31, A01.

Brown, David. 2005. "Military's Role in a Flu Pandemic." *Washington Post*, October 5, A05.

Burbank, Cynthia J. and Parson Brinckerhoff. 2009. *Strategies for Reducing the Impact of Surface Transportation on Global Climate Change: A Synthesis of Policy Research and State and Local Mitigation Strategies*. American Association of State Highway and Transportation Officials (AASHTO). Available at http://climatechange.transportation.org/pdf/nchrp_2024_59_final_report_031309.pdf (accessed March 28, 2015).

Burby, Raymond J. and Peter J. May. 1997. *Making Governments Plan: State Experiments in Managing Land Use*. Baltimore: The Johns Hopkins University Press.

Burby, Raymond J. and Peter J. May. 1998. "Intergovernmental Environmental Planning: Addressing the Commitment Conundrum." *Journal of Environmental Planning and Management* 41(1): 95–110.

Burnstein, Paul. 1991. "Policy Domains: Organization, Culture, and Policy Outcomes." *Annual Review of Sociology* 17: 327–350.

"Bush Pandemic Flu Strategy: Detection, Treatment, Response." 2005. *Environmental News Service*, November 1. Available at www.ens-newswire.com/ens/nov 2005/2005-11-01-01.html (accessed March 28, 2015).

"Bush Unveils $7.1 Billion Plan to Prepare for Flu Pandemic." 2005. *CNN*, November 2. Available at http://articles.cnn.com/2005-11-01/health/us.flu.plan_1pandemic-strain-vaccine-makers-flu-pandemic?_s=PM:HEALTH. (accessed March 28, 2015).

Butler, Declan. 2012. "Death-rate Row Blurs Mutant Flu Debate." *Science*, February 13. Available at www.nature.com/news/death-rate-row-blurs-mutant-flu-debate-1.10022 (accessed March 28, 2015).

Butler, Declan. 2012b. "Flu Surveillance Lacking." *Nature*, March 28. Available at www.nature.com/news/flusurveillance-lacking-1.10301 (accessed March 28, 2015).

Calef, David and Robert Goble. 2007. "The Allure of Technology: How France and California Promoted Electric and Hybrid Vehicles to Reduce Urban Air Pollution." *Policy Sciences* 40: 1–34.

Calmes, Jackie and Donald G. McNeil Jr. 2009. "H1N1 Widespread in 46 States as Vaccines Lag." *New York Times*, October 24, A1.

Cameron, Charley. 2012. "New Report Finds 3.7 Million U.S. Residents at Risk from Rising Sea Levels." *Inhabitat*, March 14. Available at http://inhabitat.com/new-report-finds-3-7-million-u-s-residents-at-risk-from-rising-sea-levels/(accessed March 28, 2015).

Caraley, Demetrious James. 2004. *American Hegemony: Preventing War, Iraq, and Imposing Democracy*. New York: American Academy of Political Science.

Carlson, Ann E. 2009. "Iterative Federalism and Climate Change." *Northwestern University Law Review* 103(3): 1097–1162.

Cass, Loren R. 2006. *The Failures of American and European Climate Policy: International Norms, Domestic Politics, And Unachievable Commitments.* Albany: State University of New York Press.

Center for Climate and Energy Solutions. 2008. "Legislation in the 110th Congress Related to Global Climate Change." Available at www.c2es.org/federal/congress/ 110 (accessed March 28, 2015).

Center for Climate and Energy Solutions. 2010. "Lugar Practical Energy and Climate Plan (S.3464)." Available at www.c2es.org/federal/congress/111/lugar-practical-energy-climate-plan (accessed March 28, 2015).

Center for Climate and Energy Solutions. 2012. "Climate Change Adaptation: What Federal Agencies Are Doing." February. Available at www.c2es.org/docUploads/federal-agencies-adaptation.pdf (accessed March 28, 2015).

Center for Climate and Energy Solutions. 2012b. "Climate Debate in Congress." Available at www.c2es.org/federal/congress (accessed March 28, 2015).

Center for Climate and Energy Solutions. 2012c. "Legislation in the 109th Congress Related to Global Climate Change." Available at www.c2es.org/federal/congress/109 (accessed March 28, 2015).

Center for Climate and Energy Solutions. 2012d. "109th Congress Index of Proposals." Available at www.c2es.org/federal/congress/109/bills_index (accessed March 28, 2015).

Center for Climate and Energy Solutions. 2014. "Clean Air Act Cases." Available at: www.c2es.org/federal/courts/clean-air-act-cases (accessed March 28, 2015).

Center for Climate and Energy Solutions. 2014b. "Western Climate Initiative." Available at www.c2es.org/us-states-regions/regional-climate-initiatives/western-climate-initiative (accessed March 28, 2015).

Center for Climate and Energy Solutions. 2015. "State and Local Climate Adaptation." Available at www.c2es.org/us-states-regions/policy-maps/adaptation (accessed March 28, 2015).

Center for Responsible Nanotechnology. 2005. "Nanobots Not Needed." Briefing Document, March 2. Available at www.crnano.org/BD-Nanobots.htm (accessed March 27, 2015).

Centers for Disease Control and Prevention. 2008. "Avian Influenza A Virus Infections of Humans." Available at www.cdc.gov/flu/avian/gen-info/avian-flu-humans.htm/. (accessed March 28, 2015).

Centers for Disease Control and Prevention. 2009. "CDC Estimates of 2009 H1N1 Influenza Cases, Hospitalizations, and Deaths in the United States, April-November 14, 2009." December 10. Available at www.cdc.gov/h1n1flu/estimates/April_November_14.htm (accessed March 28, 2015).

Centers for Disease Control and Prevention. 2009b. "WHO Pandemic Declaration." Available at www.cdc.gov/h1n1flu/who/ (accessed March 28, 2015).

Centers for Disease Control and Prevention. 2010. "Key Facts About Avian Influenza (Bird Flu) and Highly Pathogenic Avian Influenza A (H5N1) Virus." Available at www.cdc.gov/flu/avian/gen-info/facts.htm (accessed March 28, 2015).

Centers for Disease Control and Prevention. 2011. "Nanotechnology." Available at www.cdc.gov/niosh/topics/nanotech/ (accessed March 27, 2015).

Centers for Disease Control and Prevention. 2011b. "Preparedness 101: Zombie Apocalypse." May 16. Available at http://blogs.cdc.gov/publichealthmatters/2011/05/preparedness-101-zombie-apocalypse/ (accessed March 28, 2015).

Centers for Disease Control and Prevention. 2014. "Interim U.S. Guidance for Monitoring and Movement of Persons with Potential Ebola Virus Exposure." December 24. Available at www.cdc.gov/vhf/ebola/exposure/monitoring-and-movement-of-persons-with-exposure.html (accessed March 28, 2015).

Centers for Disease Control and Prevention. 2015. "Countries With Widespread Transmission." March 14. Available at www.cdc.gov/vhf/ebola/outbreaks/2014-west-africa/case-counts.html (accessed March 28, 2015).

"Chairman Rush, Waxman Release H.R. 5820, The Toxic Chemicals Safety Act." 2010. Committee on Energy and Commerce, July. Available at 22.http://democrats.energycommerce.house.gov/index.php?q=news/chairmen-rush-waxman-release-hr-5820-the-toxic-chemicals-safety-act (accessed March 27, 2015).

Chappell, Linda M. 2010. "Regulatory Impact Analysis for the Final Prevention of Significant Deterioration and Title V Greenhouse Gas Tailoring Rule." U.S. Environmental Protection Agency, Office of Air Quality Planning and Standards, Health and Environmental Impacts Division. Final Report, May. Available at www.epa.gov/ttn/ecas/regdata/RIAs/riatailoring.pdf (accessed March 28, 2015).

Chesto, Jon 2014. "Deval Patrick's Renewable Revolution." *Boston Business Journal*, September 26. Available at www.bizjournals.com/boston/print-edition/2014/09/26/deval-patrick-s-renewable-revolution.html?page=all (accessed March 28, 2015).

"China Accused of SARS Cover-up." 2003. *BBC News*, April 9. Available at http://news.bbc.co.uk/2/hi/health/2932319.stm (accessed March 28, 2015).

"Chirac: Iraq War Undermined U.N." CNN, September 23. Available at http://articles.cnn.com/2003-09-23/us/sprj.irq.annan_1_iraq-policy-iraq-war-invasion?_s=PM:US (accessed March 28, 2015).

City of Chicago. *Chicago Climate Action Plan*. Available at www.chicagoclimate-action.org/pages/adaptation/11.php (accessed March 28, 2015).

City of Long Beach California Department of Health and Human Services. 2012. "Avian/Pandemic Influenza Information Page." Available at www.longbeach. gov/health/influenza.asp (accessed March 28, 2015).

"City Residents Vote to Tax Selves for Carbon Use." 2006. *MSNBC News*, November 10. Available at www.nbcnews.com/id/15651688/ns/us_news-environment/t/city-residents-vote-tax-selves-carbon-use/#.VRNcC0trWlJ (accessed March 28, 2015).

Clifton, Eli. 2012. "Nine Years Since the Beginning of the Iraq War." *Think Progress*, March 19. Available at http://library.williams.edu/citing/styles/chicago2. php (accessed March 28, 2015).

Climate Change Adaptation Task Force. 2010. *Progress Report of the Interagency Climate Change Adaptation Task Force: Recommended Actions in Support of a National Climate Change Adaptation Strategy.* October 5, The White House Council on Environmental Quality. Available at www.whitehouse.gov/sites/ default/files/microsites/ceq/Interagency-Climate-Change-Adaptation-Progress-Report.pdf (accessed March 28, 2015).

Climate Change Adaptation Task Force. 2011. *Federal Actions for a Climate Resilient Nation: Progress Report of the Interagency Climate Change Adaptation Task Force.* October 28. Available at www.whitehouse.gov/sites/default/ files/microsites/ceq/2011_adaptation_progress_report.pdf (accessed March 28, 2015).

Climate Institute. 2010. "Oceans & Sea Level Rise." Available at www.climate. org/topics/sea-level/ (accessed March 28, 2015).

"Clinton Urges Americans to Act on Y2K Problem." 1998. *CNN*, July 14. Available at www.cnn.com/ALLPOLITICS/1998/07/14/clinton.y2k/ (accessed March 27, 2015).

"CNN Takes In-depth Look at Bird Flu." 2005. *CNN*, November 2. Available at http://articles.cnn.com/2005-11-02/health/birdflu.tv_1_h5n1-bird-flu-human-to human?_s=PM:HEALTH. (accessed March 28, 2015).

Coalition for American Jobs. 2010. "Stationary Source GHG Regulation Under PSD." March 5. Available at http://coalitionforamericanjobs.com/2010/03/fact-sheet-stationary-source-ghg-regulation-under-psd/ (accessed March 28, 2015).

Cobb, Clifford and Craig Rixford. 1998. *Lessons Learned from the History of Social Indicators.* San Francisco: Redefining Progress.

Cobb, Roger W. and Charles D. Elder. 1977. *Participation in American Politics: The Dynamics of Agenda Building.* Baltimore: The Johns Hopkins University Press.

Cobb, Roger W. and Marc Howard Ross. 1997. "Agenda Setting and the Denial of Agenda Access: Key Concepts." In *Cultural Strategies of Agenda Denial: Avoidance, Attack, and Redefinition*, eds. Roger W. Cobb and Marc Howard Ross. Kansas: University of Kansas Press: 3–24.

Coffee, Joyce E., Julia Parzen, Mark Wagstaff, and Richard S. Lewis. 2010. "Preparing for a Changing Climate: The Chicago Climate Action Plan's Adaptation Strategy." *Journal of Great Lakes Research* 36(2): 115–117.

Coghill, Kim. 2005. "Greenwood Praises Passage of Pandemic Influenza Plan." Biotechnology Industry Organization, December 22. Available at www.bio.org/ media/press-release/greenwood-praises-passage-pandemic-influenza-plan (accessed March 9, 2015).

Colebach, Hal K. 2006. "What Work Makes Policy?" *Policy Sciences* 39(4): 309–321.

Colman, Zack. 2014. "Democrats Shift From Climate 'Change' to 'Adaptation' to Woo Republicans." *The Washington Examiner*. April 11. Available at www. washingtonexaminer.com/democrats-shift-from-climate-change-to-adaptation-to-woo-republicans/article/2547109 (accessed March 28, 2015).

Comfort, Louise K. 1999. *Shared Risk: Complex Systems in Seismic Response*. New York: Pergamon.

Congressional Budget Office (CBO). 2006. *A Potential Influenza Pandemic: Possible Macroeconomic Effects and Policy Issues*. Washington, D.C.: The Congress of the United States. Available at www.cbo.gov/sites/default/files/12-08-birdflu. pdf (accessed March 28, 2015).

Connif, Richard. 2009. "The Political History of Cap and Trade." *Smithsonian Magazine*, August. Available at www.smithsonianmag.com/air/the-political-history-of-cap-and-trade-34711212/?no-ist (accessed March 28, 2015).

Coon, Charli E. 2001. "Why President Bush is Right to Abandon the Kyoto Protocol." *The Heritage Foundation Backgrounder* 1437 (May). Available at www.grida.no/geo/GEO/Geo-2-011.htm (accessed March 28, 2015).

Cooper, B. Kurt. 2006. "High and Dry? The Public Readiness and Emergency Preparedness Act and Liability Protection for Pharmaceutical Manufacturers." *Journal of Health Law* 40(1): 2–3.

Council on Foreign Relations (CFR). 2007. "Is The Bird Flu Threat Still Real and Are We Prepared? [Rush Transcript; Federal News Service]." April 12, Washington, D.C.: Federal News Service, Inc. Available at www.cfr.org/public-health-threats/bird-flu-threat-still-real-we-prepared-rush-transcript-federal-news-service/ p13115 (accessed March 28, 2015).

Cox, Robert. 2013. *Environmental Communication and the Public Sphere*. Thousand Oaks: Sage.

Coyner, Kris. 2011. "Preventative Care Can Cut Costly Emergency Room Visits." *The Olympian*, January 28. Available at www.theolympian.com/2011/01/28/ 1522894/preventative-care-can-cut-costly.html (accessed March 28, 2015).

Craig, Christine and Laurence Hecht. 2005. "Will Vaccine Funds Be In Time for Pandemic?" *Executive Intelligence Review*, November 4. Available at www. larouchepub.com/eiw/public/2005/2005_40-49/2005_40-49/2005-44/pdf/ 32-33_43_eco.pdf (accessed March 28, 2015).

Cresanti, Robert. 2007. "Technology Administration Speech." National Institute of Standards and Technology, April 3. Available at www.nist.gov/tpo/publications/speechtransroundtablenanotech.cfm (accessed March 27, 2015).

Crichton, Michael. 2008. *Prey*. New York: Harper.

Crichton, Michael. 2009. *State of Fear*. New York: Harper.

Crowley, Geoffrey. 1998. "Assessing the Threat: Why 'Bird Flu' Isn't Likely to Turn into a Global Killer," *Newsweek* 12.

Daly, Matthew and Frederic J. Frommer. 2010. "States Could Veto Neighbor's Offshore Drilling Under Energy-climate Bill." *Cleveland.com*, May 11. Available at www.cleveland.com/nation/index.ssf/2010/05/states_could_veto_neighbors_ of.html (accessed March 28, 2015).

Dang, Alan. 2006. "Nanotech the New Asbestos: Carbon Nanotube Toxicity." *Daily Tech*, May 5. Available at www.dailytech.com/Nanotech+the+New+Asbes tos+Carbon+Nanotube+Toxicity/article2132.htm (accessed March 27, 2015).

DARA. 2012. *2nd Edition Climate Vulnerability Monitor: A Guide to the Cold Calculus of a Hot Planet*. Available at http://daraint.org/wp-content/ uploads/2012/09/CVM2ndEd-FrontMatter.pdf (accessed March 28, 2015).

Davenport, Coral and Peter Baker. 2014. "Taking Page From Health Care Act, Obama Climate Plan Relies on States." *New York Times*, June 3, AO16.

Davidson, Keay. 2005. "Big Troubles May Lurk in Super-tiny Tech/Nanotechnology Experts Say Legal, Ethical Issues Loom." *SFGate.com*, October 31. Available at http://articles.sfgate.com/2005-10-31/news/17396870_1_foresight-nanotech-institute-nanotechnology-industry-nanomaterials (accessed March 27, 2015).

Davies, J. Clarence. 2008. *Nanotechnology Oversight: An Agenda for the New Administration*. Woodrow Wilson International Center for Scholars, Project on Emerging Technologies, July. Available at www.nanotechproject.org/process/assets/files/6709/pen13.pdf (accessed March 27, 2015).

Davies, J. Clarence. 2009. *Oversight of the Next Generation of Nanotechnology*. Woodrow Wilson International Center for Scholars, Project on Emerging Technologies. Available at www.nanotechproject.org/process/assets/files/7316/pen-18.pdf (accessed March 27, 2015).

Davis, Mike. 2006. *The Monster at Our Door: The Global Threat of Avian Flu*. New York: Henry Holt.

DeLeo, Rob A. 2010. "Anticipatory-Conjectural Policy Problems: A Case Study of the Avian Influenza." *Crisis, Risk, and Hazards in Public Policy* 1(1): 147–184.

DeLeo, Rob A. 2014. "Centers for Disease Control and Prevention: Anticipatory Action in the Face of Uncertainty (1946–Present)." In *Guide to U.S. Health and Health Care Policy*, ed. Thomas R. Oliver. Washington, D.C.: CQ Press: 51–64.

Dell'Amore, Christine 2010. "Five Global Warming 'Tipping Points.'" *National Geographic* October 28. Available at http://news.nationalgeographic.com/news/2009/03/photogalleries/tipping-points-climate-change/index.html (accessed March 28, 2015).

"Democrats Work to Protect Americans From Avian Flu." 2005. United States Senate Democrats, October 5. Available at http://democrats.senate.gov/2005/10/05/democrats-work-to-protect-americans-from-avian-flu/#.VP85_0trWlI (accessed March 28, 2015).

Dennison, Richard. 2014. "A Hint of Movement in the Super Slo-Mo that is Nanoregulation at EPA Under TSCA." Environmental Defense Fund Blog, October 8. Available at http://blogs.edf.org/health/2014/10/08/a-hint-of-movement-in-the-super-slo-mo-that-is-nanoregulation-at-epa-under-tsca/ (accessed March 27, 2015).

DeNoon, Daniel. 2005. "Bird Flu Threat Rise." *WebMD Health News*, May 20. Available at www.webmd.com/cold-and-flu/news/20050520/bird-flu-threat-rises (accessed March 28, 2015).

DeNoon, Daniel. 2005b. "WHO: Threat of Bird Flu Pandemic Rises." *Fox News*, March 23. Available at www.foxnews.com/story/0,2933,157218,00.html (accessed March 28, 2015).

Dessler, Andrew E., and Edward A. Parson. 2006. *The Science and Politics of Global Climate Change: A Guide to the Debate*. Cambridge: Cambridge University Press.

Dewar, Helen and Kevin Sullivan. "Senate Republicans Call Kyoto Pact Dead." *Washington Post*, December 11, A37.

Di Justo, Patrick. 2002. "Newt Gingrich Gets Small." *Wired*, May 20. Available at www.wired.com/science/discoveries/news/2002/05/52673 (accessed March 27, 2015).

DiLoreto, John. 2009. "Nanotechnology a Driver in TSCA Reform Push." *NanoReg News*, February 26. Available at www.nanoregnews.com/article.php?id=299 (accessed March 27, 2015).

DiLoreto, John. 2011. "State Regulation of Nanotechnology: All Politics are Local." Global Chem Conference, March 22, Baltimore, MD. Available at: www.socma.com/assets/File/socma1/PDFfiles/gcrc/2011/JD-GlobalChem-Presentation-State-Regulation-of-Nanotechnology-3-22-11.pdf (accessed March 27, 2015).

Doshi, Peter. 2011. "The Elusive Definition of Pandemic Influenza." *Bulletin of the World Health Organization* 89: 532–538.

Doubleday, Robert. 2007. "Risk, Public Engagement, and Reflexivity: Alternative Framings of the Public Dimensions of Nanotechnology." *Health, Risk & Society* 9(2): 211–227.

Drexler, Eric. 1987. *Engines of Creation: The Coming Era of Nanotechnology.* Norwell: Anchor Press.

Dutzik, Tony and Emily Figdor. 2009. *The Clean Energy Future Starts Here: Understanding the American Clean Energy and Security Act.* Environment America, Fall. Available at www.environmentamerica.org/sites/environment/files/reports/clean-energy-future-starts-here.pdf (accessed March 28, 2015).

Dyson, Kenneth. 1982. "West Germany: The Search for a Rationalist Consensus." In *Policy Styles in Western Europe*, ed. Jeremy Richardson. London: George, Allen, & Unwin: 17–46.

Elias, Bart. 2008. "National Aviation Security Policy, Strategy, and Mode-Specific Plans: Background and Considerations for Congress." *Congressional Research Service*, January 2. Available at www.au.af.mil/au/awc/awcgate/crs/rl34302.pdf. (accessed March 27, 2015).

Elliott, Philip. 2009. "Obama: Swine Flu A National Emergency." *Huffington Post*, October 25. Available at www.huffingtonpost.com/2009/10/24/obama-declares-swine-flu_n_332617.html (accessed March 28, 2015).

Enserink, Martin. 2004. "WHO Adds More '1918' to Pandemic Predictions." *Science* 5704: 2025.

Environment America. 2011. *Annual Report: Recapping our work in 2011 for our members.* Boston: Available at http://environmentamerica.org/sites/environment/files/ANN_AME_FY11_WEB.pdf (accessed March 28, 2015).

Environmental and Energy Study Institute. 2014. "Fact Sheet: Climate Adaptation at the Federal Level." January 16. Available at www.eesi.org/papers/view/fact-sheet-climate-adaptation-at-the-federal-level (accessed March 28, 2015).

Environmental Defense Fund. 2012. "California is Leading the Climate Change Fight." Available at www.edf.org/climate/AB32 (accessed March 28, 2015).

Environmental Protection Agency (EPA). 2010. "Endangerment and Cause or Contribute Findings for Greenhouse Gases under Section 202(a) of the Clean Air Act. EPA's Denial of Petitions for Reconsideration." Available at www.epa.gov/climatechange/endangerment/petitions/decision.html (accessed March 28, 2015).

Environmental Protection Agency (EPA). 2010b, "Prevention of Significant Deterioration and Title V Greenhouse Gas Tailoring Rule." *Federal Register* 75(106): 31513–31608.

Environmental Protection Agency (EPA). 2012. "Greenhouse Gas Emissions." Available at www.epa.gov/climatechange/ghgemissions/gases/co2.html (accessed March 28, 2015).

Environmental Protection Agency (EPA). 2012b. "Region 9 Strategic Plan, 2011–14." Available at www.epa.gov/region9/strategicplan/air.html (accessed March 28, 2015).

Environmental Protection Agency (EPA). 2012c. "Summary of the Clean Air Act." www.epa.gov/lawsregs/laws/caa.html (accessed March 28, 2015).

Environmental Protection Agency (EPA). 2014. "Adaptation Overview" Available at www.epa.gov/climatechange/impacts-adaptation/adapt-overview.html (accessed March 28, 2015).

Environmental Protection Agency (EPA). 2014b. "Carbon Pollution Emission Guidelines for Existing Stationary Sources: Electric Utility Generating Units, A Proposed Rule." *Federal Register* June 18. Available at www.federalregister.gov/articles/2014/06/18/2014-13726/carbon-pollution-emission-guidelines-for-existing-stationary-sources-electric-utility-generating#h-9 (accessed March 28, 2015).

Environmental Protection Agency (EPA). 2014c. "Reducing Acid Rain." October 28. Available at www.epa.gov/airquality/peg_caa/acidrain.html (accessed March 28, 2015).

Epstein, Gerald L. 2009. "Restart the Congressional Office of Technology Assessment." *Science Progress*, March 31. Available at www.fas.org/ota/2009/03/31/restart-the-congressional-office-of-technology-assessment/ (accessed March 28, 2015).

Etheridge, Elizabeth W. 1992. *Sentinel for Health: A History of the Centers for Disease Control.* Berkeley: University of California Press.

Fauci, Anthony S. 2006. "Pandemic Influenza Threat and Preparedness." *Emerging Infectious Diseases* 12(1): 73–77.

Faust, Halley S. "Prevention vs. Cure: Which Takes Precedence?" 2005. *Medscape Public Health & Prevention* 3(May).

Fisher, Erik and Roop L. Mahajan. 2006. "Contradictory Intent? U.S. Federal Legislation on Integrating Societal Concerns into Nanotechnology Research and Development." *Science and Public Policy* 33(1): 5–16.

Food and Drug Administration. 2007. *Nanotechnology Task Force Report 2007.* July 27. Available at www.fda.gov/ScienceResearch/SpecialTopics/Nanotechnology/UCM2006659.htm (accessed March 27, 2015).

Food and Drug Administration. 2011. "FDA Opens Dialogue on 'Nano' Regulation." Consumer Health Information, June. Available at http://nanotechnology.wmwikis.net/file/view/Nanotech_0611.pdf. (accessed March 27, 2015).

Food and Drug Administration. 2015. "FDA's Approach to Regulation of Nanotechnology Products." March 13. Available at www.fda.gov/ScienceResearch/SpecialTopics/Nanotechnology/ucm301114.htm (accessed March 27, 2015).

Foreman, Christopher H. Jr. 1994. *Plagues, Products, & Politics: Emergent Public Health Hazards and National Policymaking.* Washington, D.C: The Brookings Institution Press.

Friedman, Dan. 2014. "CDC Issues New Guidelines for Medical Workers Who Treated Ebola Patients in West Africa." *New York Daily News*, October 27. Available at www.nydailynews.com/life-style/health/cdc-issues-new-guidelines-medical-workers-treated-ebola-patients-article-1.1989314 (accessed March 28, 2015).

Frohock, Fred M. 1979. *Public Policy: Scope and Logic.* Englewood Cliffs: Prentice Hall.

Fuerth, Leon. 2011. "Operationalizing Anticipatory Governance." *Prism* 2(4): 31–46.

Galluci, Maria. 2013. "6 of the World's Most Expensive Adaptation Plans." *Inside Climate News*, June 20. Available at http://insideclimatenews.org/news/20130620/6-worlds-most-extensive-climate-adaptation-plans (accessed March 28, 2015).

Gargan, Edward A. 1997. "Avian Flu Strain Spreads to Humans." *New York Times*, December 9, 9.

Garrett, Laurie. 2005. "The Next Pandemic?" *Foreign Affairs* 84(4): 3–23.

Gay Stolberg, Sheryl 2009. "Vaccine Shortage Is Political Test for White House." *New York Times*, October 28, A04.

General Accounting Office. 2000. *Year 2000 Computing Challenge: Lessons Learned Can Be Applied to Other Management Challenges*. Washington. Available at www.gao.gov/assets/240/230628.pdf (accessed March 27, 2015).

Gerber, Brian J. 2007. "Disaster Management in the United States: Examining Key Political and Policy Challenges." *Policy Studies Journal* 35(2): 227–238.

Gerber, Brian J. and Grant W. Neeley. 2005. "Perceived Risk and Citizen Preferences for Governmental Management of Routine Hazards." *Policy Studies Journal* 33(3): 395–418.

Giddens, Anthony. 2011. *The Politics of Climate Change, Second Ed.* Oxford: Polity.

Gilman Duane, Betsey. 2004. "ABSA Influenza Pandemic Plan Position Paper." *ABSA News*. Available at www.absa.org/abj/abj/040904influenza.pdf (accessed March 28, 2015).

"Global Warming Bill Faces Stiff GOP Opposition." 2008. *CNN*, July 2. Available at: http://articles.cnn.com/2008-06-02/politics/senate.greenhousegas_1_global-warming-bill-energy-costs-democratic-backed-bill?_s=PM:POLITICS (accessed March 28, 2015).

Godschalk, David, Adam Rose, Elliot Mittler, Keith Porter, and Carol Taylor West. 2009. "Estimating the Value of Foresight: Aggregate Analysis of Natural Hazard Mitigation Benefits and Costs." *Journal of Environmental Planning and Management* 52(6): 739–756.

Goggin, Malcolm, Ann Bowman, James Lester, and Laurence O'Toole, Jr. 1990. *Implementation, Theory and Practice: Toward a Third Generation*. New York: HarperCollins.

Goode, Darren and Andrew Restuca. 2014. "Greens Take 2014 Fight to States." *Politico*, September 30. Available at www.politico.com/story/2014/09/environmental-groups-spending-states-2014-elections-111435.html (accessed March 28, 2015).

Goodman, Sara. 2010. "Sen. Lautenberg Introduces Chemicals Reform Bill, Saying Current Regulation 'Is Broken'" *New York Times*, April 15. Available at www.nytimes.com/gwire/2010/04/15/15greenwire-sen-lautenberg-introduces-chemicals-reform-bil-25266.html (accessed March 27, 2015).

Gore, Al. 2007. *The Assault on Reason*. New York: Penguin Books.

Gorman, Christine. 2005. "Avian Flu: How Scared Should We Be?" *Time*, October 17: 30–34.

Greenberg, George D. Jeffrey A. Miller, Lawrence B. Mohr. 1977. "Developing Public Policy Theory: Perspectives from Empirical Research." *The American Political Science Review* 71(4): 1532–1543.

Gribbin, Billy. 2011. "Energy Tax Prevention Act Passes House, Fails in Senate." Americans for Tax Reform, April 7. Available at http://atr.org/energy-tax-prevention-act-passes-house-a6032 (accessed March 28, 2015).

Guber, Deborah Lynn. 2013. "A Cooling for Climate Change? Party Polarization and the Politics of Global Warming." *American Behavioral Scientist* 57(1): 93–115.

Guston, David H. and Daniel Sarewitz. 2002. "Real-Time Technology Assessment." *Technology in Society* 23(4): 93–109.

Haddow, George, Jane Bullock, and Damon P. Coppola. 2007. *Introduction to Emergency Management, 3rd Ed.* Burlington: Butterworth-Heinemann.

Hamburger, Tom. 2014. "Fossil-Fuel Lobbyists, Bolstered by GOP Wins, Work to Curb Environmental Rules." *Washington Post,* December 7. Available at www.washingtonpost.com/politics/fossil-fuel-lobbyists-bolstered-by-gop-wins-work-to-curb-environmental-rules/2014/12/07/3ef05bc0-79b9-11e4-9a27-6fdbc612bff8_story.html (accessed March 28, 2015).

Hamilton, Kirsty. 1998. *The Oil Industry and Climate Change.* A Greenpeace Briefing. August: 37.

Hansen, James E. 2006. "Can We Still Dangerous Human-Made Climate Change?" *Social Research* 73(3): 949–971.

Harris, Gardiner. 2006. "U.S. Stockpiles Antiviral Drugs, but Democrats Critical of Pace." *The New York Times,* March 2, A21.

Hart, Paul 't and Arjen Boin. 2001. "Between Crisis and Normalcy: The Long Shadow of Post-Crisis Politics." In *Managing Crises: Threats, Dilemmas, Opportunities,* eds. Uriel Rosenthal, Arjen Boin, and Louise Comfort. Springfield: Charles C. Thomas: 28–46.

Hart, Paul 't, Uriel Rosenthal, and Alexander Kouzmin. 1993. "Crisis Decision Making: The Centralization Thesis Revisited." *Administration & Society* 25(1): 12–44.

Hart Research Associates. 2009. *Nanotechnology, Synthetic Biology, & Public Opinion: A Report of Findings.* Conducted on behalf of Project on Emerging Nanotechnologies, The Woodrow Wilson International Center for Scholars, September 22. Available at www.nanotechproject.org/process/assets/files/8286/nano_synbio.pdf (accessed March 27, 2015).

Healy, Andrew and Neil Malhotra. 2009. "Myopic Voters and Natural Disaster Policy." *American Political Science Review* 103(3): 387–406.

Healy, Bernadine. 2006. "The Young People's Plague." *U.S. News & World Report* 140(16): 63.

Heckathorn, Douglas D. and Steven M. Master. 1990. "The Contractual Architecture of Public Policy: A Critical Reconstruction of Lowi's Typology." *Journal of Politics* 52(4): 1101–1123.

Helland, Asgeir. 2004. *Nanoparticles: A Closer Look at the Risks to Human Health and the Environment.* Master's Thesis in Environment Management and Policy, International Institute for Industrial Environmental Economics. October.

Hellerman, Caleb. 2006. "Bird Flu Vaccine Eggs All in One Basket." *CNN,* March 20. Available at http://articles.cnn.com/2005-12-08/health/pdg.bird.flu.vaccine_1_vaccine-targets-sanofi-pasteur-pandemic-strain?_s=PM:HEALTH (accessed March 28, 2015).

Hess, David J., David A. Banks, Bob Darrow, Joseph Datko, Jaime D. Ewalt, Rebecca Gresh, Matthew Hoffmann, Anthony Sarkis, and Logan D.A. Williams. 2010. *Building Clean-Energy Industries and Green Jobs: Policy Innovations at the State and Local Government Level.* Science and Technology Studies Department, Rensselaer Polytechnic Institute.

Hill, Michael and Peter Hupe. 2009. *Implementing Public Policy, 2nd Edition.* London: Sage.

Hinnfors, Jonas. 1997. "Still the Politics of Compromise? Agenda Setting Strategy in Sweden." *Scandinavian Political Studies* 20(2): 159–177.

Hiskes, Richard P. 1988. "Emergent Risks and Convergent Interests: Democratic Policymaking for Biotechnology." *Policy Studies Journal* 17(1): 73–82.

Ho, Shirley S., Dominique Brossard, and Dietram A. Scheufele. 2007. "Public Reactions to Global Health Threats and Infectious Diseases." *Public Opinion Quarterly* 4 (Winter): 671–692.

Hoch, Maureen. 2010. "New Estimates Puts Gulf Oil Leak at 205 Million Gallons." *PBS Newshour*, August 2. Available at www.pbs.org/newshour/rundown/2010/08/new-estimate-puts-oil-leak-at-49-million-barrels.html (accessed March 28, 2015).

Hodge, James G., Lawrence Gostin, and Jon S. Vernick. 2007. "The Pandemic and All-Hazards Preparedness Act: Improving Public Health Emergency Response." *The Journal of the American Medical Association* 297(15): 1708–1711.

Hoffman, Matthew M. 2008. "EPA Takes First-Ever Regulatory Actions Aimed at Potential Nanomaterial Risks." Goodwin Proctor Alert, November 6. Available at www.goodwinprocter.com/Publications/Newsletters/ClientAlert/2008/1106_EPA-Takes-First-Ever-Regulatory-Actions-Aimed-at-Potential-Nanomaterial-Risks.aspx?device=print (accessed March 27, 2015).

Hogue, Cheryl. 2010. "Mixed Receptions for Chemical Bill." *Chemical & Engineering News* 88(31): 1.

Hogwood, Brian W. 1987. *From Crisis to Complacency? Shaping Public Policy in Britain*. London: Oxford University Press.

Homeland Security Council. 2006. *National Strategy for Pandemic Influenza Implementation Plan*. The White House. May. Available at www.flu.gov/planning-preparedness/federal/pandemic-influenza-implementation.pdf (accessed March 28, 2015).

Howlett, Michael and Benjamin Cashore. 2009. "The Dependent Variable Problem in the Study of Policy Change." *Journal of Comparative Policy Analysis* 11(1): 29–42.

Howlett, Michael and M. Ramesh. 2003. *Studying Public Policy: Policy Cycles and Policy Subsystems, 2nd Edition*. London: Oxford University Press.

Hsia-Khung, Katherin and Erica Morehouse. 2015. *Carbon Market in California: A Comprehensive Analysis of the Golden State's Cap-and Trade Program*. Environmental Defense Fund. Available at www.edf.org/sites/default/files/content/carbon-market-california-year_two.pdf (accessed March 28, 2015).

Huang, Yanzhong. 2010. "Comparing the H1N1 Crises and Responses in the US and China." Centre for Non-Traditional Security Studies Working Paper Series. Working Paper No. 1, November: 8 Available at www.rsis.edu.sg/NTS/resources/research_papers/NTS%20Working%20Paper1.pdf. (accessed March 28, 2015).

Huber, Peter W. 1986. "The Bhopalization of American Tort Law." *Hazards: Technology and Fairness*. Washington, D.C.: National Academy Press.

Hull, Dana. 2013. "13 Things to Know About California's Cap-And-Trade Program." *San Jose Mercury News*, February 22. Available at www.mercurynews.com/ci_22092533/13-things-know-about-california-cap-trade-program(accessed March 28, 2015).

Hunt, Geoffrey and Michael D. Mehta. 2006. *Nanotechnology: Risk, Ethics, and Law*. London: Routledge.

Hyatt, Michael S. 1999. *The Y2K Personal Survival Guide: Everything You Need to Know to Get From This Side of the Crisis to the Other*. Washington, D.C.: Regnery Publishing.

Infectious Disease Society of America (IDSA). 2011. "Vaccine Development." June 16. Available at http://biodefense.idsociety.org/idsa/influenza/panflu/biofacts/panflu_vax.html (accessed March 28, 2015).

Inhofe, James. 2004. *The Facts and Science of Climate Change*. Washington, D.C.: U.S. Senate. Available at www.epw.senate.gov/repwhitepapers/ClimateChange. pdf (accessed March 28, 2015).

Inhofe, James. 2014. "Letter to the Honorable Gina McCarthy." December 1. Available at www.inhofe.senate.gov/download/?id=be69bbb0-fe44-4542-98d4-72cedca51897&download=1 (accessed March 28, 2015).

Innes, Judith Eleanor. 1990. *Knowledge and Public Policy: The Search for Meaningful Indicators*. New Brunswick: Transaction Books.

Intergovernmental Panel on Climate Change (IPCC). 2000. *IPCC Special Report on Emissions Scenarios*. Summary for Policymakers. Available at www.ipcc.ch/pdf/special-reports/spm/sres-en.pdf (accessed March 28, 2015).

Intergovernmental Panel on Climate Change (IPCC). 2007. *Climate Change 2007: Synthesis Report*. Geneva: IPCC. Available at www.ipcc.ch/pdf/assessment-report/ar4/syr/ar4_syr.pdf (March 28, 2015).

Intergovernmental Panel on Climate Change (IPCC). 2007b. "Working Group 1: The Scientific Basis." Summary for policymakers. Geneva. Available at www.ipcc.ch/ipccreports/tar/wg1/index.php?idp=5 (accessed March 28, 2015).

Intergovernmental Panel on Climate Change (IPCC). 2007c. "6.3 Responses to Climate Change." Robust findings, Geneva. Available at www.ipcc.ch/publications_and_data/ar4/syr/en/mains6-3.html (accessed March 28, 2015).

Jackson, David. 2014. "Obama Clears the Air on Climate Deal With China." *USA Today*, November 12. Available at www.usatoday.com/story/news/nation/2014/11/12/obama-china-xi-jinping-climate-change-agreement-coal/18901537/ (accessed March 28, 2015).

Jay, Alexa. 2011. "House Republicans Prepare to Attack EPA Greenhouse Gas Rules." *Climate Science Watch*, January 10. Available at www.climatescience-watch.org/2011/01/10/house-republicans-prepare-to-attack-epa-greenhouse-gas-rules/ (accessed March 28, 2015).

Jervey, Ben. 2013. "Year After Sandy, Rebuilding for Storms and Rising Seas." *National Geographic*. Available at http://news.nationalgeographic.com/news/2013/10/131026-hurricane-sandy-anniversary-sea-level-rise-adaptation/ (accessed March 28, 2015).

John, Peter. 1999. "Ideas and Interests; Agendas and Implementation: An Evolutionary Explanation of Policy Change in British Local Government Finance." *British Journal of Politics and International Relations* 1(1): 39–62.

Johnson, Brad. 2010. "Citing Katrina Myth, Obama Claimed 'Oil Rigs Today Don't Generally Cause Spills.'" *Think Progress*, April 28: 456. Available at http://thinkprogress.org/climate/2010/04/28/174651/obama-katrina-spill/ (accessed March 28, 2015).

Johnson, Renee J. and Michael J. Scicchitano. 2000. "Uncertainty, Risk, Trust, and Information: Public Perceptions of Environmental Issues and Willingness to Take Action." *Policy Studies Journal* 28(3): 633–647.

Jones, Bryan D. and Frank R. Baumgartner. 2005. *The Politics of Attention: How Government Prioritizes Problems*. Chicago: The University of Chicago Press.

Jordan, Andrew J. 2001. "National Environmental Ministries: Managers or Ciphers of European Environmental Policy?" *Public Administration* 79(3): 643–663.

Jordan, Andrew J. 2002. *The Europeanisation of British Environmental Policy: A Departmental Perspective*. London: Palgrave.

Jordan, Grant and Jeremy Richardson. 1987. *British Politics and the Policy Process: An Arena Approach*. London: Unwin Hyman.

Kamalick, Joe. 2011. "US Congress Move to Block EPA Greenhouse Role." *ICIS. com*, February 3. Available at www.icis.com/Articles/2011/02/03/9431933/insight-us-congress-moves-to-block-epa-greenhouse-role.html (accessed March 28, 2015).

Karanjia, Vahbiz. 2011. "United States: Nanosteps Toward Regulating Nanotechnology, Part 2." Minority Corporate Counsel Association, July 19. Available at www.mondaq.com/unitedstates/x/139436/Environmental+Law/Nanosteps+Towards+Regulating+Nanotechnology+Part+II (accessed March 27, 2015).

Karl, Thomas R., Gerald A. Meehl, Christopher D. Miller, Susan J. Hassol, Anne M. Waple, and William L. Murray. 2008. *Weather and Climate Extremes in a Changing Climate*. U.S. Climate Change Program, June. Available at http://downloads.globalchange.gov/sap/sap3-3/sap3-3-final-all.pdf (accessed March 28, 2015).

Katz, Rebecca. 2011. *Essentials of Public Health Preparedness*. Sudbury: Jones & Bartlett Learning.

Kay, W.D. and Christopher J. Bosso. 2003. "A Nanotech Velvet Revolution? Issues for Science Inquiry." *STEP Ahead* 3(2): 2–4.

Keiner, Suellen. 2008. *Room at the Bottom? Potential State and Local Strategies for Managing the Risks and Benefits of Nanotechnology*. Woodrow Wilson International Center for Scholars, Project on Emerging Nanotechnologies, March. Available at www.nanotechproject.org/process/assets/files/6112/pen11_keiner.pdf (accessed March 27, 2015).

Keiper, Adam. 2003. "The Nanotechnology Revolution." *The New Atlantis* Summer: 17–34.

Kettl, Donald F. 2007. *System Under Stress: Homeland Security and American Politics, 2nd Ed*. Washington, D.C.: CQ Press.

Khong, Yuen Foong. 1992. *Analogies at War: Korea, Munich, Dien Bien Phu, and the Vietnam Decisions of 1965*. Princeton: Princeton University Press.

Kingdon, John W. 2003. *Agendas, Alternatives, and Public Policies, 2nd Ed*. New York: Addison-Wesley Educational Publishers.

Klinberg, Eric. 2013. "Adaptation: How Can Cities Climate Proof?" *The New Yorker*, January 7: 33.

Knox, Richard. 2009. "Swine Flu Vaccine Shortages: Why?" *NPR*, October 26. Available at www.npr.org/templates/story/story.php?storyId=114156775 (accessed March 28, 2015).

Krauthammer, Charles. 2009. "Preventative Care Isn't the Magic Bullet for Health Care Costs." *Washington Post*, August 14. Available at www.washingtonpost.com/wpdyn/content/article/2009/08/13/AR2009081302898.html (accessed March 28, 2015).

Kunreuther, Howard. 2006. "Has the Time Come for Comprehensive Natural Disaster Insurance?" In *On Risk and Disaster: Lessons from Hurricane Katrina*, eds. Ronald J. Daniels, Donald F. Kettle, and Howard Kunreuther. Philadelphia: University of Pennsylvania Press: 175–202.

Lakoff, Andrew. 2007. "From Population to Vital System: National Security and the Changing Object of Public Health." ARC Working Paper, No. 7. Available at http://anthropos-lab.net/wp/publications/2007/08/workingpaperno7.pdf (accessed March 28, 2015).

Lallanilla, Marc. 2005. "Spanish Flu of 1918: Could It Happen Again?" *ABC News*, October 5. Available at: http://abcnews.go.com/Health/AvianFlu/story?id=1183172 (accessed March 28, 2015).

Lambright, Henry W. and Jane A. Heckley. 1985. "Policymaking for Emerging Technology: The Case of Earthquake Prediction." *Policy Sciences* 18(3): 227–240.

Landy, Marc. 2010. "Climate Adaptation and Federal Megadisaster Policy: Lessons from Katrina." Resources for the Future, February. Available at www.rff.org/RFF/Documents/RFF-IB-10-02.pdf (accessed March 28, 2015).

Langer, Laura and Paul Brace. 2005. "The Preemptive Power of State Supreme Courts: Adoption of Abortion and Death Penalty Legislation." *Policy Studies Journal* 33(3): 317–340.

Larsen, John. 2006. "Global Warming Legislation in the 109th Congress." World Resources Institute, November 3. Available at www.wri.org/stories/2006/11/global-warming-legislation-109th-congress (accessed March 28, 2015).

Larson, Erik. 1998. "The Flu Hunters." *Time*, February 23: 56.

Lawrence, Deborah and Karen Vandecar. 2015. "Effects of Tropical Deforestation on Climate and Agriculture." *Nature Climate Change* 5: 27–36.

Lazarus, Richard J. 2010. "Super Wicked Problems and Climate Change: Restraining the Present to Liberate the Future." *Environmental Law and Policy Annual Review* 40: 10749–10756.

Leahy, Stephen. 2004. "Grim Signs Mark Global Warming." *Wired*, November 10. Available at www.wired.com/science/discoveries/news/2004/11/65654 (accessed March 28, 2015).

Lefton, Rebecca and Daniel J. Weiss. 2010. "Oil Dependence is a Dangerous Habit." Center for American Progress, January 13. Available at www.americanprogress.org/issues/2010/01/oil_imports_security.html (accessed March 28, 2015).

Levit, Fred. 1998. "Asian 'Bird Flu' Isn't Similar to 1918." *New York Times*, January 5, A18.

Limbaugh, Rush. 2009. "Obama's Swine Flu Vaccine Fiasco." The Rush Limbaugh Show, November 3. Available at www.rushlimbaugh.com/daily/2009/11/03/obama_s_swine_flu_vaccine_fiasco (accessed March 28, 2015).

Lindquist, Eric Katrina N., Mosher-Howe, and Xinsheng Liu. 2010. "Nanotechnology.... What is it Good For? (Absolutely Everything): A Problem Definition Approach." *Review of Policy Research* 27(3): 255–271.

Lister, Sarah A. 2005. *Pandemic Influenza: Domestic Preparedness Efforts*. November 10. Congressional Research Service. Washington, D.C.: Library of Congress.

Lister, Sarah A. 2007. *Pandemic Influenza: Appropriations for Public Health Preparedness and Response*. January 23. Congressional Research Service. Washington, D.C.: Library of Congress. Available at https://fas.org/sgp/crs/misc/RS22576.pdf (accessed March 8, 2015).

Lister, Sarah A. and C. Stephen Redhead. 2009. *The 2009 Influenza Pandemic: An Overview*. August 6. Congressional Research Service. Washington, D.C.: Library of Congress. Available at http://fpc.state.gov/documents/organization/128854.pdf (accessed March 28, 2015).

Little, Amanda. 2003. "The Climate Bill Lost Out, but the Environment May Yet Prove the Winner." *Grist*, November 5. Available at http://grist.org/article/thrill/ (accessed March 28, 2015).

Lizza, Ryan. 2010. "As the World Burns." *The New Yorker*, October 11. Available at www.newyorker.com/reporting/2010/10/11/101011fa_fact_lizza (accessed March 28, 2015).

Loris, Nicholas and Ben Lieberman. 2009. "Cap and Trade (American Clean Energy and Security Act): Talking Points." The Heritage Foundation, Distributed by the Sumter County Republican Executive Committee. Available at www. sumterrepublicans.com/images/talkingpoints/Talking%20Points%20%20 American%20Clean%20Energy%20and%20Security%20Act.pdf (accessed March 28, 2015).

Lomax, Simon. 2010. "EPA Advisory Panel Splits Over U.S. Technology Rules for Carbon Dioxide." *Bloomberg*, September 28. Available at http://mobile.bloomberg. com/news/2010-09-28/epa-advisory-panel-splits-over-u-s-technology-rules-for-carbon-dioxide (accessed March 28, 2015).

Lowi, Theodore J. 1964. "American Business, Public Policy, Case Studies, and Political Theory." *World Politics* 16(4): 687–691.

Lowi, Theodore J. 1972. "Four Systems of Policy, Politics, and Choice." *Public Administration Review* 33(4): 298–310.

Lux Research. 2014. "Nanotechnology Update: Corporations Up Their Spending as Revenues for Nano-enabled Products Increase." State of the Market Report. February 17. Available at https://portal.luxresearchinc.com/research/report_excerpt/ 16215 (accessed March 27, 2015).

Mackey, Brendan and Song Li. 2007. "Stand Up for the Earth Community: Win the Struggle Against Global Warming." *Pacific Ecologist* (Summer): 10–13.

Maher, Brendan. 2010. "Crisis Communicator." *Nature* 463(14): 150–152.

Mair, Michael, Beth Maldin, and Brad Smith. 2006. "Passage of S. 3678: The Pandemic and All-Hazards Preparedness Act." Center for Biosecurity, University of Pittsburgh Medical Center. Available at www.upmchealthsecurity.org/our-work/ pubs_archive/pubs-pdfs/archive/2006-12-20-allhazardsprepact.pdf (accessed March 12, 2015).

Manjoo, Farhad. 2009. "Apocalypse Then." *Slate*, November 11. Available at www.slate.com/articles/technology/technology/features/2009/apocalypse_then/ was_y2k_a_waste.html. (accessed March 27, 2015).

Manzer, Ronald. 1984. "Public Policy-Making as Practical Reasoning." *Canadian Journal of Political Science* 17(3): 577–594.

Massey, Nathaniel. 2012. "1988 vs. 2012: How Heat Waves and Droughts Fuel Climate Perception." Environmental Change in the News, University of Maryland. Available at www.climateneeds.umd.edu/climatewire-08-28-12/article-02. php. (accessed March 28, 2015).

Matthews, James T. 2006. "Egg-Based Production of Influenza Vaccine: 30 Years of Commercial Experience." *The Bridge* 36(3): 17–24.

Matzner, Franz and Jim Presswood. 2010. "An Energy Bill Without a Carbon Cap Could Do More Harm than Good." National Resources Defense Council. Available at http://docs.nrdc.org/legislation/files/leg_10030901a.pdf (accessed March 28, 2015).

May, Peter J. 1986. "Politics and Policy Analysis." *Political Science Quarterly* 101(1): 109–125.

May, Peter J. 1991. "Addressing Public Risks: Federal Earthquake Policy Design." *Journal of Policy Analysis and Management* 10(2): 263–285.

May, Peter J. and Chris Koski. 2013. "Addressing Public Risks: Extreme Events and Critical Infrastructures." *Review of Policy Research* 30(2): 139–159, 139.

May, Peter J. and Thomas A. Birkland. 1994. "Earthquake Risk Reduction: An Examination of Local Regulatory Efforts." *Environmental Management* 18(6): 923–937.

May, Peter J., Ashley E. Jochim, and Joshua Sapotichne. 2011. "Constructing Homeland Security: An Anemic Policy Regime." *Policy Studies Journal* 39(2): 285–307.

Maynard, Andrew. 2009. "Nanotechnology Safety Research Funding Up." *2020 Science*, May 21. Available at http://2020science.org/2009/05/21/nanotechnology-safety-research-funding-on-the-up/ (accessed March 27, 2015).

Maynard, Andrew. 2010. "US Government Kicks Nanotechnology Safety Research Up a Gear." *2020 Science*, February 18. Available at: http://2020science.org/2010/02/18/us-government-kicks-nanotechnology-safety-research-up-a-gear/ (accessed March 27, 2015).

McCarthy, James E. and Larry Parker. 2010. *EPA Regulation of Greenhouse Gases: Congressional Responses and Options*. Congressional Research Service. June 8. Washington, D.C.: Library of Congress. Available at www.fas.org/sgp/crs/misc/R41212.pdf (accessed March 28, 2015).

McCombs, Maxwell. 2004. *Setting the Agenda: The Mass Media and Public Opinion*. Cambridge: Polity.

McCool, Daniel C. 1995. "Policy Typologies." In *Public Policy Theories, Models, and Concepts: An Anthology*, ed. Daniel C. McCool. Englewood Cliffs: Prentice Hall: 174–181.

McCray, Patrick W. 2005. "Will Small be Beautiful? Making Policies for Our Nanotechnology Future." *History and Technology* 21(2): 177–203.

McGinn, Robert E. 1990. *Science, Technology, and Society*. New Jersey: Prentice Hall.

McNeil, Donald G. 2010. "U.S. Reactions to Swine Flu Apt and Lucky." *The New York Times*, January 2, A1.

Menaham, Gila. 1998. "Policy Paradigms, Policy Networks and Water Policy in Israel." *Journal of Public Policy* 18(3): 283–310.

Mikhail, Natalie J. 2003. "Bush Approves Billions for Nanotechnology Research." *The Badger Herald*, December 4. Available at http://badgerherald.com/news/2003/12/04/bush_approves_billio.php (accessed March 27, 2015).

Milburn, Colin. 2008. *Nanovision: Engineering the Future*. Durham: Duke University Press.

Mileti, Dennis. 1999. *Disasters by Design: A Reassessment of Natural Hazards in the United States*. Washington, D.C.: Joseph Henry Press.

Miller, Georgia. 2008. "Contemplating the Implications of a Nanotechnology 'Revolution.'" In *The Yearbook of Nanotechnology in Society: Presenting Futures*, eds. Erik Fisher, Cynthia Selin, and Jameson M. Wetmore. New York: Springer: 215–225.

Miller, Kristen L. Janet L. Kaminski Leduc, and Kevin E. McCarthy. 2012. "Sea-Level Rise Adaptation Policy in Various States." Connecticut Coastal Zone Management, October 26. Available at www.cga.ct.gov/2012/rpt/2012-R-0418.htm (accessed March 28, 2015).

Mills, James M. Robins, Carl T. Bergstrom, and Marc Lipsitch. 2006. "Pandemic Influenza: Risk of Multiple Introductions and the Need to Prepare for Them." *PLoS Medicine* 3(6): 769–773. Available at http://octavia.zoology.washington.edu/publications/MillsEtAl06.pdf (Accessed March 8, 2015).

Minnesota Department of Public Health. 2012. "Avian Flu and Pandemic Flu: The Difference—and the Connection." Available at: www.health.state.mn.us/divs/idepc/diseases/flu/avian/avianpandemic.html (accessed March 28, 2015).

Mintrom, Michael. 2000. *Policy Entrepreneurs and School Choice*. Washington, D.C.: Georgetown University Press.

Monica, John C., Jr., Patrick T. Lewis, and John C. Monica. 2006. "Preparing for Future Health Litigation: The Application of Products Liability Law to Nanotechnology." *Nanotechnology Law & Business* 3(1): 54–63.

Mooney, Chris. 2005. "Some Like It Hot." *Truthout*, May/June. Available at http://archive.truthout.org/article/chris-mooney-some-like-it-hot (accessed March 28, 2015).

Mooney, Chris. 2014. "38 Federal Agencies Reveal Their Vulnerabilities to Climate Change—and What They're Doing About It." *Washington Post*, Wonkblog, October 31. Available at www.washingtonpost.com/blogs/wonkblog/wp/2014/10/31/38-federal-agencies-reveal-their-vulnerabilities-to-climate-change-and-what-theyre-doing-about-it/ (accessed March 28, 2015).

More, Charles. 2000. *Understanding the Industrial Revolution*. London: Routledge.

Moreno, Robert B. and Peter Zalzal. 2010. "Greenhouse Gas Dissonance: The History of EPA's Regulations and the Incongruity of Recent Legal Challenges." *UCLA Journal of Environmental Law and Policy* 30: 121–156.

Morin, Richard. 2005. "Bush Approval Rating at All-Time Low." 2005. *Washington Post*, September 12. Available at www.washingtonpost.com/wpdyn/content/article/2005/09/12/AR2005091201158_pf.htm (accessed March 28, 2015).

Moser, Susanne C. 2009. *Good Morning, America! The Explosive U.S. Awakening to the Need for Adaptation*. California Energy Commission and National Oceanic and Atmospheric Administration (NOAA). Available at www.preventionweb.net/files/11374_MoserGoodMorningAmericaAdaptationin.pdf (accessed March 28, 2015).

Moss, David A. 2002. *When All Else Fails: Government as the Ultimate Risk Manager*. Cambridge: Harvard University Press.

Müller, Benito. 2001. "Fatally Flawed Inequity: Kyoto's Unfair Burden on the United States & the Chinese Challenge to American Emission Dominance." Presented at World Bank Climate Change Day, Washington D.C., June 14, 2001 and Climate Strategies Review, Brussels, June 19. Available at www.oxfordclimatepolicy.org/publications/documents/ffi.pdf (accessed March 28, 2015).

"Nanotechnology Risks: How Buckyballs Hurt Cells." 2008. *Science Daily*, May 27. Available at www.sciencedaily.com/releases/2008/05/080527091910.htm (accessed March 27, 2015).

National Association of Clean Air Agencies. 2012. "Background and History of EPA Regulation of Greenhouse Gas (GHG) Emissions Under the Clean Air Act & National Association of Clean Air Agencies' Comments on EPA GHG Regulatory and Policy Proposals." February 17. www.4cleanair.org/Documents/BackgroundandHistoryEPARegulationGHGsFeb2012post.pdf (accessed March 28, 2015).

National Institute for Occupational Safety and Health. 2013. *Protecting the Nanotechnology Workforce: NIOSH Nanotechnology Research and Guidance Strategic Plan, 2013–2016*. Department of Health and Human Services, Center for Disease Control and Prevention. Available at www.cdc.gov/niosh/docs/2014-106/pdfs/2014-106.pdf (accessed March 27, 2015).

National Institutes of Health, Office of Research Services, Division of Occupational Health and Safety. 2014. "Nanotechnology Safety and Health Program." October. Available at www.ors.od.nih.gov/sr/dohs/Documents/Nanotechnology%20Safety%20and%20Health%20Program.pdf (accessed March 27, 2015).

National Nanotechnology Initiative. 2014. "NNI Supplement to the President's 2015 Budget." March 25. Available at www.nano.gov/node/1128 (accessed March 27, 2015).

National Research Council. 2009. *Review of the Federal Strategy for Nanotechnology-Related Environmental, Health, and Safety Research*. Washington, D.C.: The National Academies Press.

National Science and Technology Council. 2000. *National Nanotechnology Initiative: The Initiative and Its Implementation Plan*. Committee on Technology, Subcommittee on Nanoscale Science, Engineering and Technology. July. Available at www.nsf.gov/crssprgm/nano/reports/nni2.pdf (accessed March 27, 2015).

Naylor, Brian. 2009. "Obama Flu Response Relied On Bush Plan." *NPR*, May 7. Available at www.npr.org/templates/story/story.php?storyId=103908247 (accessed March 28, 2015).

Neary, Ian. 1992. "Japan." In *Power and Policy in Liberal Democracies*, ed. Martin Harrop. Cambridge: Cambridge University Press.

Neustadt, Richard E. and Ernest R. May. 1988. *Thinking in Time: The Uses of History for Decision Makers*. New York: Free Press.

Neustadt, Richard E. and Harvey V. Fineberg. 1978. *The Swine Flu Affair: Decision-Making on a Slippery Disease*. Washington, D.C.: U.S. Department of Health, Education, and Welfare.

Nohrstedt, Daniel. 2008. "The Politics of Crisis Policymaking: Chernobyl and Swedish Nuclear Energy Policy." *Policy Studies Journal* 36(2): 257–278.

Nohrstedt, Daniel and Christopher M. Weible. 2010. "The Logic of Policy Change After Crisis: Proximity and Subsystem Interaction." *Risk, Hazards, & Crisis in Public Policy* 1(2): 1–32.

"Obamas Get Their Swine Flu Shots." 2009. *New York Times*, December 21. Available at www.nytimes.com/2009/12/22/us/politics/22brfs-OBAMASGETTHE_BRF.html (accessed March 28, 2015).

Oberdörster, Eva. 2004. "Manufactured Nanomaterials (Fullerines, C60) Induce Oxidative Stress in the Brain of Juvenile Largemouth Bass." *Environmental Health Perspectives* 122(10): 1058–1062.

O'Donnell, Sean T. and Jacqueline A. Isaacs. 2010. "A World of Its Own? Nanotechnology's Promise—and Challenges." In *Governing Uncertainty: Environmental Regulation in the Age of Uncertainty*, ed. Christopher J. Bosso. Washington, D.C.: Resources for the Future: 12–27.

Office of the Press Secretary. 2007. "Press Briefing on National Strategy for Pandemic Influenza Implementation Plan One Year Summary." The White House, July 17. Available at http://georgewbush-whitehouse.archives.gov/news/releases/2007/07/20070717-13.html (accessed March 28, 2015).

Olsen, Johan, Paul Roness, and Harald Saetren. 1982. In *Policy Styles in Western Europe*, ed. Jeremy Richardson. London: George, Allen, & Unwin.

O'Neill, Xana. 2009. "The Little Boy Behind the Swine Flu Crisis." *NBC Miami*, April 28. Available at www.nbcmiami.com/news/archive/NATL-The-Little-Boy-Behind-the-Swine-Flu-Crisis-.html (accessed March 28, 2015).

O'Rourke, Thomas. 2007. "Critical Infrastructure, Interdependencies, and Resilience." *The Bridge* 37(1): 22–30.

Oreskes, Naomi and Erik M. Conway. 2010. *Merchants of Doubt: How a Handful of Scientists Obscured the Truth on Issues from Tobacco to Global Warming*. New York: Bloomsbury Press.

Organic Consumers Association. 2009. "Studies Show Nanoparticles Used in Sunscreens and Makeup can Harm the Environment." March 26. Available at www.organicconsumers.org/scientific/studies-show-nanoparticles-used-sunscreens-and-makeup-can-harm-environment (accessed March 27, 2015).

Osterholm, Michael T. 2005. "Preparing for the Next Pandemic." *The New England Journal of Medicine* 352: 1839–1842.

Palacios-Huerta, Ignacio. 2014. *In 100 Years: Leading Economists Predict the Future*. Boston: The MIT Press.

Parker, Charles F. and Eric K. Stern. 2005. "Bolt From the Blue or Avoidable Failure? Revisiting September 11 and the Origins of Strategic Surprise." *Foreign Policy Analysis* 10(3): 301–331.

Parker, Larry John Blodgett, and Brent D. Yacobucci. 2011. "U.S. Global Climate Change Policy: Evolving Views of Cost, Competiveness, and Comprehensiveness." February 24. Congressional Research Service. Washington, D.C.: Library of Congress. Available at www.fas.org/sgp/crs/misc/RL30024.pdf (accessed March 28, 2015).

Parkinson, Clair L. 2010. *Coming Climate Crisis? Consider the Past Beware the Big Fix*. Plymouth: Rowan & Littlefield.

Parry, Jane. 2007. "Ten Years of Fighting Bird Flu." *Bulletin of the World Health Organization* 85(1): 3–6.

Parson, Wayne. 1995. *Public Policy: An Introduction to the Theory and Practice of Policy Analysis*. Cheltenham: Edward Elgar.

Patashnik, Eric M. 2008. *Reforms at Risk: What Happens After Major Policy Changes Are Enacted*. Princeton: Princeton University Press.

Patashnik, Eric M. and Julian E. Zelizer. 2013. "The Struggle to Remake Politics: Liberal Reform and the Limits of Policy Feedback in the Contemporary American State." *Perspectives on Politics* 11(4): 1071–1087.

Paton, Douglas and David Johnston. 2001. "Disasters and Communities: Vulnerability, Resilience, and Preparedness." *Disaster Prevention and Management: An International Journal* 10(4): 270–277.

Pattberg, Philipp and Johannes Stripple. 2008. "Beyond the public and private divide: remapping transnational climate governance in the 21st century." *International Environmental Agreements* 8: 367–388.

Pear, Robert and Gardiner Harris. 2009. "Obama Seeks to Ease Fears on Swine Flu." *New York Times*, April 28, 1.

Peters, Guy 2012. *American Public Policy: Promise & Performance, 8th Ed.* Washington, D.C.: CQ Press.

Petru, Alexis. 2014. "Two Years After Sandy: NYC Plans for Transportation Resiliency." *Triple Pundit*. September 29. Available at www.triplepundit.com/2014/09/two-years-sandy-nyc-plans-transportation-resiliency/ (accessed March 28, 2015).

PEW Center on Global Climate Change. 2008. "Economy-wide Cap-and-Trade Proposals in the 110th Congress." December. Available at www.c2es.org/docUploads/Chart-and-Graph-120108.pdf (accessed March 28, 2015).

Pielke, Roger A. 2000. "Policy history of the US Global Change Research Program: Part I. Administrative development." *Global Environmental Change* 10: 9–25.

Pielke, Roger A. 2000b. "Policy history of the US Global Change Research Program: Part II. Legislative process." *Global Environmental Change* 10: 133–144.

Pielke, Roger Jr., Gwyn Prins, Steve Rayner, and Daniel Sarewitz. 2007. "Lifting the taboo on adaptation." *Nature* 445(8): 597–598.

Pierson, Paul. 2004. *Politics in Time: History, Institutions, and Social Analysis*. Princeton: Princeton University Press.

Plautz, Jason. 2013. "Hearing Lays Path for TSCA Reform, but Boxer Role Unclear." *E&E Daily*, August 1. Available at www.eenews.net/stories/1059985441 (accessed March 27, 2015).

Plautz, Jason. 2015. "Obama's Budget Goes All-In on Climate Change." National Journal, February 2. Available at www.nationaljournal.com/energy/obama-budget-goes-all-in-on-climate-change-20150202 (accessed March 28, 2015).

Poland, Gregory A. 2006. "Vaccine against Avian Influenza—A Race against Time." *The New England Journal of Medicine* 354(13): 1411–1413.

Pollitt, Christopher. 2008. *Time, Policy, Management: Governing With the Past.* Oxford: Oxford University Press.

Posner, Paul. 2005. "The Politics of Preemption: Prospects for the States." *PS: Political Science and Politics* 38(3): 371–374.

Pralle, Sarah B. 2006. *Branching Out, Digging In: Environmental Advocacy and Agenda Setting.* Washington, D.C.: Georgetown University Press.

Pralle, Sarah B. 2006b. "The 'Mouse That Roared': Agenda Setting in Canadian Pesticide Politics." *Policy Studies Journal* 34(2): 171–194.

Pralle, Sarah B. 2009. "Agenda Setting and Climate Change." *Environmental Politics* 18: 781–789.

Prather, Gordon. 2002. "Clear Skies, But No Carbon Dioxide Cap." *WND Commentary*, February 16. Available at www.wnd.com/2002/02/12813/ (accessed March 28, 2015).

Prugh, Russell. 2011. "Proposed Legislation Seeking to Block EPA Greenhouse Gas Regulation Picks Up Speed." Marten Law, February 17. Available at www.marten law.com/newsletter/20110217-epa-greenhouse-gas-regulation (accessed March 28, 2015).

Quiggin, John. 2008. "Uncertainty and Climate Change Policy." *Economic Analysis & Policy* 38(2): 203–210.

Rabe, Barry. 2004. *Statehouse and Greenhouse: The Emerging Politics of American Climate Change Policy.* Washington, D.C.: Brookings Institution Press.

Rabe, Barry. 2008. "States on Steroids: The Intergovernmental Odyssey of American Climate Policy." *Review of Policy Research* 25(2): 105–128.

Ramseur, Jonathan L. 2014. *The Regional Greenhouse Gas Initiative: Lessons Learned and Issues for Policy Makers.* Congressional Research Service. November 14. Washington, D.C.: Library of Congress. Available at www.fas.org/sgp/crs/misc/R41836.pdf (accessed March 28, 2015).

Rejeski, David. 2008. "Gearing Up for the Reauthorization of the Nanotechnology R&D Act." *Nanotechnology Now*, May 16. Available at www.nanotech-now. com/columns/?article=195 (accessed March 27, 2015).

Reynolds, Glenn Harlan. 2003. "Nanotechnology and Regulatory Policy: Three Futures." *Harvard Journal of Law & Technology* 17(1): 179–209.

Richardson, Jeremy J., Gunnel Gustafsson, and Grant A. Jordan. 1982. "The Concept of Policy Style." In *Policy Styles in Western Europe*, ed. Jeremy Richardson, London: George, Allen, & Unwin: 1–16.

Riebeek, Holli. 2005. "The Rising Cost of Natural Hazards." The National Aeronautics and Space Administration, March 28. Available at http://earthobservatory. nasa.gov/Features/RisingCost/rising_cost.php (accessed March 28, 2015).

Ripley, Randall B. and Grace A. Franklin. 1991. *Congress, the Bureaucracy, and Public Policy.* Belmont: Wadsworth.

Roberts, Karlene. 2009. "Managing the Unexpected: Six Years of HRO-Literature Reviewed." *Journal of Contingencies and Crisis Management* 17(1): 50–54.

Rochefort, David A. and Roger W. Cobb. 1994. "Problem Definition: An Emerging Perspective." In *The Politics of Problem Definition: Shaping the Policy Agenda*, eds. David A. Rochefort and Roger W. Cobb. Kansas: University of Kansas Press: 1–31.

Roco, Mihail C. 2003. "Broader Societal Issues of Nanotechnology." *Journal of Nanoparticle Research* 5: 181–189.

Roco, Mihail C. 2011. "The Long View of Nanotechnology Development: The National Nanotechnology Initiative in 10 Years." In *Nanotechnology Research Directions for Societal Needs in 2020: Retrospective and Outlook*, eds. Mihail C. Roco, Chad A. Mirkin, and Mark C. Hersam. New York: Springer: 1–28.

Roe, Emery and Paul R. Schulman. 2008. *High Reliability Management: Operating on the Edge*. Stanford: Stanford University Press.

Roller, Elisa and Amanda Sloat. 2002. "The Impact of Europeanisation on Regional Governance: A Study of Catalonia and Scotland." *Public Policy and Administration* 17(2): 68–86.

Rose, Adam, Keith Porter, Nicole Dash, Jawhar Bouabid, Charles Huyck, John Whitehead, Douglass Shaw, Ronald Eguchi, Craig Taylor, Thomas McLane, L. Thomas Tobin, Philip T. Ganderton, David Godschalk, Anna S. Kiremidjian, Kathleen Tierney, and Carol Taylor West. 2007. "Benefit-Cost Analysis of FEMA Hazard Mitigation Grants." *Natural Hazards Review* 8(4): 97–111.

Rose, Richard. 1993. *Lesson-Drawing in Public Policy: A Guide to Learning Across Time and Space*. Chatham: Chatham House Publishers.

Roseboro, Ken. 2006. "From GMO to Nano: A Familiar Debate Over a New Technology." Organic Consumers Association, September. Available at: www.organicconsumers.org/articles/article_1946.cfm (accessed March 27, 2015).

Rosenberg, Daniel. 2010. "It's Official: Strong TSCA Reform Bill Debuts in the House." *Switchboard*, July 23. Available at http://switchboard.nrdc.org/blogs/drosenberg/its_official_strong_tsca_refor.html. (accessed March 27, 2015).

Rosenberg, Daniel. 2011. "Decades of Delay: TSCA Turns 35—Chemical Industry Still Stifles Protection from Toxic Chemical." Switchboard, October 18. Available at: http://switchboard.nrdc.org/blogs/drosenberg/decades_of_delay_tsca_turns_35.html (accessed March 27, 2015).

Rosenthal, Elisabeth. 2006. "On the Front: A Pandemic is Worrisome But 'Unlikely.'" *New York Times*, March 2006, F1.

Ross, Ashley D. 2014. *Local Disaster Resilience: Administrative and Political Perspectives*. New York: Routledge.

Ruin, Olof. 1982. "Sweden in the 1970s: Policy-Making Becomes More Difficult." *Policy Styles in Western Europe*, ed. Jeremy Richardson. London: George, Allen, & Unwin: 141–167.

Sabatier, Paul A. 1991. "Toward Better Theories of the Policy Process." *PS: Political Science and Politics* 24(2): 147–156.

Sabatier, Paul A. and Christopher M. Weible. 2007. "The Advocacy Coalition Framework: Innovations and Clarifications." In *Theories of the Policy Process, 2nd Edition*, ed. Paul A. Sabatier. Colorado: Westview Press: 190–220.

Sadeghpour, Nura. 2013. "NIOSH Recommends New Level of Exposure for Nanomaterials" National Institute for Occupational Safety and Health, April 24. Available at www.cdc.gov/niosh/updates/upd-04-24-13.html (accessed March 27, 2015).

Sainz-Snatamaria, Jamie and Sarah E. Anderson. 2013. "The Electoral Politics of Disaster Preparedness." *Risks, Hazards & Crisis in Public Policy* 4(4): 234–249.

Salaam-Blyther, Tiaji. 2011. *Centers for Disease Control and Prevention Global Health Programs: FY2001-FY2012 Request*. June 27. Congressional Research Service. Washington, D.C.: Library of Congress. Available at www.fas.org/sgp/crs/misc/R40239.pdf (accessed March 28, 2015).

Sanchez, Christine. 2010. "New TSCA Reform Bill Threatens Domestic Chemical Manufacturing and Innovation." Society of Chemical Manufactures & Affiliates,

July 23. Available at www.socma.com/pressroom/index.cfm?subSec=3&sub=71 &articleID=2528 (accessed March 27, 2015).

Sandler, Ronald. 2006. "The GMO-Nanotech (Dis)Analogy?" *Bulletin of Science, Technology & Society* 26(1): 57–62.

Sandler, Ronald and W.D. Kay. 2006. "The National Nanotechnology Initiative and the Social Good." *Journal of Law, Medicine, & Ethics* 34(4): 675–681.

Santorum, Rick. 2012. "Blown and Tossed by the Winds of Political Correctness." *Redstate.com*, March 10. Available at www.redstate.com/rjsantorum/2012/ 03/10/blown-and-tossed-by-the-winds-of-political-correctness/ (accessed March 28, 2015).

Sarewitz, Daniel and Roger Pielke Jr. 2000. "Breaking the Global Warming Deadlock." *The Atlantic*, July. Available at www.theatlantic.com/past/issues/2000/07/ sarewitz.htm (accessed March 28, 2015).

Sargent, John F. 2011. *Nanotechnology: A Policy Primer*. Congressional Research Service, January 19. Available at https://fas.org/sgp/crs/misc/RL34511.pdf (accessed March 27, 2015).

Sargent, John F. 2011b. *Nanotechnology and Environmental, Health, and Safety: Issues for Consideration*. Congressional Research Service. January 20. Available at www.fas.org/sgp/crs/misc/RL34614.pdf (accessed March 27, 2015).

Sargent, John F. 2011c. *The National Nanotechnology Initiative: Overview, Reauthorization, and Appropriations Issues*. Congressional Research Service, March 25. Available at: http://assets.opencrs.com/rpts/RL34401_20110325.pdf (accessed March 27, 2015).

Schattschneider, Elmer E. 1960. *The Semisovereign People: A Realist's View of Democracy in America*. Hinsdale: Dryden Press.

Schelmetic, Tracy E. 2010. "EPA Releases Best Available Control Technology Guidelines for Polluters." *Green Technology World*, November 11. Available at http://green.tmcnet.com/topics/green/articles/116361-epa-releases-best-available-control-technology-guidelines- (accessed March 28, 2015).

Schipper, E. Lisa F. 2006. "Conceptual History of Adaptation in the UNFCC Process." *Reciel* 15(1): 82–92.

Schneider, Andrew. 2010. "Amid Nanotech's Dazzling Promise, Health Risks Grow." *AOL News*, March 24. Available at www.aolnews.com/2010/03/24/ amid-nanotechs-dazzling-promise-health-risks-grow/ (accessed March 27, 2015).

Schneider, Andrew. 2010b. "Obsession With Nanotech Growth Stymies Regulators." *AOL News*, March 24. www.aolnews.com/2010/03/24/obsession-with-nanotech-growth-stymies-regulators/ (accessed March 27, 2015).

Schram, Martin. 2009. "Will Swine Flu be Obama's Katrina?" *Washington Times*, August 29. Available at www.washingtontimes.com/news/2009/aug/29/will-swine-flu-be-obamas-katrina/ (accessed March 28, 2015).

Schreurs, Miranda A. 2010. "Climate Change Politics in the United States: Melting the Ice." *Analyse & Kritik* 32(1): 177–189.

Schulman, Paul R. 1980. *Large-Scale Policy Making*. New York: Elsevier.

Schulman, Paul R. 1993. "The Analysis of High Reliability Organizations: A Comparative Framework." In *New Challenges to Understanding Organizations*, ed. Karlene H. Roberts, London: Macmillan: 33–54.

Scoones, Ian and Paul Forster. 2010. "Unpacking the International Response to Avian Influenza: Actors, Networks and Narratives." In *Avian Influenza: Science, Policy, and Politics*, ed. Ian Scoones. Washington, D.C.: Earthscan: 19–64.

Selin, Cynthia. 2010. "Anticipation." In *Encyclopedia of Nanoscience and Society*, ed. David H. Guston. Thousand Oaks: Sage. Available at http://knowledge. sagepub.com/view/nanoscience/n10.xml (accessed March 27, 2015).

"S.F. Bay Area Passes Carbon Tax." 2008. *Environmental Leader*, May 22. Available at www.environmentalleader.com/2008/05/22/sf-bay-area-passes-carbon-tax/ (accessed March 28, 2015).

Shaw, Gwyneth. 2011. "NIOSH: Nano-Titanium Dioxide "A Potential Occupational Carcinogen." *The New Haven Independent*, April 27. Available at www. merid.org/Content/News_Services/Nanotechnology_and_Development_News/ Articles/2011/Apr/27/A_NIOSH.aspx. (accessed March 27, 2015).

Shear, Michael D. and Rob Stein. 2009. "Administration Officials Blame Shortage of H1N1 Vaccine on Manufacturers, Science." *Washington Post*, October 27, A11.

Shull, Steven A. 1983. *Domestic Policy Formation: Presidential-Congressional Partnership?* Westport: Greenwood Press.

Shute, Nancy. 2005. "A Man With An Antiflu Plan." *U.S. News & World Report*. 139(18): 32.

Siegel, Marc. 2006. "A Pandemic of Fear." *Washington Post*, March 26, B07.

Silva, Gabriel A. 2007. "Nanotechnology Approaches for Drug and Small Molecule Delivery Across the Blood Brain Barrier." *Surgical Neurology* 67: 113–116.

Simons, John. 1999. "Clinton's Y2k Czar Won't Have Time for Champaign." *The Wall Street Journal*, December 30. Available at www.wsj.com/articles/ SB946511229746246519 (accessed March 27, 2015).

Simonson, Stewart. 2010. "Reflections on Preparedness: Pandemic Planning in the Bush Administration." *Saint Louis University Journal of Health Law & Policy* 4(1): 5–33. Available at www.slu.edu/Documents/law/SLUJHP/Simonson_Article.pdf (accessed March 28, 2015).

Sims Bainbridge, William. 2002. "Public Attitudes Towards Nanotechnology." *Journal of Nanoparticle Research* 4(6): 561–570.

Smith, Kevin B. 2002. "Typologies, Taxonomies, and the Benefits of Policy Classification." *Policy Studies Journal* 30(2): 379–395.

Starling, Grover. 1988. *Strategies for Policy Making*. Chicago: The Dorsey Press.

State of California. 2006. "State of California's Actions to Address Global Climate Change." Available at www.climatechange.ca.gov/climate_action_team/reports/2006 report/2005-12-08_STATE_ACTIONS_REPORT.PDF (accessed March 28, 2015).

Steinberger, Peter J. 1980. "Typologies of Public Policy: Meaning Construction and Their Policy Process." *Social Science Quarterly* 61(1): 185–197.

Stimers, Paul. 2008. "The Implications of Recent Nanomaterials Toxicity Studies for the Nanotech Community." *Nanotechnology Law & Business* 5(3): 313–318.

Stobbe, Mike. 2011. "CDC's Zombie Apocalypse Advice An Internet Hit." *Huffington Post*, May 20. Available at www.huffingtonpost.com/2011/05/21/ zombie-apocalypse-advice-cdc_n_865078.html (accessed March 28, 2015).

Stone, Deborah. 2002. *Policy Paradox: The Art of Political Decision Making*. New York: W.W. Norton.

Stone, Diane. 2008. "Global Public Policy, Transnational Policy Communities, and their Networks." *Policy Studies Journal* 36(1): 19–38.

Strikas, Raymond A., Gregory S. Wallace, and Martin G. Myers. 2002. "Influenza Pandemic Preparedness Action Plan for the United States: 2002 Update." *Clinical Infectious Diseases* 5(September 1): 590–596.

"Study: Swine Flu Resembles 1918 Virus." 2009. *Fox News*, July 13. Available at www.foxnews.com/story/0,2933,532020,00.html. (accessed March 28, 2015).

Sullivan, Chris. 2008. "Making Data Breach Prevention a Matter of Policy." *SC Magazine*, July 9. Available at www.scmagazine.com/making-data-breach-prevention-a-matter-of-policy/article/112213/ (accessed March 27, 2015).

Sussman, Fran, Nisha Krishnan, Kathryn Maher, Rawlings Miller, Charlotte Mack, Paul Stewart, Kate Shourse, and Bill Perkins. 2014. "Climate Change Adaptation Cost in the US: What Do We Know?" *Climate Policy* 14(2): 242–282.

Talbert, Gerald F. 2012. "Conservation Marketplace." National Association of Conservation Districts. Available at www.nacdnet.org/resources/reports/marketplace.phtml (accessed March 28, 2015).

Tatalovichand, Raymond and Byron W. Daynes. 1988. "Conclusion: Social Regulatory Policymaking." In *Social Regulatory Policy: Moral Controversies in America Politics*, eds. Raymond Tatalovich and Byron W. Daynes. Boulder: Westview Press: 210–225.

Taylor, Michael. 2006. *Regulating the Products of Nanotechnology: Does the FDA Have the Tools It Needs?* Woodrow Wilson Center for International Scholars, Project on Emerging Technologies. October. Available at: www.nanotechproject.org/file_download/files/PEN5_FDA.pdf. (accessed March 27, 2015). 29.

"The Hottest Years on Record." 2010. *The Economist*, December 3. Available at www.economist.com/blogs/dailychart/2010/12/climate_change (March 28, 2015).

The National Aeronautics and Space Administration. 2015. "NASA, NOAA Find 2014 Warmest Year in Modern Record." January 16. Available at www.nasa.gov/press/2015/january/nasa-determines-2014-warmest-year-in-modern-record/#.VRK1FEtrW1J (March 28, 2015).

"The OTA Legacy." 2012. Princeton University Archives of the Office of Technology Assessment. Available at www.princeton.edu/~ota/ (accessed March 28, 2015).

Theodoulou, Stella Z. 1995. "The Contemporary Language of Public Policy: A Starting Point." In *Public Policy: The Essential Readings*, eds. Stella Z. Theodoulou and Matthew A. Cahn. Upper Saddle River: Prentice Hall: 1–9.

Tierney, Kathleen. 2014. *The Social Roots of Risk: Producing Disasters, Promoting Resilience*. Redwood City: Stanford University Press.

Toffler, Alvin. 1970. *Future Shock*. New York: Bantam Books.

Trouiller, Benedicte, Ramune Reliene, Aya Westbrook, Parrisa Solaimani, and Robert H. Schiestl. 2009. "Titanium Dioxide Nanoparticles Induce DNA Damage and Genetic Instability In vivo in Mice." *Molecular Biology, Pathology, and Genetics* 69 (22): 8784–8789, 8788.

Tuckman, Jo and Robert Booth. 2009. "Four-year-old Could Hold Key in Search for Source of Swine Flu Outbreak." *Guardian*, April 27. Available at www.guardian.co.uk/world/2009/apr/27/swine-flu-search-outbreak-source (accessed March 28, 2015).

Unger, Brigitte and Frans Van Waarden. 1995. "Introduction: An Interdisciplinary Approach to Convergence." In *Convergence or Diversity? Internationalization and Economic Policy Response*, eds. Frans van Waarden and Brigitte Unger. Aldershot: Avebury: 1–36.

Union of Concerned Scientists. 2003. "Early Warning Signs of Global Warming: Spreading Disease." Available at www.ucsusa.org/global_warming/science_and_impacts/impacts/early-warning-signs-of-global-9.html (accessed March 28, 2015).

Union of Concerned Scientists. 2004. *Scientific Integrity in Policymaking: An Investigation into the Bush Administration's Misuse of Science*. March. Cambridge. Available at www.ucsusa.org/our-work/center-science-and-democracy/promoting-scientific-integrity/scientific-integrity-in.html#.VRdGx0trWlI (accessed March 28, 2015).

Union of Concerned Scientists. 2012. "Manipulation of Global Warming Science." Available at www.ucsusa.org/scientific_integrity/abuses_of_science/manipulation-of-global.html.(accessed March 28, 2015).

Union of Concerned Scientists. 2014. "The Clean Power Plan: A Climate Game Changer." Available at www.ucsusa.org/our-work/global-warming/reduce-emissions/what-is-the-clean-power-plan#.VOFp5UtrWlJ (accessed March 28, 2015).

United States Department of Health and Human Services. 2006. *Department of Health and Human Services Pandemic Planning Update*. March 13, Washington. Available at www.cap.org/apps/docs/committees/microbiology/panflu_final3_13_(2).pdf (accessed March 28, 2015).

United States Department of Health and Human Services. 2006b. *Department of Health and Human Services Pandemic Planning Update III*. A report from Secretary Michael O. Leavitt, November 13. Available at www.flu.gov/pandemic/history/panflureport3.pdf (accessed March 28, 2015).

United States Department of Health and Human Services. 2008. *Pandemic and All-Hazards Preparedness Act: Progress Report on the Implementation of Provisions Addressing High Risk Individuals*. August. Available at www.phe.gov/Prepared-ness/legal/pahpa/Documents/pahpa-at-risk-report0901.pdf (accessed March 28, 2015).

United States Department of Health and Human Services. 2009. "Obama Administration Calls on Nation to Begin Planning and Preparing for Fall Flu Season & the New H1N1 Virus." July 9. Available at www.hhs.gov/news/press/2009 pres/07/20090709a.html (accessed March 28, 2015).

United States Department of Health and Human Services. 2009b. *Pandemic Planning Update VI*. A report from Secretary Michael O. Leavitt, January 8. Available at www.flu.gov/pandemic/history/panflureport6.pdf (accessed March 28, 2015).

United States Department of Health and Human Services. 2012. "An HHS Retrospective on the 2009 H1N1 Influenza Pandemic to Advance All Hazards Preparedness." June 15. Available at www.phe.gov/Preparedness/mcm/h1n1-retrospective/Documents/h1n1-retrospective.pdf (accessed March 28, 2015).

U.S. Congress. House. 1999. Subcommittee on Basic Research. "Nanotechnology: The State of Nano-Science and Its Prospects for the Next Decade." June 22. 106th Cong., 1st sess. Washington, D.C.: Government Printing Office.

U.S. Congress. House. 2003. "H.R. 766, Nanotechnology Research and Development Act of 2003." Committee on Science. March 19. 108th Cong., 1st sess. Washington, D.C.: Government Printing Office.

U.S. Congress. House. 2003. "Kyoto Global Warming Treaty's Impact on Ohio's Coal-Dependent Communities." Committee on Resources. May 13. 108th Cong., 1st sess. Washington, D.C.: U.S. Government Printing Office.

U.S. Congress. House. 2005. "Energy Demand in the 21st Century: Are Congress and the Executive Branch Meeting the Challenge?" Subcommittee on Energy and Resources. March 16. 109th Cong., 1st sess. Washington, D.C.: U.S. Government Printing Office.

U.S. Congress. House. 2005. "Environmental and Safety Impacts of Nanotechnology: What Research is Needed?" Committee on Science. November 17. 109th Cong., 1s sess. Washington, D.C.: Government Printing Office.

U.S. Congress. House. 2005. "National Nanotechnology Initiative: Review and Outlook." Subcommittee on Research. May 18. 109th Cong., 1st sess. Washington, D.C.: Government Printing Office.

U.S. Congress. House. 2005. "The Threat of and Planning for Pandemic Flu." Subcommittee on Health. May 26. 109th Cong., 1st sess. Washington, D.C.: Government Printing Office: 65.

U.S. Congress. House. 2006. "Research on Environmental and Safety Impacts of Nanotechnology: What are the Federal Agencies Doing?" Committee on Science. September 21. 109th Cong., 2nd sess. Washington, D.C.: Government Printing Office.

U.S. Congress. House. 2007. Political Interference With Climate Change Science Under the Bush Administration. Committee on Government Reform. Report: ii. Available at http://earthjustice.org/sites/default/files/library/reports/house-of-representative-2007-majority-report-on-climate-change-science.pdf (accessed March 28, 2015).

U.S. Congress. House. 2008. "Planning for a Changing Climate and Its Impacts on Wildlife and Oceans: State and Federal Efforts and Needs." Subcommittee on Fisheries, Wildlife, and Oceans. June 24. 110th Cong., 2nd sess. Washington, D.C.: U.S. Government Printing Office.

U.S. Congress. House. 2008. "The National Nanotechnology Initiative Amendments Act of 2008." Committee on Science. April 16. 111th Cong., 1st sess. Washington, D.C.: Government Printing Office.

U.S. Congress. House. 2009. "Planning for a Changing Climate—Smart Growth, Public Demand, and Private Opportunity." Select Committee on Energy Independence and Global Warming. June 18. 110th Cong., 2nd sess. Washington, D.C.: U.S. Government Printing Office.

U.S. Congress. House. 2009. "Preparing for Climate Change: Adaptation Policies and Program." Subcommittee on Energy and Environment. March 25. 111th Cong., 1st sess. Washington, D.C.: U.S. Government Printing Office.

U.S. Congress. House. 2009. "Revisiting the Toxic Substances Control Act of 1976." Subcommittee on Commerce, Trade, and Consumer Protection. February 26. 111th Cong., 1st sess. Washington, D.C.: Government Printing Office.

U.S. Congress. House. 2009. "The Administration's Flu Vaccine Program: Health, Safety, and Distribution." Committee on Oversight and Government Reform. September 29. 111th Cong., 1st sess. Washington, D.C.: Government Printing Office.

U.S. Congress. House. 2014. "Nanotechnology: Understanding How Small Solutions Drive Big Innovation." Subcommittee on Commerce, Manufacturing, and Trade. July 29. 113th Cong., 1st sess. Washington, D.C.: Government Printing Office.

U.S. Congress. Senate. 2003. "Nanotechnology." Subcommittee on Science, Technology, and Space. September 17. 107th Cong., 2nd sess. Washington, D.C.: Government Printing Office.

U.S. Congress. Senate. 2003. "S. 189, 21st Century Nanotechnology Research and Development Act." Committee on Commerce, Science, and Transportation. May 1. 108th Congress, 1st sess. Washington, D.C.: Government Printing Office.

U.S. Congress. Senate. 2004. "Electricity Generation." Committee on Energy and Natural Resources. April 27. 108th Cong., 2nd sess. Washington, D.C.: Government Printing Office.

U.S. Congress. Senate. 2005. "U.S.–E.U. Regulatory Cooperation on Emerging Technologies." Subcommittee on European Affairs. May 11. 109th Cong., 1st sess. Washington, D.C.: Government Printing Office.

U.S. Congress. Senate. 2009. "H1N1 Flu: Monitoring the Nation's Response." Committee on Homeland Security and Governmental Affairs. 2009. October 21. 111th Cong., 1st sess. Washington, D.C.: Government Printing Office.

U.S. Congress. Senate. 2009. "Swine Flu: Coordinating the Federal Response." Committee on Homeland Security and Government Affairs. April 29. 111th Cong., 1st sess. 3. Washington, D.C.: Government Printing Office.

U.S. Green Building Council. 2009. "Highlights: American Clean Energy and Security Act of 2009 (H.R. 2454)." Available at www.usgbc.org/ShowFile. aspx?DocumentID=6070 (accessed March 28, 2015).

Van Putten, Jan. 1982. "Policy Styles in the Netherlands: Negotiation and Conflict." In *Policy Styles in Western Europe*, ed. Jeremy Richardson. London: George, Allen, & Unwin: 168–196.

Vassey, Brett A. 2009. "Testimony for Legislative Hearing on S. 1733, Clean Energy Jobs and American Power Act, U.S. Senate Committee on Environment and Public Works." October 28. Available at http://epw.senate.gov/public/index. cfm?FuseAction=Files.View&FileStore_id=30d85d3d-d032-4293-a9da-d239e06d040a (accessed March 28, 2015).

Volpe, Rosalind. 2010. "Industry Comments on EPA OPP Proposed Nanopesticide Policy." Silver Nanotechnology Working Group. Presentation to the Office on Management and Budget.

Votaw, James G. 2014. "NIOSH Releases Nanomaterial Safety Research Plans." *Environmental Leader*, May 22. Available at www.environmentalleader. com/2014/05/22/niosh-releases-nanomaterial-safety-research-plans/ (accessed March 27, 2015).

Wade, Larry L. and Robert L. Curry. 1970. *A Logic of Public Policy: Aspects of Political Economy*. Belmont: Wadsworth.

Wallace, David L. and Justin A. Schenk. 2014. "EPA Targets Nanotechnology: Hi-Ho, Nanosilver Away?" *Nanotechnology Law & Business* 207 (Fall 2014): 207–218.

Walsh, Bryan. 2007. "Indonesia's Bird Flu Showdown." *Time*, May 10. Available at http://content.time.com/time/health/article/0,8599,1619229,00.html (accessed March 28, 2015).

Walsh, Bryan. 2008. "How to Win the War on Global Warming." *Time*, April 17. Available at: www.time.com/time/specials/2007/article/0,28804,1730759_17313 83_1731363,00.html (accessed March 28, 2015).

Webby, Richard J. and Robert G. Webster. 2003. "Are We Ready for Pandemic Influenza?" *Science* 302(5650): 1519–1522.

Webster, Robert G. and Elena A. Govorkova. 2006. "H5N1 Influenza—Continuing Evolution and Spread." *New England Journal of Medicine* 355: 2174–2177.

Weible, Christopher M. 2008. "Expert-Based Information and Policy Subsystems: A Review and Synthesis." *Policy Studies Journal* 36(4): 615–635.

Weick, Karl E. and Kathleen M. Sutcliffe. 2007. *Managing the Unexpected: Resilient Performance in an Age of Uncertainty, 2nd Ed.* San Francisco: Jossey Bass.

Weiss, Jurgan and Mark Sarro. 2009. "The Economic Impact of AB 32 on California Small Businesses." Prepared for the Union of Concerned Scientists. December. Available at www.ucsusa.org/sites/default/files/legacy/assets/documents/global_ warming/AB-32-and-CA-small-business-report.pdf (accessed March 28, 2015).

Weiss, Rick. 2005. "Nanotechnology Regulation Needed, Critics Say." *The Washington Post*, December 5, A08.

Welch, Gilbert H., Lisa Schwartzl, and Steve Woloshin. 2011. *Over-Diagnosed: Making People Sick in the Pursuit of Health*. Boston: Beacon Press.

White House. 2001. "NATIONAL NANOTECHNOLOGY INITIATIVE: LEADING THE WAY TO THE NEXT INDUSTRIAL REVOLUTION." Office of the Press Secretary, January 21. Available at http://clinton4.nara.gov/WH/New/html/20000121_4.html (accessed March 27, 2015).

White House. 2013. *The President's Climate Action Plan.* Executive Office of the President. June. Available at www.whitehouse.gov/sites/default/files/image/president27sclimateactionplan.pdf (accessed March 28, 2015).

White House. 2015. *Investing in America's Future: The Budget for Fiscal Year 2016.* Washington. Available at www.whitehouse.gov/sites/default/files/omb/budget/fy2016/assets/investing.pdf (accessed March 28, 2015).

White House Homeland Security Council. 2005. *National Strategy for Pandemic Influenza.* November 5. Washington. Available at www.flu.gov/planning-preparedness/federal/pandemic-influenza.pdf (accessed March 28, 2015).

Wildavsky, Aaron. 1988. *Searching for Safety.* New Brunswick: Transaction Books.

Wilson Scott, Scott and Spencer S. Hsu. 2009. "A Bush Team Strategy Becomes Obama's Swine Flu Playbook." *The Washington Post,* May 1. Available at www.washingtonpost.com/wp dyn/content/article/2009/04/30/AR2009043003910.html (accessed March 28, 2015).

Woodrow Wilson International Center for Scholars. 2002. "Congressional Clearinghouse on the Future." October 1. Available at www.wilsoncenter.org/article/congressional-clearinghouse-the-future (accessed March 28, 2015).

World Council of Churches. 2005. "A Tiny Primer on Nano-scale Technologies and the Little Bang Theory." January 15. Available at www.oikoumene.org/en/resources/documents/wcc-programmes/justice-diakonia-and-responsibility-for-creation/science-technology-ethics/nano-scale-technologies.html (accessed March 27, 2015).

World Economic Forum. 2008. *Global Risks 2008: A Global Risk Network Report.* January. Available at https://members.weforum.org/pdf/globalrisk/report2008.pdf. (accessed on March 27, 2015).

World Health Organization (WHO). 2011. "Avian Influenza." April. Available at www.who.int/mediacentre/factsheets/avian_influenza/en/ (accessed March 28, 2015).

World Health Organization (WHO). 2011b. "H5N1 Avian Influenza: Timeline of Major Events." December 13. Available at www.who.int/influenza/human_animal_interface/avian_influenza/H5N1_avian_influenza_update.pdf (accessed March 28, 2015).

World Health Organization (WHO). 2012. "Cumulative Number of Confirmed Human Cases of Avian Influenza A(H5N1) Reported to WHO." May 2. Available at www.who.int/influenza/human_animal_interface/H5N1_cumulative_table_archives/en/index.html (accessed March 28, 2015).

World Health Organization (WHO). 2014. "Climate Change and Infectious Diseases." Available at www.who.int/globalchange/climate/summary/en/index5.html (accessed March 28, 2015).

World Health Organization (WHO). 2014b. "WHO Risk Assessment Of Human Infections With Avian Influenza A(H7N9) Virus" February 23. Available at www.who.int/influenza/human_animal_interface/influenza_h7n9/RiskAssessment_H7N9_23Feb20115.pdf?ua=1 (accessed March 28, 2015).

World Health Organization (WHO). 2015. "Cumulative Number of Confirmed Cases for Avian Influenza (AH5N1) Reported to WHO, 2003–2015." Available at www.who.int/influenza/human_animal_interface/EN_GIP_20150126CumulativeNumberH5N1cases.pdf?ua=1 (accessed March 28, 2015).

Wood, Robert S. 2006. "The Dynamics of Incrementalism: Subsystems, Politics, and Public Lands." *Policy Studies Journal* 34(1): 1–16.

Wyatt, Kristen. 2009. "Senators Tour US Park, Hear About Global Warming." *Guardian*, August 24. Available at www.guardian.co.uk/world/feedarticle/8672910 (accessed March 28, 2015).

Yacobucci, Brent D. and Robert Bamberger. 2010. *Automobile and Light Truck Fuel Economy: The CAFÉ Standards*. Congressional Research Service. April 23. Washington, D.C.: Library of Congress. Available at http://assets.opencrs.com/rpts/R40166_20100423.pdf (accessed March 28, 2015).

Yan, Holly and Greg Botelho. 2014. "Ebola: Some U.S. States Announce Mandatory Quarantines—Now What?" *CNN*, October 27. Available at www.cnn.com/2014/10/27/health/ebola-us-quarantine-controversy/ (accessed March 28, 2015).

Yeatman, William. 2015. "The Clean Power Plan's Dirty Secret." *U.S. News and World Report*, January 26. Available at www.usnews.com/opinion/articles/2015/01/26/obamas-epa-clean-power-plan-has-a-dirty-secret (accessed March 28, 2015).

Yeo, Sophie. 2014. "Hawaii Passes Climate Change Adaptation Law." Responding to Climate Change, June 12. Available at www.rtcc.org/2014/06/12/hawaii-passes-climate-change-adaptation-law/ (accessed March 28, 2015).

Youde, Jeremy. 2008. "Who's Afraid of Chicken? Securitization and Avian Flu." *Democracy and Security* 4(2): 148–169.

Zahran, Sammy, Samuel D. Brody, Arnold Vedlitz, Himanshu Grover, and Caitlyn Miller. 2008. "Vulnerability and Capacity: Explaining Local Commitment to Climate-Change Policy." *Government and Policy* 26(3): 544–562.

Zajac, Andrew. 2010. "FDA Urged to be Proactive, Not Reactive, in Preventing Food Safety Problems." *Los Angeles Times*, June 8. Available at http://articles.latimes.com/2010/jun/08/nation/la-na-fda-20100609 (accessed March 27, 2015).

Zengerle, Patricia. 2009. "Obama Declare Swine Flu a National Emergency." *Reuters*, October 24. Available at http://mobile.reuters.com/article/topNews/idUSTRE59N19E20091024 (accessed March 28, 2015).

Zerbe, Noah. 2007. "Risking Regulation, Regulating Risk: Lessons from the Transatlantic Biotech Dispute." *Review of Policy Research* 24(5): 407–423.

# Index

Page numbers in *italics* denote tables, those in **bold** denote figures.